THE WESTERN INTELLECTUAL TRADITION

THE WESTERN INTELLECTUAL TRADITION

From Leonardo to Hegel

J. BRONOWSKI
BRUCE MAZLISH

HARPER TORCHBOOKS
Harper & Row, Publishers
New York, Hagerstown, San Francisco, London

THE WESTERN INTELLECTUAL TRADITION:

FROM LEONARDO TO HEGEL

Printed in the United States of America

This book was originally published in 1960
by Harper & Brothers

First HARPER COLOPHON edition published 1975 by
Harper & Row, Publishers, Incorporated
New York and Evanston

Library of Congress catalog card number: 59-12671

ISBN: 0-06-133001-9

88 89 90 29

Contents

Part III. The Great Revolutions: From Smith to Hegel, 1760–1830

Preface

EVERY THOUGHTFUL man who hopes for the creation of a contemporary culture knows that this hinges on one central problem: to find a coherent relation between science and the humanities. In education in particular, this problem faces us in two forms. We have to give the future scientist an abiding sense of the value of literature and the arts; and at the same time we have to give to those whose preoccupation lies with the liberal arts a glimpse of the methods, the depth, and the inspiration of science. These are living problems all the way from the school desk through the university and beyond, into the daily life of all thoughtful men. They are, however, focused most sharply in the universities, where the traditional division between science and the humanities cannot be rooted out of the timetable in one generation. It is not accidental, therefore, that this book was conceived and begun in the Department of Humanities in the Massachusetts Institute of Technology, which is actively devoted to making an intellectual fusion between the sciences and the humanities.

The specific field with which the authors were concerned there was the field of history. Here, more even than in some other fields, there is a cleavage between the conventional presentation of history, and the recent but still specialized interest in the historical growth of science and of techniques. By contrast, the authors set out from the beginning to see all history, certainly all intellectual history, as a unity. This book is the result of this endeavor; and its aim happens to have been put very cogently for us, since the book was written, by Sir Charles Snow in an article "Challenge to the Intellect" in *The Times Literary Supplement* of August 15, 1958.

What is needed is that in the general history books the development of science should take its place along with political and economic developments. It is only just that this should be done even from the historian's point of view, for the world we live in is as much the product of science as of politics and economics. The steam engine helped to shape the modern world at least as much as Napoleon and Adam Smith, but only rarely do the historians admit the fact. There are few living historians who can write history in this way; but this is one way in which history must be written if the worlds of science and the humanities are not to drift still farther apart.

The vision of an integrated history, written in this way, brought the authors together—one an Englishman, the other an American. The book owes its origin, in fact, to the appointment in 1953 of the English author to be Carnegie Visiting Professor at the Massachusetts Institute of Technology, where the American author was teaching in the Department of Humanities. We are both indebted to the Carnegie Foundation and to the Massachusetts Institute of Technology for creating this opportunity, which has resulted in a lasting friendship and collaboration on the themes of this book.

We have many debts and obligations to acknowledge. Our overriding obligation is, of course, to the thinkers whom we discuss in the following pages, because their work has shaped our understanding and influenced our outlook in so many ways. On similar grounds, we owe thanks to the students of the Massachusetts Institute of Technology, and to other groups of students in England and America, whose arguments have influenced not only the presentation but often the essential integration of our ideas.

We are most indebted to the constant interest and support we have had from John E. Burchard, Dean of the School of Humanities and Social Sciences, and from Howard R. Bartlett, Head of the Department of Humanities. The authors were first brought together by Professor Giorgio de Santillana, and his stimulus has remained lively at every stage since then, to his final reading of our manuscript. Others who have given us valuable advice are Professor Herbert Rowen, now Visiting Professor at the University of California at Berkeley, Professor Hanna Gray of Harvard University, and Professors Roy Lamson and Charles Gray of the Massachusetts Institute of Technology. We express our particular thanks to Professor William Bottiglia, also of the Massachusetts Institute of Technology, for his acute comments and help on the chapters which deal with French thinkers and writers. And of course, we must thank those many friends who sacrificed hours of leisure and mental ease in order to expose themselves to our enthusiastic arguments about this major conception or that scholarly detail in the book.

In general, we have given the source of translations which are not our own; where no such acknowledgment is made, it can be assumed that we are responsible for the translation. Although we have supported or qualified most important points by bibliographical references, this has not been the main purpose of our bibliographies. Their main purpose is to guide the interested reader to the best works in which he can pursue in detail any subject which attracts him in principle. For this reason, we have usually confined ourselves to works in English and in French, on the assumption that our readers will find their way through books in these languages more easily than others; but we have occasionally drawn attention to important works in other languages, chiefly German and Spanish. The bibliographies we give are not complete, but they are critical bibliographies: they present what we believe to be the most important works on each subject, with short appraisals of them. No complete and critical bibliography for intellectual history exists at present. Naturally, many general readers will prefer reading straight through the book, with only a few glances at the footnotes; this, of course, the text allows, for it was written as an entity in itself.

As befits a work on the Western Intellectual Tradition, this book was drafted and redrafted in many parts of Europe—in England, Spain, France, and Italy—as well as in America. So scattered but intense an effort created constant problems of organization and of presentation. In England, these were resolved by Mrs. Elise Charles, who typed and coördinated much of the material, and who prepared the Index of Names. In America, Mrs. Ruth Dubois and the secretaries at the Massachusetts Institute of Technology typed most of the material for the press. We must not end without acknowledging the labors of many unnamed typists in Europe, who struggled delightfully with English as a strange language, and with the authors as very strange thinkers.

J. Bronowski
Bruce Mazlish

Introduction

I

THIS BOOK is a study of the development of ideas from the Renaissance to the opening of the nineteenth century. It differs in several ways, in content and in presentation, from other books on the history of ideas. We ought to draw attention at the outset to three differences which are fundamental, and which amount to differences of principle.

First, this is an intellectual history in the largest sense. It is not confined to ideas in some one field—in politics, say, or in philosophy. We are interested in the whole spectrum of the mind, and an important feature of our book is its stress on the interplay of ideas from different fields. In particular, we give more attention than is usual to the ideas of science, to the movements of literary style, and to the innovations of the arts.

Second, we believe that the history of ideas must be coupled with a concrete knowledge of the events that took place at the same time. The study of ideas is an evolutionary study, and the reader should feel behind it the context of events, the physical environment within which ideas have evolved. He should see the influence of events on ideas, and of ideas on events. Therefore, our book gives more attention than is usual to the history of the times which it covers. Particularly, it stresses the impact of technical and social inventions, which go hand in hand with new ideas, and in which the intellectual vigor of a community displays itself in practice.

Third, we think that the right way to present an idea is in the speech of the men to whom it came as a revelation. Therefore, we have constantly shown ideas as held by people—by single men or by groups of men. In general, each of our chapters focuses on the outlook of a man or of a group of men who epitomize a way of thinking: such men as Descartes and Bentham, groups of men such as the early humanists and the dissenting manufacturers in the Lunar Society. These men are not paraded as heroes or as dramatic figures in themselves. We see them and we show them as the living embodiment of the thought of an age, and the interest of their lives is that in them the conflicts of the age take on the sharp edge of struggles of conscience.

These three principles, in our view, give reality to the history of ideas. Ideas are not dead thoughts, even when they are no longer contemporary; for they remain steps in the evolution of contemporary ideas. We have wanted to present the ideas of each age not as fossils but as evolving organisms, and not as butterflies in a box but as the vital processes of the human mind. This is a history of the life of ideas: active, mobile, and changing. We have therefore presented it and written it as integrated history.

II

We have tried to show both the movement and the conflict of ideas. Even one man may hold ideas which go oddly together, as the ideas of Newton did; or which change sharply between youth and age, as those of Hobbes did: or which at bottom are inconsistent, as those of Rousseau were. In a community the conflict of ideas has a special force: established ideas held rigidly for social reasons or for reasons of interest are at odds with new ideas, and the character of the community derives from the struggle and balance between these. We have wanted to show the creative value of the conflict of ideas.

Yet, the evolution of an idea also has an inner logic, which it follows through every conflict. Consider, for example, how the idea of progress has changed and yet has retained its direction through the four centuries which we study. We have been at pains to trace this logic of ideas through their changes; and we draw attention, as one example, to our extended discussion of the idea of man's unity with nature, which we follow from the simple materialism of Hobbes to Kant's subtle theory of scientific knowledge.

In the four centuries which this book covers, the world has been transformed from medieval to modern. They are centuries of change, in every detail of life; and the history of their ideas, as we have said, is necessarily a history of movement. The movement is created by that which gives life to ideas: by the interplay of all the interests of the mind, by the pressure of events, and by the expression of personalities. It is this sense of movement, of ideas felt as the mind in action, which we have wanted to communicate. To us, the Age of Enlightenment—to take a single example—is not a restful abstraction. It is a complex of people and groups with conflicting ideas which yet have a common direction; and this is why we see it and present it under a new name, as the Age of Reasoned Dissent. This is our approach to the intellectual history of all the periods that we cover.

III

The integration of intellectual history which we have attempted has advantages, but it also creates some difficulties for the authors. We ought to explain the difficulties of which we are aware, but which we could not outflank.

The central difficulty is to keep so large an undertaking in a manageable frame. There is a limit to the detail which a book can hold in focus, and this limit is strained when the book presents ideas and events together. In particular, in our method we must make a choice of men and groups of men to single out for discussion. Everyone will agree that we ought to include Erasmus, but ought we to have included Robert Owen? Ought we to have left out Loyola? Everyone will agree that we ought to include the Royal Society, but ought we to have included the Lunar Society? Ought we to have left out the Académie Royale des Sciences of France?

We cannot hope to include every great mind in four hundred years, or every important group of minds; indeed, however many we include, we cannot hope that it will be agreed that none of equal importance has been left out. In short, we cannot hope that the choice we have made will content every scholar. Yet, we should explain the method of our choice, and particularly because the method of the book is intimately related to the method of choice.

Every history is a map: it leaves out some features of reality, and singles out others which are thought to display its essential structure. When a historical map is drawn, as ours is, from specific men and groups outward, it has a special character. The process of mapping becomes the surveyor's process of triangulation, and the reference points are important less in themselves than as marks for triangulation—that is, as points from which to take a bearing. This is the principle which has guided us whenever we have had to make a choice between men: which of them, we have asked, best expresses the essence of his time.

Some scholars may hold that the nature of our presentation, what we have called the method of triangulation, must make our survey of thought unsystematic. Certainly our presentation is discontinuous, and in outward appearance yields a series of essays rather than a single story. We believe that the appearance is superficial, and that our choice of themes ensures that the reader constructs a coherent map of intellectual and general history which has a greater sweep and a clearer perspective than other methods give. This is an attempt to write a new kind of intellectual history, and it depends for its effect on the total pattern built up by its parts.

IV

A history which is to communicate to its readers the whole sensibility of an age ought to be written, like the Authorized translation of the Bible, by a group of like-minded people—and it should, like the Authorized translation, then give the effect of a single mind. We cannot claim to be more than two men, but we are specialists in more fields, and in different fields, than the historian of ideas usually is. One of us is a mathematician with an interest in the philosophy of science and in literature. The other is an intellectual historian with qualifications in literature and in philosophy. As a result, we have been able to support one another, and to present (out of our own discussions) a deeper interpretation, as well as a wider background, than could be expected of most single writers.

Some features of our book grow out of our interests, and we ought to point to them. Thus, we give a deeper treatment of the inventions on which our civilization rests than is usual: our discussion of the evolution of mechanisms to keep time, with the wider implications of time keeping, is an example. At the same time, we do not consider that the only worth-while inventions are mechanical devices. We discuss the invention of the inductive method in science, and the philosophical questions which it raises, as a natural part of the working of modern civilization. We discuss the invention of political and social ideas in the same direct spirit. To us, Machiavelli's political method is an invention of the same kind as the mechanical inventions of Leonardo, and both spring from the same soil in the Renaissance.

We adopt a method of exact textual criticism which is more familiar in literature, and apply it to texts which it is not usual to read so closely. One patent advantage is that this presents to the reader many passages written by the men we have chosen, in which he can see at first hand how they spoke and thought. But more deeply than this, we regard the scrutiny of texts which are not principally literary—of political, scientific, economic, and philosophical texts—as a personal encounter with the style and attitude of an age. To us, the style of a period is a vivid expression of its totality, in which we read, as it were, the thumbprint of history—or, to change the metaphor, we discover the character of an age from its handwriting. When the Royal Society encouraged its Fellows to write with a simplicity in which "men delivered so many *things* almost in an equal number of *words,*" it was giving the stamp of the new science and the very climate of empiricist philosophy. Or, to take another example: we discuss the style in which the Declaration of Independence is written, because this throws light directly both on Franklin and on Jefferson as men, and at the same time it expresses the intellectual unity which America had at that time.

In short, a style can become a method; and the style of early science in the seventeenth century is characteristic of the new method. Consider, for example, Galileo's purpose in putting his criticism of traditional astronomy into the form of popular dialogues. This method, the teasing question and answer of the dialogue, readily becomes the method of doubt which Descartes developed; and it is characteristic that Pascal combined the two in his *Provincial Letters.* The thread runs from Pascal (through Montesquieu) into Voltaire's *Candide,* in which dialogue and doubt frankly become satire. We have tried to capture the spirit of such literary traditions and transitions as a link to the ideas which they intertwine.

V

The period which we cover goes from the Renaissance to the opening of the nineteenth century. It is natural to begin with the Renaissance, because then the modern concept of man as an individual, to be judged in himself, was formed. The very method of this book, the discussion of ideas as held personally by men, derives from the Renaissance. And the Renaissance is also the beginning of that secularization of society which, for example, is expressed in the word "state." There were no states, in this sense, before the Renaissance: the idea and the institution grew up together with the word. In a book dealing with ideas of man and state, the Renaissance is the natural starting point.

The natural end point to our book might be the present day, but we felt that, at least for the present, this would be to overstrain our frame. We have chosen instead to treat as a culmination of the Renaissance impulse to individuality the period of the great revolutions at the end of the eighteenth century. With the Industrial Revolution and the American and French Revolutions, the ideas of three centuries in some ways had their potential exhausted: something was closed, and something new was opened. The rise of Napoleon and the long disaster of his wars showed that liberty was no longer a self-contained and self-sustaining idea. On the scale of the modern state, which dwarfs the tyrannies of Sforza and Cromwell, a new political age had begun. The experience of the French Revolution and of Napoleon, the philosophy of a Hegel, marks the transition from dynastic empires and tradition to nation states and charismatic authority, from revolution to the possibilities of dictatorship; and because these movements and ideas contain the seeds of all totalitarian states, of the right and of the left, that age seems to us a natural close to the book—and a natural window onto our own times.

VI

We have divided the four centuries of the book into three swaths. They are not those into which the period is usually divided, but they seem to us more expressive of the totality of their ideas. The first swath we have called "The Expanding World." It runs from Leonardo in fifteenth-century Italy to the late climax of the Elizabethan age in England, say to the decade marked by the publication of Shakespeare's folio in 1623. This period, then, covers the Renaissance, the Reformation, and the rise of science and early industry.

The next swath of history begins at the first major revolution in the world, the resistance to Charles I from about 1630. It runs through the settled calm of the merchant empires at the turn of the seventeenth century, and ends in the slow decline of their prosperity and stability —to about the time of the troubles of the East India Company and of the French exchequer about 1760. This we have called "The Age of Reasoned Dissent." Much of it is more familiar as the Age of Reason, and we have coupled with that concept the neglected but powerful movement of dissent which runs from the Puritans to the rising nonconformists in religion and in politics in eighteenth-century England and America.

The third swath in our history we have called "The Great Revolutions," and here what is characteristically modern in our society becomes concrete. We have given more detail and attention to the Industrial Revolution than is usual, because the important social and economic ideas that have influenced us have their roots in it—the ideas of Adam Smith, for example, and of Jeremy Bentham. Theirs are still powerful ideas, though changed by the setting of industry; so, too, both the American Revolution and the French Revolution threw up ideas which are still omnipresent. The sense that civilization has become a burden, the search for a fresh inspiration in nature, is one of these ideas, which links the two revolutions intellectually. We show this carried into the simple socialism of Robert Owen, which has given the English labor movement a different character from that of the continent of Europe. The other shaping idea is the romantic conservatism of Burke, which finds a rational basis for its opposition to the Revolution in the rejection of rationalism as a foundation for politics. When to this is added the work of Hegel, we have the historical process elevated to the rank of a goddess; and it is a nice irony of the goddess that she, who pleased Burke so much, at last becomes the inspiration of Karl Marx. History grows to an absolute value—a mysterious power and a mystic study; and it is appropriate that a book of history should stop then, when the writing of history first becomes a natural expression of creative minds.

VII

We have explained the general grounds of our choice in including some men and groups, and leaving out others. Why have we passed over the Counter Reformation? Why have we ignored Spinoza and some other philosophers? No doubt we have been guided by our interests, but in the end the answer must be found in the perspective of the whole book. We believe that the choice that we have made gives a coherent and living map of the intellectual history which has formed the ideas by which we live today.

The spirit in which we have written this book is empirical. To both of us history has been an instrument of vision; in our own studies, we have learned by its means to see as differently as one sees through a microscope. Both of us have experienced the study of history as a liberation—a liberation from accepted ideas, and a perspective into their evolution which brings them sharply into focus. History liberates because it refines our understanding of men, of ideas, and of events. This intellectual history is written in that spirit.

PART I

THE EXPANDING WORLD:

From Leonardo to Galileo, 1500-1630

CHAPTER 1

Leonardo and His Times

THE IDEA of the Renaissance is a singularly complicated one. Historians differ sharply as to when it began and when it ended. Some see its beginnings as early as the twelfth or thirteenth centuries; others prolong the Middle Ages until as late as the seventeenth century. Some historians date the Renaissance from the fall of Constantinople to the Turks in 1453, on the grounds that this drove Greek scholars westward into Mediterranean Europe; others hold that the Renaissance was really set in motion by the rapid printing of books from movable type, which was introduced about 1451 and became common fifty years later.

These differences in dating arise, naturally, from different conceptions of the character of the Renaissance. For example, did it involve, among other things, the birth of "modern man" and the emergence of individualism? Should the rise of empirical science and the close attention to nature be regarded as part of it? And, on a more trivial level, did it include a new love of mountain climbing? Obviously, one's decision as to what the Renaissance *was* affects one's idea of when it *occurred*.

Our purpose in this chapter, however, is not to attempt a frontal attack on the entire problem of the Renaissance;[1] it is, rather, to look

[1] The best surveys of this large debate are to be found in Wallace Ferguson, *The Renaissance in Historical Thought* (Boston, 1948), and in the bibliography given by Federico Chabod, *Machiavelli and the Renaissance* (London, 1958). Jacob Burckhardt gave the classical picture of the Renaissance as the birthtime of modern man in *Civilization of the Renaissance in Italy* (1860). According to Burckhardt, Italy of the fourteenth and fifteenth centuries witnessed: the emergence of a secular concept of the state, "the state as a work of art"; a stress on the develop-

upon the period around 1450–1500 as one of transition between the classical beginnings (leaving this date vague) of the Renaissance and its popular spread by the printed book. We thus make a separation: between the aristocratic Renaissance, with, for example, its reading

---

ment of the individual, i.e., a new attention to fame, glory, and the expression of personality; a discovery of the world based partly on the new voyages of exploration and partly on the new work in natural science; and a discovery of man, involving a new psychology and a new concept of humanity. The society which stood behind these novel attitudes was strongly influenced by the revival of antiquity, but this was merely one element in the new world it was creating—a world marked by increased violence and increased doubt and skepticism. Throughout, Burckhardt's emphasis was on the "newness" either of the particular elements or of their synthesis; his tendency to neglect the historical origins of the "new" attitudes derived in part from the fact that he was attempting to depict a *Zeitgeist* rather than to write a narrative history. Thus, one of the major attacks on the Burckhardt thesis has centered on his view of the Renaissance as a sudden flowering; and much recent scholarship has attempted to show the links between the fourteenth and fifteenth centuries in Italy and the centuries which preceded them. The continuity of the Middle Ages and the Renaissance is stressed and the "newness" of Renaissance characteristics is played down (See Chabod's bibliography, pp. 215–216, for details). The extreme of this has been to credit earlier "Renaissances" (for example, a Renaissance in the twelfth century) with most of the foundations of modern Western civilization. Another significant attack on the Burckhardt thesis puts forward the notion that humanism was really a reactionary movement and fundamentally opposed to the new emerging science. Along these lines, for example, Leonardo Olschki contends that the new science was not the work of the humanists—remote from reality—but of practical men: craftsmen, engineers, and artists.

It might be well, here, to indicate briefly our position in these matters. First, it is not the purpose of our book—nor could we hope to do it in the framework that we have set up—to write a history of the Renaissance or even to argue the various positions concerning the concept of the Renaissance. We believe that a history of the Italian Renaissance *would* show the continuity between the Renaissance and the Middle Ages and, therefore, would lessen the emphasis on the "newness" of the period and its culture. This is a correct view, from that perspective of the historian. On the other hand, we believe it a mistake to go so far in this direction as to lose all sight of the originality and fundamental modernity of the culture produced in Italy at this time; we take the position that the Renaissance *is* the origin of the modern world, without overlooking the fact that the Renaissance is, in turn, and quite naturally, based on the Middle Ages. We believe, too, that there were at least two stages in the Renaissance—what we have called an aristocratic, idealistic stage and a more popular, empirical one—and that what is called humanism was involved in a very complex way in both. Insofar as our book starts with Leonardo and the period of the Renaissance, we tend, for obvious reasons, not to talk about the historical links with medieval culture; we are quite clearly stressing certain dominant traits which we regard as dominant because we are looking toward the modern "future" and not back to the medieval past.

of the Greeks and Romans in manuscript and its taste for a curious Platonic idealism, as discussed in the Platonic Academy at Florence; and another kind of Renaissance which followed or supplanted it—a popular, empirical, less traditional and hierarchical, and more scientific and forward-looking Renaissance.[2] To show concretely this change in the aspect of the Renaissance, we take a single figure, Leonardo da Vinci.

I

Leonardo lived from 1452 to 1519. He was born near the small town of Vinci, which lies between Pisa and Florence in North Italy. His father was a young lawyer; we cannot be sure who his mother was, but it is likely that she was a village girl, called Caterina. Leonardo's father did not marry his mother; instead, he married into a good family of Florence. This marriage was childless, and so were two later marriages; only in his last marriage, more than twenty years afterwards, did Leonardo's father have children again—about a dozen.

Leonardo was thus his father's only child all through his formative childhood and youth. He was taken into the family, in the house of his grandparents, when very young; and his mother was probably a servant in the house.

What effect did Leonardo's strange upbringing have on his life? What did his age think about his illegitimacy? The answer is that illegitimacy, itself, was a commonplace of the time. Men were proud of making their own way, and cardinals, condottieri, and well-known artists, such as Leonardo's forerunner, Leon Battista Alberti, boasted that they were born out of wedlock. Indeed, these men frequently wielded a power which was as illegitimate as their birth. Theirs was an age in which power was often personal, and usurpers existed at the head of many states. As the great historian of the Renaissance, Jacob Burckhardt, said: "The fitness of the individual, his worth and capacity, were of more weight than all the laws and usages which prevailed elsewhere in the West."

[2] Hiram Haydn makes a division between a Renaissance and a Counter-Renaissance; see his book *The Counter-Renaissance* (New York, 1950). Theodore Spencer discusses something similar to this change in his book *Shakespeare and the Nature of Man* (New York, 1942), where he sees the basic pattern of Elizabethan thinking questioned by Copernicus in the cosmological, Montaigne in the natural, and Machiavelli in the political order.

It is, therefore, strange that Leonardo's life suggests that he took his irregular birth and boyhood amiss. Something withdrawn in Leonardo's character, his remote and secret air, his lack of male sensuality seem to be the marks of a divided childhood. There is in his actions always some awkwardness, an unspoken but deliberate opposition, which calls up a picture of a silent and willful boy in a home where his mother is not in her rightful place.

II

When he was about 14, Leonardo was taken by his father to Florence and apprenticed to the distinguished artist Andrea del Verrocchio. Florence was then, about 1465, ruled by the heads of the banking family of the Medici.

For such men, grown rich by trade, Verrocchio and others were making works of art: painting and sculpture, tableware and ornaments. An artist's workshop was a shop, and it was as much his business to make a golden chafing dish as to paint an altarpiece, and to design a chalice as to make improper drawings for the boudoir of a cardinal's mistress. Verrocchio was, in fact, famous as a goldsmith as well as a painter.

This is the setting in which Leonardo became a man: tall, strong, handsome, well known for having a fine singing voice, and endlessly gifted. When he completed his apprenticeship, about 1472, he was the leading painter in Verrocchio's workshop. We know that Verrocchio himself gave up painting, and had probably done so by this time. The story goes that he did so because the young Leonardo, in helping him with a commission, painted a more lifelike angel than his master.

Such a story is told of other painters too; for example, of Raphael; it is a characteristic Renaissance story. Here is an age in love with the unexpected, which wants to discover the wonder of childhood and the wonder child. Genius must burst on the ordinary air, native and untaught; and it must instantly convert all those who behold it.

This story is appropriate to its age for another reason. The Renaissance artist could do many things, but his workshop ran better if he did what he could do best. The age was discovering the advantages in the division of labor. If Verrocchio had an apprentice who painted well, it was to everyone's advantage; and Verrocchio would gladly leave him the painting and take over something that no apprentice could do.

Yet, when this has been said of the story, we have still to say that it may actually be true of Leonardo. For we have the painting which bears it out: it is the *Baptism* by Verrocchio. Among the rather stiff figures in it, there is one unlike the others: a curly-haired angel that has come to life, so that it is no longer an angel but a child. No previous figure in Italian painting has this tender yet distant, this dedicated touch. And more than the angel, it is the grass and the rocks that make us guess the hand, at once warm and inhuman, of the young Leonardo: the accurate vision, the loving detail of a man who is fascinated to watch a blade of grass push up through the earth and begin to grow.

III

In the official documents of the time, Leonardo in his twenties is mentioned twice: once as having completed his training, and once as having been accused with other young artists of making improper advances to a male model. Although he was acquitted at a trial, many of his contemporaries still entertained suspicions that he was homosexual. Later in his life, the only entries in his diary and the only drawings which show a personal passion concern a man: his pupil Giacomo Andrea Salai, whose curly head and lost looks he went on drawing for twenty-five years with contemptuous tenderness.

In one particular passage in his notebooks, Leonardo interrupts a scientific description of the flight of the vulture to recall, suddenly, a childhood dream. This is the dream: "It seems that it had been destined before that I should occupy myself so thoroughly with the vulture, for it comes to my mind as a very early memory, when I was still in the cradle, a vulture came down to me, opened my mouth with his tail and struck me many times with his tail against my lips." Sigmund Freud interprets this dream as evidence that Leonardo was a latent homosexual. The exegesis of Freud is brilliant and original in the extreme; it has the same flavor as a detective story in which one unlikely clue after another is linked together to provide the convincing, though improbable, solution. However, arresting as is Freud's interpretation, it has been seriously challenged and must be held to be based on tenuous evidence.[3]

We are on firmer ground when we say that Leonardo was left-handed. The shading in the pictures of a right-handed man runs from the bottom left-hand corner to the top right-hand corner; in Leonardo's pictures, it runs the other way, from the bottom right-hand corner to the top left-hand corner. The many notebooks he left behind are in mirror writing, clearly formed but running from right to left, and this is one reason why they were unread for more than two centuries after his death. Nevertheless, it is possible that Leonardo did the more delicate parts of his drawings and paintings with his right hand and only used his left hand for the coarser daily work, because he had injured his right hand in Florence.

Leonardo's life and work, alike, are characterized by a kind of impatience and self-assertion, which also typify the age of which he is such a dominating figure. He was self-taught and self-willed. He did everything, almost truculently, in his own way. Some of his pictures have perished by ill luck, but others more simply because he insisted on mixing the paints with unusual ingredients or on drying them in odd ways. He never followed the career that was expected of him, and seemed almost to need to be at odds with everything that he could do best. In a profound sense, Leonardo was a perverse man.

[3] The quotation is from Sigmund Freud, *Leonardo da Vinci: A Psychosexual Study of an Infantile Reminiscence,* tr. by A. A. Brill (New York, 1916), p. 34. The original statement by Leonardo is in the *Codex Atlanticus.* For a criticism of Freud's entire interpretation, see Meyer Schapiro's excellent article, "Leonardo and Freud: An Art Historical Study," *Journal of the History of Ideas* (April, 1956). Schapiro's main points are that Freud has used a faulty German translation of some of the key words in the original Italian, and that he has been misled by insufficient knowledge of the art history of the period into attributing to Leonardo's psyche what were merely standard characteristics of the art of the period.

General treatments of Leonardo's life, in English, are Vasari's *Life,* reprinted in *Leonardo da Vinci* (London, Phaidon Press, 1943), R. Langton Douglas' *Leonardo da Vinci: His Life and His Pictures* (Chicago, 1944), which contains a bibliography and plates, and Antonina Vallentin's *Leonardo da Vinci: The Tragic Pursuit of Perfection,* tr. by E. W. Dickes (New York, 1938), which, although it is written as fiction, is based on solid reading in the known material about Leonardo. Much of *Leonardo da Vinci: The Artist and the Man,* by the Swedish writer Osvald Sirén (New Haven, 1916), is outdated, and he is overly imaginative about Leonardo's life, but his discussion of the paintings is of interest, as are the plates adjoined to the book. The last chapter, "A Sketch of Leonardo's Personality, Character, and Views of Life," is also worthy of attention.

IV

In 1481, Verrocchio went to Venice to work on his great statue of the condottiere Colleoni on horseback, which still stands there. There was no plain reason why Leonardo should have left Florence. Yet he did. Already famous as an artist, he wrote the following letter to Lodovico Sforza, the usurper of Milan, nicknamed the Moor.

Most Illustrious Lord, Having now sufficiently considered the specimens of all those who proclaim themselves skilled contrivers of instruments of war, and that the invention and operation of the said instruments are nothing different from those in common use: I shall endeavour, without prejudice to any one else, to explain myself to your Excellency, showing your Lordship my secrets, and then offering them to your best pleasure and approbation to work with effect at opportune moments on all those things which, in part, shall be briefly noted below.

(1) I have a sort of extremely light and strong bridges, adapted to be most easily carried, and with them you may pursue, and at any time flee from the enemy; and others, secure and indestructible by fire and battle, easy and convenient to lift and place. Also methods of burning and destroying those of the enemy.

(2) I know how, when a place is besieged, to take the water out of the trenches, and make endless variety of bridges, and covered ways and ladders, and other machines pertaining to such expeditions.

(3) Item. If, by reason of the height of the banks, or the strength of the place, and its position, it is impossible, when besieging a place, to avail oneself of the plan of bombardment, I have methods for destroying every rock or other fortress, even if it were founded on a rock, etc.

(4) Again, I have kinds of mortars; most convenient and easy to carry; and with these I can fling small stones almost resembling a storm; and with the smoke of these cause great terror to the enemy, to his great detriment and confusion.

(5) Item. I have means by secret and tortuous mines and ways, made without noise, to reach a designated [spot], even if it were needed to pass under a trench or a river.

(6) Item. I will make covered chariots, safe and unassailable, which, entering among the enemy with their artillery, there is no body of man so great but they would break them. And behind these, infantry could follow quite unhurt and without any hindrance.

(7) Item. In case of need I will make big guns, mortars, and light ordnance of fine and useful forms, out of the common type.

(8) Where the operation of bombardment might fail, I would contrive catapults, mangonels, trebuchets and other machines of marvellous efficacy and not in common use. And in short, according to the variety of cases, I can contrive various and endless means of offence and defence.

(9) And if the fight should be at sea I have many kinds of machines most efficient for offence and defence; and vessels which will resist the attack of the largest guns and powder and fumes.

(10) In time of peace I believe I can give perfect satisfaction and to the equal of any other in architecture and the composition of buildings public and private; and in guiding water from one place to another.

Item. I can carry out sculpture in marble, bronze, or clay, and also I can do in painting whatever may be done, as well as any other, be he who he may.

Again, the bronze horse may be taken in hand, which is to be the immortal glory and eternal honour of the prince your father of happy memory, and of the illustrious house of Sforza.

And if any of the above-named things seem to any one to be impossible or not feasible, I am most ready to make the experiment in your park, or in whatever place may please your Excellency—to whom I commend myself with the utmost humility.[4]

This sober and prophetic list of inventions to make war was accepted by the Moor, and Leonardo went to the Court of Milan, carrying with him a silver lute which he had made for himself in the shape of a horse's head. He remained at that treacherous, turbulent court through the most creative years in his life, until Lodovico Sforza was deposed in 1499.

Why did Leonardo leave Florence? One reason was that Verrocchio had gone. So had the other known Florentine painters: Botticelli, Ghirlandaio, Perugino, and Cosimo Rosselli. They had gone to Rome at the invitation of Pope Sixtus IV to paint competitively in the Sistine Chapel. So Leonardo may have felt that Florence was no longer the center for artists which it had once been.

It may be that Leonardo went for no better reason than that Lodovico Sforza was looking for an artist to make a statue of his father on horseback. Verrocchio, as we have said, had gone to Venice to finish the statue of a mounted condottiere. Leonardo was not a sculptor, but he was a man who had to do everything, and do it better. The father of Lodovico Sforza had been a greater condottiere than Col-

[4] Letter from Leonardo to Sforza, to be found in Jean-Paul Richter, *The Literary Works of Leonardo da Vinci,* 2 vols., (London, 1883), Vol. II, pp. 395–398.

leoni, and perhaps Leonardo could not resist the itch to set up his statue against Verrocchio's.

Perhaps so; but there was also a deeper reason. Florence under the Medici was a city of tradition. Here was the classical Renaissance: the beautiful libraries, the Greek and Roman texts in manuscript, the Platonic Academy set up earlier by Cosimo de' Medici. It was graceful, literary, and derivative; its golden dream was of the past.

This first form of the Renaissance was now changing to another, and Leonardo personifies the transition. It is the transition from a classical to a popular, from an idealistic to an empirical Renaissance; from a worship of past humanism to a fierce belief in the human present. The Medici had given up the medieval ideal of an unapproachable godhead; but man and nature were still remote ideals to them, to be found mainly in books. The self-made men of the new Renaissance, however, wanted to grasp man and nature through the senses, physically, in handfuls. As Leonardo wrote in his notebooks: "He who has access to the fountain does not go to the water-pot."

By the standards of Florence, Leonardo was in fact an uneducated man. He taught himself Latin later, in his forties in Milan, but he never learned Greek. To the aristocratic bankers of Florence, proud of their schools, Leonardo was an unscholarly and unlettered painter.

Things were different in Milan. There the court had for some years been ruled by a family of condottieri, the Sforzas. They lived by their wits and by their popularity. The first Sforza had left his son three maxims: "Let other men's wives alone; do not strike your followers; and do not ride a hard-mouthed horse." The grandson, Lodovico, still boasted that he was a self-made man, and claimed kinship with all who, like himself, stood on their personal merits—with scholars, poets, artists, and musicians. Leonardo, who went to Milan when it was just becoming the center of the new printing in Italy, was attracted to this ruffian who was the son of his own deeds.

Leonardo, too, was interested in what was new. He was not willing to look back: his look was outward and forward into nature. He looked at her with two passions: a passion for the exact, which turned him toward mathematics, and a passion for the actual, which urged him to experiment. These two strands, the logical and the experimental, have remained the two sinews of the scientific method ever since.

Leonardo had already begun anatomical studies in Florence, and carried out elaborate dissections. With his extraordinary camera of an eye that could stop the hawk in flight and fix the rearing horse, he saw everything, and saw it precisely. It was the detail, the articulation of nature which fascinated him. Leonardo was the first to see that the detail has a meaning; and he wanted to find the meaning. Step by step he wanted to observe, to discover, and to invent. There is a mounting impatience, in Leonardo's notes as well as in his drawings, with all that merely catches an effect—a movement or a likeness—of light, of water, of face or flower. He wanted to understand and uncover, layer by layer, the muscle under the skin and the bone under the muscle. He was in a rage to know.

There is a beautiful portrait, which Leonardo painted soon after he came to Milan, of a mistress of Lodovico Sforza holding in her arms his emblem, a stoat. She was probably Cecilia Gallerani, so that the stoat with her is also a Greek pun on her name. Yet, behind that smooth Renaissance wit, the picture is more cruel than either of the delicate and handsome, yet stupid, heads that it paints, because it matches the skull behind the girl's temples with the stoat's, and the bones of her hand with his paw. The man who painted it is less an artist than an explorer; and the portrait is less a likeness than a work of discovery—an emblematic research into anatomy and character together. Indeed, if one looks closely at the painting, there is a sense in it in which the skull of the stoat and the skull of the girl, both looking the same way, are so alike that the whole theory of evolution is, as it were, contained in the picture.

In Milan, Leonardo was free to follow his interest in science wherever it led him. These were the mature and active years of his life, from the age of 30 until he was nearly 50. The scope of his researches was prodigious. There are, first, his anatomical drawings, for example of the hollows and blood vessels in the head. These are so exact that, even today, it is striking to compare them point by point with X-ray photographs and with photographs taken with radioactive tracers. It is worth citing at length a characteristic quotation from the notebooks, to show how absorbed and meticulous Leonardo was in this work, and how intently he looked for the mechanism behind what he saw.

And this old man, a few hours before his death, told me that he had lived a hundred years, and that he did not feel any bodily ailment other than weakness, and thus while sitting upon a bed in the hospital of Santa Maria Nuova at Florence, without any movement or sign of anything amiss, he passed away from this life.

And I made an autopsy in order to ascertain the cause of so peaceful a death, and found that it proceeded from weakness through failure of blood and of the artery that feeds the heart and the other lower members, which I found to be very parched and shrunk and withered; and the result of this autopsy I wrote down very carefully and with great ease, for the body was devoid of either fat or moisture, and these form the chief hindrance to the knowledge of its parts.

The other autopsy was on a child of two years, and here I found everything the contrary to what it was in the case of the old man.

The old who enjoy good health die through lack of sustenance. And this is brought about by the passage to the mesaraic veins becoming continually restricted by the thickening of the skin of these veins; and the process continues until it affects the capillary veins, which are the first to close up altogether; and from this it comes to pass that the old dread the cold more than the young, and that those who are very old have their skin the colour of wood or of dried chestnut, because this skin is almost completely deprived of sustenance.

And this network of veins acts in man as in oranges, in which the peel becomes thicker and the pulp diminishes the more they become old. And if you say that as the blood becomes thicker it ceases to flow through the veins, this is not true, for the blood in the veins does not thicken because it continually dies and is renewed.[5]

What Leonardo looked for when he drew men and animals and plants was the structure, because to him this was how Nature showed her meaning: she expressed the purpose in the structure. He was looking for the mechanism which moves the creature.

Instrumental or mechanical science is the noblest and above all others the most useful, seeing that by means of it all animated bodies which have movement perform all their actions; and the origin of these movements is at the centre of their gravity, which is placed in the middle with unequal weights at the sides of it, and it has scarcity or abundance of muscles and also the action of a lever and counter lever.

[5] *Extracts from Leonardo da Vinci's Note-Books*, ed. by Edward McCurdy (London, 1908), pp. 78–79.

It is only a step from this to inventing a flying machine, and a dozen other machines. Some of these, of course, were impracticable, because that age did not command (and did not understand) the mechanical energy needed, for example, to fly—though Leonardo did invent a screw helicopter and a parachute which work. But most of his machines which control an action or a process are thoroughly practical. There are in his notebooks machines for grinding needles and mirrors and for cutting screws and files, there is a rolling mill, and a special lathe, all of which work and all of which are original. There are lock gates and excavators and girder bridges; there are instruments for measuring wind and water; there is the clock pendulum. And there are others, particularly engines of war, of whose originality we are less certain, yet to which Leonardo certainly added some neat device from his own springing invention.

V

Leonardo's life at the court of Milan kept him busy with other schemes, too. He designed buildings and canals, he made surveys and maps, he was expected to mount the elaborate masques with which the courts of the Renaissance decorated a public occasion. He is said to have made a mechanical lion to welcome the King of France, in order that it would spill a shower of lilies from its breast at his feet.

Behind all this activity, however, stood one shadow: the unfinished commission for a monument to Lodovico Sforza's father. As always, Leonardo was full of plans and preparations. He made numberless sketches of a horse—for, to him, that was the center of the monument, which he simply called "the horse." The horse on which Verrocchio had mounted Colleoni was one of the first to raise one foot from the ground; Leonardo was determined to have a horse rearing with both forelegs in the air. This scheme posed many mechanical difficulties, for example, in the casting: at once, Leonardo began to sketch devices for casting and transporting the horse. Meanwhile, the years passed, and the statue remained a series of sketches.

Something had to be done, after ten years, in 1493 at the ceremony to usher in the wedding of Lodovico's niece to the Emperor Maximilian. Leonardo made a model of the statue, full size, in clay. With that, the bronze to cast the statue was got together too, at last. But the metal had to be sent off again next year to found cannon in order to help Lodovico's allies. Lodovico was now intriguing with one city-state after another, and was inviting the French and even the Turks to help him. The French came and marched up and down Italy; but the tyrant whom they deposed at last was Lodovico the Moor himself. When the French archers conquered Milan, they used the clay horse as a target, and gaily shot it to pieces.

Leonardo fled from Milan when the French came in 1499. Characteristically, he left in the company of Luca Paccioli, who had made important advances in algebra and in the mathematics of gaming, and who was now writing a book on geometry for which Leonardo was drawing the diagrams. Nothing remained of the full years in Milan but the *Last Supper,* which began to molder on its damp convent wall before Leonardo died. The picture, which has been repainted so many times since that it is impossible to conceive that any single touch of paint that Leonardo put on remains, was one of the first paintings of the *Last Supper* in which Judas sat on the same side of the table as Christ. The decision to bring Judas around to the same side as Christ produces a heightened dramatic and artistic effect; it accords, psychologically, with what we conceive to have been the real relationship of the betrayed and his betrayer.

VI

With the fall of the court of Milan, Leonardo's life seemed to fall to pieces. He wandered from one unhappy task to another. In 1502 he became military engineer to Cesare Borgia for a short time, in one of the most treacherous of Borgia's campaigns (in fact, the one so brilliantly described by Machiavelli). Then Florence recovered from the long hysteria during which the monk Girolamo Savonarola had expelled the Medici and burned their wonderful treasures.[6] Thereupon Leonardo and his young rival Michelangelo were commissioned by Florence to paint two patriotic pictures, neither of which was finished. Working now and again, arguing about the money, Leonardo seems to have spent his leisure time for four years in painting the third wife of an obscure merchant. The picture is the *Mona Lisa,* which still faintly glows through a green sea of varnish. Freud makes the interesting suggestion that all the smiling women—St. Anne, the

[6] Cf. Chap. 6, "The Reformation."

Virgin, Mona Lisa—who occur in Leonardo's pictures were attempts to capture the peculiar air of tenderness and humility, which the memory of his mother, the servant-girl Caterina, called to his mind.

From time to time, Leonardo went back to Milan to plan, of all things, a monument to the condottiere Trivulzio in the service of the French who had defeated Lodovico Sforza. But nothing came of this either. Then, in 1513, the son of Lorenzo (Medici) the Magnificent was elected to be the Pope Leo X. He invited Leonardo to Rome, where Raphael and Michelangelo were now painting, and offered him a commission. It is said that Leonardo at once began to make the varnish for the picture, and that Leo X said, sadly and rightly, "This man will never do anything, for he begins to think of the end before the beginning."

At last, in 1516, Leonardo was offered a retreat from this aimless and restless life. The king invited him to France, and there he remained in the country seat of Charles d'Amboise until he died in 1519. Though the setting was peaceful, his mind was not at peace. The later notebooks are filled with whirling patterns of storm and flood, clouds and waterspouts, in a vision of apocalyptic ruin. And on page after page, one phrase is scribbled: "Tell me if anything at all was done. . . ." "Tell me if anything at all was done. . . ."

Did Leonardo fail? His contemporaries did not think so; and not fifty years after his death, in 1568, he was a hero to his first biographer, Giorgio Vasari. True, he did much less than he might have done, and much of what he did was destroyed or forgotten. Yet he contained in himself, and brought to life, the aspirations of an age. In this sense, he was the Renaissance man.

He was, first, the boy wonder: the personification of the belief in the native genius of man. His story is that of the painter who, still in his teens, enters the studio of a distinguished artist and immediately outshines him. As the genius leaping perfect into the world, he embodied and made real the Renaissance feeling that every individual carries in him unlimited potentialities and requires, not an elaborate indoctrination or a lifetime of monastic devotion to one kind of work, but simply the proper environment in which, like a flower, he can unfold. It was this feeling of the way all human achievement is contained in the individual which was one main point in Renaissance belief.

Second, Leonardo was the man of the people: what he saw he saw for himself, with little attention to the learning of ancient predecessors.[7] Indeed, the books of the great Greek and Latin authors were to him rather second-hand, already the water-pot which we quoted earlier, because he believed that the things about which they talked were discoverable in nature herself. His belief marks a critical change in the Renaissance. The humanists before him, by 1450, had super-

[7] Pierre Duhem, *Études sur Léonard de Vinci*, 3 vols., (Paris, 1906–1913), shows, however, that Leonardo was not totally unlearned and, in fact, used and copied the writings of many ancient and medieval thinkers. It was, therefore, a question of emphasis; and Leonardo emphasized the fountain as against the water-pot.

On the general subject of Leonardo and science, see John H. Randall, "Leonardo da Vinci and Modern Science," *Journal of the History of Ideas* (April, 1953), which attempts to evaluate the new and the old elements in Leonardo and emphasizes the difference between Leonardo's belief in the fundamental "identity between the microcosm of man and the macrocosm of nature" and the seventeenth-century scientist's divorce of man from nature. Specialized works on Leonardo as scientist are *The Mechanical Investigations of Leonardo da Vinci* (London, 1925), by Ivor B. Hart, which pays special attention to Leonardo's aeronautical as well as mechanical work, and contains a translation of his "On the Flight of Birds." Hart's introduction also contains a good analysis of why Leonardo wrote up his notes as he did. See, too, J. Playfair McMurrich, *Leonardo da Vinci the Anatomist* (Baltimore, 1930), and K. D. Keele, *Leonardo da Vinci on Movement of the Heart and Blood*, with a foreword by Charles Singer (London, 1952).

On the general subject of Leonardo's thought, see Gabriel Séailles, *Léonard de Vinci, l'artiste et le savant* (Paris, 1892), and Paul Valéry's famous *Introduction à la méthode de Léonard de Vinci* (Paris, 1919; there is an English tr., 1926). Edward McCurdy, *The Mind of Leonardo da Vinci* (London, 1928), promises more in its title than it delivers. Kenneth Clark, in the 1952 edition of his book, *Leonardo da Vinci: An Account of His Development as an Artist* (Cambridge), has a suggestive introductory note of four pages on the necessary connection of Leonardo's thought and art.

Leonardo's work as an artist is admirably discussed in Clark's book, which also has an excellent collection of plates at the end. Bernard Berenson has an important essay on Leonardo in his *The Study and Criticism of Italian Art*, Third Series (London, 1917). Walter Pater's *The Renaissance* (London, 1893), contains a famous essay on Leonardo, originally published as a separate article in 1869; as Kenneth Clark comments, however, Pater's essay is really on the Leonardesque rather than on Leonardo.

For illustrations of Leonardo's art the following are recommended. *Leonardo: Des Meisters Gemälde und Zeichnungen*, herausgegeben und eingeleitet von Heinrich Bodmer, Klassiker der Kunst series (Stuttgart, 1931), is a fine collection of reproductions. *The Drawings of Leonardo da Vinci*, with an introduction and notes by A. E. Popham (New York, 1945), contains almost all of Leonardo's drawings of aesthetic value; however, the reproductions are not always as fine as can be found scattered about in other, less complete, collections. Cf., for example, *Drawings of Leonardo da Vinci*, with an introduction by C. Lewis Hind (London, no date given).

seded in importance the medieval scholastics and their speculations and returned to the classic pagan authors—to Greek and Latin literature. Leonardo was one of the first to break with this break; his interest was not in the authority of the ancients but in a direct appeal to nature.

This brings us to our third point. Leonardo was the discoverer: the man who saw in the detail of nature the meaning which had been missed for centuries. He approached the world through his drawings, and showed nature's lineaments beneath the surface of his pigments. His new vision pierced to the structure of things which lay hidden behind outward appearances; he was concerned with the bone underneath the muscle as much as with the rendering of the skin color.

Was Leonardo a scientist? He found nothing that we should now call a scientific theory, because he lacked the gift to isolate those abstract concepts—gravitation, momentum, energy—in which science seeks the unity under the chaos of natural phenomena. His mind leaped to the concrete and the particular.

Yet, to an age still dominated by the traditional categories of Aristotle and St. Thomas Aquinas, he brought the right mind. When almost all thinking was still guided by universal and a priori plans of nature, he made a single profound discovery. He discovered that Nature speaks to us in detail, and that only through the detail can we find her grand design.

This is the discovery at the base of modern science, all the way from atomic structure to genetics. In the nature of things, this discovery had to be made by an artist. The Renaissance painters before Leonardo had already taken the first step in it; they had shown that the detail of nature marks one scene from another and gives meaning to each. What Leonardo did was to take this discovery from the studio into the laboratory. He made the artist's eye for meaningful detail become part of the essential equipment of the scientist.

Leonardo's insight was always instant and astonishing. Before Copernicus, he wrote, "The sun does not move." He thought of sound and perhaps of light as waves. He understood, before Galileo, that perpetual motion is impossible. He read the rings in trees and the history of fossil shells.

Stubborn, prodigal, and perverse with his gifts, balked, Leonardo lived in the richest and in the most menacing age of Europe. There was suddenly disclosed to the men of his generation a store of wealth and power in the world which they were too stunned and intoxicated to use well. For this the condottieri fought and marched, for this popes bribed and princes poisoned, and for this artists fawned and played away their lives at the courts of their unruly masters. Leonardo, too, was fascinated and dominated by power in others; he lay under the spell that has bound men for 500 years, so that they cannot tear themselves away from the loved and brutal image of the gangster and the tyrant.

Leonardo left fewer than 20 paintings, not a whole statue, machine, or book, and 5000 pages of notes and sketches which lay unread for 250 years. His way of painting had a lasting influence, his dissections some, and his inventions none. Raphael and Dürer learned from him, he was the friend of Machiavelli and Paccioli and the contemporary of Martin Luther and Christopher Columbus. In that heady, modern-seeming age, he was the prototype of the inveterate explorer of the unknown, the inspired man of genius who gazed in a new way at the microcosm within and the macrocosm without.

CHAPTER 2

The City-States of Italy

I

THE WORLD of Leonardo is familiar to us in a way in which the worlds of St. Thomas Aquinas and of Geoffrey Chaucer, for instance, are not. We recognize it as modern. We recognize the problems and attitudes of the close of the fifteenth century as similar to our own. The Renaissance is the birth of modern history, and the Italian Peninsula is its birthplace.

The hierarchy of the feudal system was largely absent here. In the early fifteenth century, the life of St. Joan in Northern France was still shaped by feudal custom, by church law, and by traditional status. In Italy, fifty years later, Leonardo was bound by none of these. The old habits and links in society were gone, and, in the process of forming new ways, every part of society was elbowing every other for power.

The concentrations of power in the Italian Renaissance were the city-states. The most important of these were Milan and Venice in the north, Florence, and Rome and Naples toward the south. As Jacob Burckhardt put it, in these cities (with the exception of Rome and, perhaps, Naples) the state became a work of art. It ceased to be dominated by the authority of the church. Indeed, in the struggle between the city-states, the papacy itself became transformed into a secular power.

The new state was no longer molded and guided by custom, but by men. Therefore, the state as a work of art created (because it had to be created by) a new art: the art of statesmanship. The first modern book on this subject was written in Leonardo's last years by Niccolò Machiavelli.

The dealings of states with one another, too, were no longer regulated by church and feudal authority. From the intrigues and alliances by which the city-states tried to outwit one another grew a second art: the art of diplomacy.[1] In these struggles for power, the concept of the balance of power was evolved, and with it the modern notion of a community of states.

II

Venice at the end of the fifteenth century had, it is estimated, about 200,000 inhabitants.[2] Milan and Naples had about the same number, and Rome rather fewer. Only Florence was much smaller, with a population of about 100,000.

These cities are, of course, dwarfed by modern cities, which we do not begin to think large until they have topped at least half a million. All the city-states together did not make up the population of a city like Boston today. Nevertheless, they were by the standards of the time sizable cities. In the setting of their time, their organization—for example, the system of canals in Venice—was a large achievement. We must remember that the population of Europe as a whole was then, at most, a tenth of what it is today.

Of the population of one of the city-states, perhaps 1000 to 2000 called themselves noble. Leonardo and other artists may have been leaders of the popular Renaissance, but it was this small aristocracy for whom they worked. It was not, however, a permanent aristocracy of birth, for it was constantly entered by the more successful merchants and bankers. This rising class of traders lay just below the aristocracy and formed the reservoir for it and for the counsels of government. Below this lay the larger classes of artisans and then of laborers.

[1] Cf. Garrett Mattingly, *Renaissance Diplomacy* (Boston, 1955).

[2] *Cambridge Modern History*, Vol. I, p. 257. See further, on the population question, Pasquale Villari, *The Life and Times of Niccolò Machiavelli*, tr. by Linda Villari (New York, Charles Scribner's Sons, no date given), p. 4. In the *New Cambridge Modern History*, Vol. I (*The Renaissance, 1493–1520*), H. C. Darby, in Chap. 2, cites lower population figures than ours; he suggests 100,000 for Venice, Naples, and Milan and 50,000 for Florence (pp. 46–47). Most authorities, however, prefer the higher figures we have given.

The city-states had prospered from a combination of causes, but one cause was central to the rest: they traded. Italy lay at the hub of trade: one could travel east to Asia Minor and the Far East and west to Spain, south to Africa and north to Europe. Above all, the rising trade between the East and Europe made her rich.

This trade created the merchant bankers, who risked and made their wealth in the cargoes of ships. It was a real risk, and money tied up in a cargo lay idle for a long time; to cover both these drawbacks, the profits were spectacular. As a result, the mathematics of interest, of insurance, and of gambling began in the Renaissance.

The medieval church had insisted that all money is equal and equally barren—the money tied up and risked in a cargo as much as money in the pocket. In the trading cities, this was plainly not true. Therefore the church laws against usury broke down. The feeling against the lending of money at interest died hard, as the story of the *Merchant of Venice* shows; but, as this story also shows, the man who had ships at sea could not afford to quarrel with the man who kept his capital liquid. The calling of banker became respectable.

Money made in trade became the new backbone of power in Italy. It provided the material conditions for the Renaissance, and the psychological climate as well. Making money was no longer regarded as somewhat disgraceful. The Medici in Florence carried proudly in their crest the golden balls which now hang over pawnbrokers' shops. The bankers of Venice, with soft eyes and dark beards, were painted with their jeweled hands counting money.

The Renaissance was also the beginning of modern capitalism. Capitalism can be defined as a system in which the individual seeks, consciously and rationally, by economic means to attain financial profit as his end. Thus, on the scale of individual business activity, rational methods of bookkeeping were introduced and involved banking operations were carried out. On the scale of the city-state, calculated economic imperialism marked the new age.

How wealthy were these city-states? In the year 1500 the treasury of Venice collected in export and import taxes, in guild taxes, and from its salt monopoly, a sum estimated at 1,145,580 ducats.[3] We can calculate, from this and other figures, that the national income of Venice must have been then about 10 million ducats a year. (And there was

[3] *Cambridge Modern History,* Vol. I, p. 278.

a national debt of the same size.) Over a population of 200,000, this gives an income of 50 ducats a head—that is, of about 200 ducats for each wage earner. The ducat was worth about 2½ gold dollars.[4] Thus the average family income in Venice in 1500 was about 500 gold dollars a year. This may be equivalent, at present prices, to 1500 to 2000 dollars a year.

Such figures, adjusted from the values of one age to another, must not be taken too literally. But they leave no doubt that Venice and the city-states were rich by the standards of any age. And in an age in which wealth was less evenly distributed than today, such an "average income" meant that the actual fortunes of individuals were fantastic. When Lodovico Sforza married his niece Bianca Maria to the Emperor Maximilian, he gave her as a dowry 400,000 ducats—about a million gold dollars.

It was with such sums in their hands that the great families of bankers and of warlords built the splendors of these cities. Florence and Rome are beautiful today because men with such fortunes displayed them in fine buildings and churches. One of the greatest of all the condottieri, Francesco Sforza, father of Lodovico, built the forbidding castle which, in its later form, still stands as an island of stone and green parkland in the center of the industrial city of Milan today. The display of wealth was a part of the mechanism of power. Luxury was less a reward than an expression of success; and the itch for luxury as a mark of social status was one of the drives of the Renaissance. Much of the trade on which the city-states thrived was, in fact, in luxuries.[5]

III

The banker often started as a trader. This was probably the case with the Medici. Purchasing goods in one town and selling them in

[4] Jacob Burckhardt, *The Civilization of The Renaissance in Italy* (London, 1945), p. 46 and especially the note by L. Goldscheider on p. 53 as to the present-day value of the ducat. Other estimates value the golden ducat at about 1½ gold dollars, and the silver ducat of Italy at 50 cents in gold. Carlo M. Cipolla, in *Money, Prices and Civilization in the Mediterranean World* (Princeton, 1956), however, suggests a higher value; that the ducat was pure gold of about 3½ to 4½ grams, against less than 1 gram in the U.S. dollar; see especially pp. 22–23. The florin of Florence was about equal in value to the ducat of Venice.

[5] See Werner Sombart, *Luxury and Capitalism* (New York, 1938, mimeographed), for the notion that luxury was one of the main causes of the rise of capitalism.

another, where a different sort of money prevailed, they began to transfer credit rather than coin at the conclusion of these transactions. Gradually, they entered upon branch banking: Venice, Rome, Pisa, Milan, and even Bruges and London became seats of their financial power. Princes of state and church became their debtors.

To the enterprises of banking and trading, the Medici added silk and wool workshops in their home town of Florence. When deposits of alum, necessary to fix dyes, were found at a site near Rome, the Medici entered into an agreement with the papacy and others to exploit them. Thus, a complex of banking, trade, and manufacturing was built up, and out of it a golden treasure made to run. A hint at Medici wealth is given in the statement that, from 1434 to 1471, they paid for charities, public buildings, and taxes no less than 663,755 gold florins; and Lorenzo the Magnificent "was delighted that the money had been so well spent."

Two figures stand large behind the age of Leonardo. They are its symbols in the adventure of power: the banker and the professional soldier—the Merchant of Venice and Othello.

For there was another way to get money: war. Soldiers could often win money and power over money for the man who led them. And the bankers and traders of the rivaling city-states were prepared to put their money out to fight for them, to protect them and to extend their city's prerogatives. There were statesmen—Machiavelli was one of them—who wanted to create a soldiery of citizens to hold what the city owned. But in a time of trade it was more natural to traffic in war, and to hire a body of mercenaries. Feudalism had left behind it many noble families with no other vocation than warfare, and they and their bands now came into the market.

Thus, there was created the unlovely but romantic trade of the hired soldier. He might be of any nationality: the Borgias were Spaniards, and the papal guard to this day is Swiss. He might fight for any nation: Trivulzio was an Italian, but he was fighting for the French when he defeated Lodovico Sforza and sacked Milan. In the twilight of the Renaissance, in 1527, German and Spanish mercenaries in the pay of the Emperor Charles V of the Holy Roman Empire sacked Rome with fearful cruelty.

We can see the career of the successful soldiers of fortune epitomized in the history of the Sforzas. The first of the family, Jacopo, made his own way in the bands of different condottieri until he became a leader himself. He was a man of rough strength and peasant looks who inspired confidence in those who paid or financed him and loyalty in those who fought for him. But, in fact, he rose because he gave his mind wholly to his profession. He and other condottieri once again turned war into an organized technique of skill, which Machiavelli later linked to the Roman military tradition and formalized in a book called *The Art of War*. At the same time, Jacopo Sforza shrewdly calculated each step in his career. The marriages of his relatives were planned to a nicety, and when he was himself ready for a princely marriage, he married off his mistress, Lucia, without a pang.

His son by Lucia was Francesco, perhaps the most famous of the condottieri. Like his father, he knew the value of confidence and loyalty, and knew also when the moment had come to abandon them. He was a skillful general and planner, and he made himself indispensable to the Duke of Milan and married his daughter. He himself thus became duke, and the Milanese were proud of his rule. Pope Pius II has left this description of Francesco Sforza as Duke of Milan.

> In the year 1459, when the Duke came to the congress at Mantua, he was 60 [really 58] years old; on horseback he looked like a young man; of a lofty and imposing figure, with serious features, calm and affable in conversation, princely in his whole bearing, with a combination of bodily and intellectual gifts unrivalled in our time, unconquered on the field of battle—such was the man who raised himself from a humble position to the control of an empire. His wife was beautiful and virtuous, his children were like the angels of heaven; he was seldom ill, and all his chief wishes were fulfilled. And yet he was not without misfortune. His wife, out of jealousy, killed his mistress; his old comrades and friends, Troilo and Brunoro, abandoned him and went over to King Alfonso; another, Ciarpollone, he was forced to hang for treason; he had to suffer it that his brother Alessandro set the French upon him; one of his sons formed intrigues against him, and was imprisoned; the March of Ancona, which he had won in war, he lost again the same way. No man enjoys so unclouded a fortune, that he has not somewhere to struggle with adversity. He is happy who has but few troubles.[6]

This was the man whose monument Leonardo was to have made. Alas, the children like the angels of heaven grew up to be tyrants in bitter rivalry. His oldest son was murdered after a senseless rule, and

[6] Quoted in Burckhardt, *op. cit.*, pp. 25–26.

the dukedom was usurped by a younger son, Lodovico, who then kept the rightful heir in prison and probably poisoned him.

Lodovico Sforza, dark enough to be nicknamed the Moor, had about him that sinister air of doom which often marks the last of a great and short line. He was as shrewd, as ruthless, and, for a time, as popular as his grandfather and his father. He was also as ambitious. At the head of a large city-state, he set himself to manipulate the politics of all Italy. At his height, he could boast that even the pope, Alexander, the father of Cesare Borgia, was his chaplain, and that marriage had made the Emperor Maximilian his condottiere. But what had been boldness in his forebears had now, with the stakes so high, become an overbearing recklessness. In the intrigues between the city-states, Lodovico called in the French as allies, and they did not scruple to betray him. Lodovico Sforza, ruler of one of the most brilliant, intellectual, and corrupt courts in Europe, was overthrown in 1499.

IV

War could be a costly trade. Lodovico Sforza had and used a yearly income of about 700,000 ducats—about 1¾ million gold dollars; and in his last years he came to fear his subjects because he knew that they murmured against his taxes and his wars. About the same time, in 1495, the Venetians, when they prepared for war, raised 15,000 horsemen: a sizable army, indeed. The 6 Venetian state fleets made a total of about 330 ships, and their crews numbered in all about 36,000.

War was knit into the fabric of the great city-states.[7] Then, as now, techniques devised for peaceful mercantile purposes were often diverted to military uses. For example, Venice had already standardized her ships so that spare parts could be stocked at her outposts in the eastern Mediterranean, and these could be used as much for her war as for her trading fleet. In the same way, much scientific and mathematical work, for example in ballistics, was prompted by the needs and the fascination of war. So, too, the crafts and skills of the sculptor could be employed in the casting of cannon as well as of statues, and the architect, like Michelangelo, turned his designing mind to fortifications instead of cathedrals.

[7] Cf. Richard Ehrenberg, *Capitalism and Finance in the Age of the Renaissance*, tr. from the German (London, 1928).

Yet war is an unproductive calling; and so, of course, is trade, in itself. The Renaissance was dominated and, in many ways, kept in motion by the merchant banker and the condottiere. But neither of them created wealth. There was indeed little wealth created at all, and most of that was produced in the form of foodstuffs. The Renaissance destroyed the narrow view of wealth of the medieval church, but it did not put in its place the modern view. Instead, it had the mercantilist view, that the total store of wealth in a society is constant and is measured by the gold it owns.

The lack of real wealth as measured in production, and the hankering after gold, took their revenge in the coming century from 1500 on. When the Americas were discovered, large quantities of treasure were imported into Europe. The import of gold thus outran the rate at which the community was producing new goods. The result was that Europe suffered from its first catastrophic inflation: an inflation produced not by printing paper but by minting gold, which in itself is every bit as worthless as paper. As a result, between 1500 and 1600, prices in Europe rose so steeply that at the end of the century they were between three times and four times as high as at the beginning.[8]

Problems such as inflation and monetary balance, war and trade, link our states to the city-states of the Renaissance. In general, the period of Leonardo and Machiavelli bore witness to the modern form of problems in politics and economics and in ethics: those concerning the relations of state to state, and of state to individual. Our "solutions" are different from those of the fifteenth century, but they are no more final. The problems are not yet solved, and part of the interest of history is in seeing how other times and places tried to solve them. The history of the 400 years between us and the Renaissance is the story of the attempts of man and society to come to grips with the problems then raised for the first time in modern guise.

[8] Cf. Earl Hamilton, *American Treasure and the Price Revolution in Spain, 1501–1650* (Cambridge, 1934).

CHAPTER 3

Machiavelli

I

NICCOLÒ MACHIAVELLI, who introduced a revolution in political thinking, is generally described as being unprepossessing in appearance and somewhat timid in conversation. Born in 1469, he lived in Florence under the rule of the Medici, till 1494, and then of Savonarola. The downfall of Savonarola opened the way for an effective republic. At this point, in 1498, Machiavelli achieved his first important post in the Florentine diplomatic service. As a young man of 29, he became secretary to the second chancery (concerned with home, war, and certain foreign affairs).

Although his post did not give him a very strong hand in the formulation of policy, it did give him a direct view of the political processes of the Italian city-states. And, as these political processes varied from oligarchy in Venice to monarchy in Naples and to democracy in Florence, they provided a veritable school of politics for an attentive observer.

Machiavelli's secretariat also involved him in a number of "foreign" missions. On one of these, he met the cruel, warlike Cesare Borgia, then only 26 years old, who was employing, as an engineer, Leonardo da Vinci. Later, Machiavelli was to make Borgia the hero of his book *The Prince.*

The Prince might never have been written, however, if the Medici had not been restored to Florentine leadership. With the fall of the republic in 1512, Machiavelli lost his post. Indeed, he was even tortured. Exiled from Florence, he retired to his little estate at San

Casciano; and, in this enforced leisure, he composed *The Prince.* In a wonderful letter to his friend, Vettori, he describes how,

> On the threshold I slip off my day's clothes with their mud and dirt, put on my royal and curial robes, and enter, decently accoutered, the ancient courts of men of old, where I am welcomed kindly and fed on that fare which is mine alone, and for which I was born: where I am not ashamed to address them and ask them the reasons for their action, and they reply considerately; and for two hours I forget all my cares, I know no more trouble, death loses its terrors: I am utterly translated in their company. And since Dante says that we can never attain knowledge unless we retain what we hear, I have noted down the capital I have accumulated from their conversation and composed a little book, *De Principatibus,* in which I probe as deeply as I can the consideration of this subject, discussing what a principality is, the variety of such states, how they are won, how they are held, how they are lost.[1]

Machiavelli wrote *The Prince* rapidly in 1513 and dedicated it to Lorenzo de' Medici (not Lorenzo the Magnificent, but his grandson) as a bid for Lorenzo's favor. Unread and neglected by Lorenzo, the volume circulated in manuscript, and was only published posthumously, in 1532.

[1] We use here the translation from Ralph Roeder, *Man of the Renaissance* (New York, 1933), pp. 284–285. For another version of this same letter, see the chapter on Machiavelli in Giorgio de Santillana's admirable book, *The Age of Adventure* (New York, 1956). For further evidence of Machiavelli's correspondence, see Orestes Ferrara, *The Private Correspondence of Nicolo Machiavelli* (Baltimore, 1929); in his commentary on the correspondence, Ferrara places the emphasis on "the human side" of Machiavelli. For some of Machiavelli's letters in an English translation, see Allan H. Gilbert, *The Prince and Other Works* (Chicago, 1941), which also includes "Reform in Florence," Castruccio Castracani," "On Fortune," and "Ten Discourses on Livy."

Jean Félix Nourrisson, *Machiavel* (Paris, 1875), has an interesting account of how *The Prince* was plagiarized and printed in 1523 by Augustino Nifo, nine years before the real *Prince* appeared. *The Life and Times of Niccolò Machiavelli,* by Pasquale Villari, tr. by Linda Villari, 2nd ed. (New York, Charles Scribner's Sons, no date given), is primarily concerned with an evaluation of Machiavelli's doctrines and their "scientific" worth. Villari believes that the Florentine was the first theorist of the modern state and the first truly scientific thinker on politics. Chap. 16 of Book I and Chaps. 2, 3, and 5 and the conclusion of Book II can bear special reading; they deal mainly with an analysis of Machiavelli's correspondence with Vettori, with the *Discourses* and with later criticisms of *The Prince.* Valeriu Marcu, *Accent on Power: The Life and Times of Machiavelli,* tr. from the German by Richard Winston (New York, 1939), is a colorful study of Machiavelli, emphasizing that the thirst for power is a constant in human affairs; the problem is to exercise power with justice and in the service of the law.

II

The Prince is the book for which Machiavelli is famous and to which the term "Machiavellism" refers. But it is by no means his only work. There are also the *Discourses,* from which, probably, the chapters for *The Prince* were taken; a *History of Florence;* the life of a condottiere named Castruccio Castracani; *The Art of War;* a first-rate play called *Mandragola;* and two or three rather poor farces.

All of these were the products of a humanist. Unlike Leonardo da Vinci, Machiavelli was a Latin scholar. Thus, his *Discourses* were based on the *Histories* of Livy, and his *Art of War* on the Latin author Vegetius. It was, therefore, as a practicing politician who was also a humanist that Machiavelli turned to the study of statecraft. As one modern writer has commented, Machiavelli represented "the culture that is born of humanism becoming aware of political problems because they are at a crisis. It is because of this that he seeks to solve them from the elements with which humanism has endowed the western mind."[2]

The elements of humanism in Machiavelli were twofold. One was a straightforward acceptance of the classical authors as models. The other was the adoption of a rigorously secular attitude to politics. In a real sense, we have here a fusion of attitudes from the aristocratic and the popular Renaissance.

III

Machiavelli may have differed from Leonardo in that he was a Latin scholar and a humanist, but he was like Leonardo in being a scientist. In fact, Machiavelli was the first social scientist, in the same sense that Leonardo was one of the first natural scientists of modern times.

[2] J. H. Whitfield, *Machiavelli* (Oxford, 1947), p. 18. Whitfield rejects the notion of Machiavelli as a "cold scientist" and seeks to interpret him as an "idealist, with dreams of improvement." Filled with highly individual and intensely held views, the book is very much worth reading. Felix Gilbert, "The Humanist Concept of the Prince and The Prince of Machiavelli," *Journal of Modern History* (Dec., 1939), is an important contribution on the subject of what is new in Machiavelli and what is merely the development of elements to be found in the "idealistic" literature of the humanists and in the "mirror-of-princes" literature of the medievalists. This article should be compared with Allan H. Gilbert, *Machiavelli's "Prince" and Its Forerunners* (Durham, N.C., 1938), which investigates the connection of *The Prince* with the mirror-of-princes literature.

He, himself, was aware that he was opening up a "new route."[3] For the first time in political thought, Machiavelli attempted, haltingly and hesitatingly, to look at his material from a scientific perspective. This meant that Machiavelli, like Leonardo, had to make a selection from the total facts of life; he had to decide what should be left out of the picture. The problem in both Renaissance art and science was "what to exclude."[4]

In Machiavelli's case, the thing to be left out of a consideration of politics was morality, or "what should be." The attention was to be firmly riveted to "what is." Thus, *The Prince* made a break with the whole tradition of what is called the "mirror-of-princes" literature; of sweet, pious advice to a ruler.[5] Over and over again, Machiavelli stressed the fact that he was not describing a good or honorable way to behave, or how society ought to be run. He was simply dealing with the question of how society *is* run and how people *do* behave (and, therefore, how a ruler must behave if he wishes to survive).

This was an extraordinary approach in what was really a prescientific age, in which only a few men like Jean Buridan, Tartaglia, and Leonardo da Vinci had looked in such a manner at natural objects. It meant that Machiavelli's choice of perspective had led him

3 These are the words Machiavelli uses in the introduction to the first book of the *Discourses*. The best edition is *The Discourses of Niccolò Machiavelli*, tr. with an introduction and notes by Leslie J. Walker, S.J. (New Haven, 1950). The notes do not run along with the text but are kept in the second volume. Father Walker's obvious prejudices do not mar his scholarly apparatus; but in the introduction he makes clear that the excuse of Machiavelli's critics, "that his writings are in conformity with the spirit of the age," is not his standpoint; in fact, that such a standpoint is morally wrong. A very convenient edition of the *Discourses*, along with *The Prince*, is the Modern Library version (New York, 1950), which has a stimulating introduction by Max Lerner.

4 Cf. Charles Singleton's article, "The Perspective of Art," *The Kenyon Review* (Spring, 1953). Mario Praz, in his British Academy lecture "Machiavelli and the Elizabethans" (London, 1928), makes the very interesting remark that "The discovery of the individuum was parallel to the discovery of the nude. . . . Machiavelli's hero is the counterpart of the nudes painted by Signorelli or sketched by Leonardo: he is a scientific being." (p. 9.) Praz's lecture is a fascinating piece of research and shows penetrating thought. It continues the investigation started by Edward Meyer in his *Machiavelli and the Elizabethan Drama* (Weimar, 1897), but emphasizes that much of the influence which Meyer attributes to Machiavelli should really be attributed to the example of Italian Senecan drama.

5 For a discussion of this literature, see Allan H. Gilbert, *Machiavelli's "Prince" and Its Forerunners*, as well as the other entries under n. 2.

to empiricism. Francis Bacon, who was both a moralist and a natural scientist, correctly recognized a forerunner and kindred spirit when he said, "We are much beholden to Machiavelli and others that wrote what men do and not what they ought to do."

The empirical and pragmatic, however, was only one side of Machiavelli's method. There was also the rational and a priori.[6] We see this best in the fact that Machiavelli tried to base his science on a number of general postulates.

The most important of his postulates concerned the nature of man. A century after Machiavelli, Galileo worked with the general postulate that "matter is unalterable" and of a like nature whether on the earth or in the heavenly bodies. In the same way, Machiavelli laid it down as a general rule that human nature is everywhere and at all times the same. The passions that moved men in former times are the same as those that move us now. It is for this reason, he believed, that examples drawn from the Romans could apply to modern times; we are dealing with the same "materials." The type of science that he wished to erect forced him to ignore the effect of culture upon men and to treat them, so to speak, out of history.

Having postulated that man's nature is everywhere the same, Machiavelli had to decide what were the characteristics of that nature. He decided that man's nature, while actually good and bad, had to be treated, for the purposes of politics, as bad. Thus, *The Prince* is filled with such statements as that "men are wicked and will not keep faith with you" and that "unless men are compelled to be good they will inevitably turn out bad."

It ought to be said that Machiavelli also saw the other side of man's nature. Although most commentators overlook this, Machiavelli (among others) anticipated Rousseau's notion that man is pure but is corrupted by civilization. This notion was implied when Machiavelli said, in the *Discourses*, that "if anyone wanted to establish a republic at the present time, he would find it much easier with the simple mountaineers, who are almost without any civilization, than with such as are accustomed to live in cities, where civilization is already corrupt." And it is stated directly in his comment that "the consti-

[6] Leonardo Olschki emphasizes a "speculative and scientific" side in *Machiavelli the Scientist* (Berkeley, 1945).

tution and laws established in a republic at its very origin, *when men were still pure* [our italics], no longer suit when men have become corrupt and bad."

Machiavelli, like Rousseau, might have tried to build a political system on the postulate of man's goodness; instead, he took as his theoretical assumption man's evilness. Perhaps Machiavelli chose the postulate of man's evilness because of his own bitter personal experiences as a Florentine diplomat. Or he may have elected evil because it was the dominant view of his times, especially as expressed in the theological notion of man's depravity.

However, the particular reason for his particular choice does not matter; the essential thing was that he needed a general rule on which to base his new science. Machiavelli faced the same necessity to make a one-eyed, general statement about man that we shall encounter later in other attempts at a social science: for example, Locke's "All men are equal by nature," or Smith's "Every man seeks his own interest." He needed to "exclude" all the extraneous, qualifying aspects of actual life so as to focus on the essential in politics. In this, he was like the physicist who looked at matter as colorless and tasteless so as to deal with it better as a mathematical entity.

Machiavelli's rationalism, however, collided with the empirical part of his method. His desire to state general rules and principles came into conflict with his attempt to pay attention only to "what was." He thought the facts confirmed his a priori postulates, but he neither drew his postulates from the facts nor tested the facts which supposedly tested the postulates. We see this in his use of Livy. Machiavelli used Livy, constantly, without any real attempt ever to check his historical "facts"; indeed, Machiavelli simply used the ancient histories as the "given." Thus he saw, not "what was," but what his postulates told him would be there. His speculative theory prevented him from looking closely at the facts, and the erroneous facts made a shaky foundation for his science.

Nevertheless, whatever criticism we make about Machiavelli's specific use of empiricism or rationalism, we must not lose sight of the fact that he grasped the essential point about modern scientific method—its combined use of empiricism and rationalism. As a result, he was more consciously "scientific" about politics than any previous thinker. Faulty, in parts, as his use of the scientific method was, he succeeded in fulfilling his boast to open a "new route."

IV

That Machiavelli was so blunt about the fact that he wished to talk about the "how" of statecraft and not whether his advice accorded with the Ten Commandments is striking. It meant, of course, the emancipation of politics from religious guidance. This was an extraordinary event. It set the tone for one of the main themes of modern history: the secularization of life and thought. We shall see this trend carried into field after field, whether it is in natural science with Galileo or in economic science with Adam Smith. Machiavelli's work is a first example of the way in which science and secularism have developed together in the West.

Machiavelli's secularism does not, of course, mean that he was antireligious, although he was anticlerical. He looked at religion as a social force rather than as a spiritual one. His "realistic" attitude toward religion produced the "objective" view that religion (i.e., Christianity) hinders the strong state by preaching meekness. At the same time, Machiavelli recognized that religion acts as a form of social cement which helps keep the state together. We read, for example, in the *Discourses,* that "as the observance of divine institutions is the cause of the greatness of republics, so the disregard of them produces their ruin; unless it be sustained by the fear of the prince, which may temporarily supply the want of religion." This is probably the first time in modern political theory that religion is looked upon primarily as a coercive and not a spiritual force, and it anticipates similar discussion by such diverse figures as Rousseau, Burke, and Marx.

Machiavelli had nothing against religion as a coercive force. In fact, his criticism of Savonarola was that the latter was an "unarmed prophet"; and he generalized his observation of Savonarola's downfall by commenting that unarmed prophets never succeed. Similarly, Machiavelli did not criticize the Roman papacy on religious grounds. Instead, he accused it of having failed completely as a source of moral inspiration and of having brought nothing but disaster to Italy, as a secular power. In short, the papacy had brought Italy to the edge of political and moral degradation by creating a power vacuum.

V

Machiavelli turned away from morality, religion, and the papacy toward the state as a thing-in-itself. We must not be misled by the title *The Prince.* While seeming to address an individual, Machiavelli had in mind the body politic. His book was based on the unwritten assumption that in advising the prince, he was advising the state; that the interest of the prince was to be identified with that of the community.

This is, of course, an unfounded assumption. If it were true, politics would be a very simple matter. There would be no need to wrestle with the problem of how a community discovers its "general will," and almost all the works on political science of the past 400 years could have been left unwritten. On the contrary, politics concerns itself largely with the tensions and conflicts of interests between individuals and the government that claims to speak for them. It is this conflict which stands behind the great political revolutions of modern times.

Machiavelli, however, took it for granted that the prince is the personification of the good of the state. This is an important point. One of the reasons why modern readers so often censure Machiavelli is that they assume they are reading merely a book of advice to a prince on the means of obtaining power; that is, a kind of textbook of tyranny. We believe that it is a misreading of *The Prince* to think of it in this way, as a short guide on "How to Become and Remain a Dictator." The advice to be unscrupulous is not necessarily advice to the prince as an individual.

Indeed, ground for the identification of the prince and the state did actually exist in Machiavelli's time. The sixteenth century was a century in which individualism was a dominant characteristic. The ability and power of the individual—whether artist or statesman—was given great respect and perhaps exaggerated importance.[7]

[7] Cf. *Il Principe,* ed. by L. Arthur Burd (Oxford, 1891), p. 24. Burd's edition, which has an introduction by Lord Acton, gives the Italian text with a stupendous array of notes in English as well as footnotes in Italian taken from Machiavelli's other works for comparative purposes. There is also a useful though somewhat outdated bibliographical note. Burd's introduction explains the purpose of *The Prince* and, in addition, discusses the early criticism of the book; it is well worth reading. Burd's aim is to give the historical background necessary for an understanding of Machiavelli; his general conclusion is that Machiavelli did not

This accompanied a disbelief in the intervention of God as the decisive factor in history. Men like Machiavelli put the stress, instead, on "natural" factors, such as the personality and power of the prince. Thus, the state was looked upon as a creation of the individual, and the identity of the two assumed.

There was, however, another reason for the identification of the prince and the state. We can best understand this if we trace the origin of the word "state." Its original meaning, derived from the Latin, was that of "position" or standing, but it gradually came to mean primarily a political standing which was superior or supreme. In medieval Italy, it became the aim of podestas and "captains of the people" to keep their *status* as public functionaries as long as possible. Gradually, the holders of such a *status* were able, during the development of Italian despotism, to identify their position with the entire administrative and bureaucratic structure of the community. When this was achieved, the prince and the state, in the political theory of the time, became one.[8] Machiavelli accepted this identification of interest between the prince and the state without examination, as a given axiom.

VI

What Machiavelli was really writing was a book advising the prince how to make his country hold her place among the other countries fighting for power. He identified the prince with the country, and held that the tenets of ordinary morality need not apply to the prince. The prince acts on behalf of the community and must be willing, therefore, to let his own conscience sleep. In reality, the moral obligation of a prince is like that of a soldier who must achieve victory at any price.

This can be a very terrible doctrine. The belief that states are somehow absolved from the ordinary decencies of conduct is, if any-

---

primarily intend his book as a source of general maxims of politics but as a solution for the particular case of Italy. Acton's introduction follows Macaulay in excusing Machiavelli on the grounds of contemporary conditions, and then degenerates into a series of learned quotations on the subject of reasons of state versus morality.

[8] Cf. Leonardo Olschki, *The Genius of Italy* (New York, 1949), pp. 166–167, and Ernest Barker, *Principles of Social and Political Theory* (Oxford, 1951), p. 90. The concept is familiar to us in the alleged saying of Louis XIV: "*L'ètat, c'est moi.*"

thing, worse than the belief that the prince is absolved from them. Yet this is close to being the doctrine that underlies Machiavelli. Nor can we disguise from ourselves that it is the doctrine generally underlying all international conduct from that moment to this.

Let us put this in colloquial terms. If a man comes to you and says, "I'm awfully sorry, but I shall have to shoot you. I have four starving children at home and I absolutely need the food in your house," we do not consider this a sufficient excuse. But if a government says it has four million inhabitants in this situation and must therefore claim the right to eliminate your state—many people still accept that as a realistic and outspoken way of carrying on international politics. Thus, in Machiavelli's time, Florentine imperialism against Pisa and the surrounding country was considered a "vital necessity," not open to question. It was considered essential that Florence have a sure supply of food for her populace as well as an open port for her trade.[9]

Machiavelli, who accepted the idea of "vital necessity," made the further assumption that the overriding aim of the state is to persist and to be powerful. Thus, he contended that Lycurgus deserved the "highest praise" for having "created a government which maintained itself for over 800 years in the most perfect tranquillity" and that Solon's work was unimportant because it lasted so short a time.[10] He ignored the different cultural contents of the two states, Athens and Sparta, and accepted the stagnated endurance of Sparta as the proof of excellence.

For Machiavelli, the existence of the state and its acquisition of power became ends in themselves. Toward these ends all other considerations must be sacrificed. One idea, which Machiavelli referred to again and again, sums this up: reason of state. A reason of state overrides everything else. It is the highest good to which Machiavelli appeals, and from which there is no appeal. A prince must be entitled to do whatever he wants provided it is for the satisfaction of the community as a whole and not for his own personal aggrandizement. He serves a "higher morality" than any ordinary code of ethics. In Machiavelli's own words: "It must be understood that a

[9] In our day, the examples might be America's claim to the Panama Canal or Russia's claim to a warm-water port in the Far East.

[10] Machiavelli's assertion that Sparta endured in "perfect tranquillity," of course, is not true.

prince . . . cannot observe all those things which are considered good in men, being often obliged, *in order to maintain the state* [our italics], to act against faith, against charity, against humanity, and against religion."

Machiavelli's assumption was clearly that the end justifies the means. "A wise mind," Machiavelli commented, "will never censure any one for having employed any extraordinary means for the purpose of establishing a kingdom or constituting a republic. It is well that, when the act accuses him, the result should excuse him." Thus all Machiavelli's thinking is pervaded by a "war psychology." In war, the chief aim is the complete destruction of the enemy state. To realize that aim anything is permitted.[11] As Machiavelli stated it, "Where the very safety of the country depends upon the resolution to be taken, no considerations of justice, humanity or cruelty, nor of glory or of shame, should be allowed to prevail."[12] Not since Thucydides and the Melian debates has the rule of power politics been so ruthlessly and candidly propounded.

VII

That Machiavelli saw the pitfalls which were opened by such reasoning is shown in his own revealing statement that

> . . . as the reformation of the political condition of a state presupposes a good man, whilst the making of himself prince of a republic by violence naturally presupposes a bad one, it will consequently be exceedingly rare that a good man should be found willing to employ wicked means to become prince, even though his final object be good; or that a bad man, after having become prince, should be willing to labor for good ends, and that it should enter his mind to use for good purposes that authority which he has acquired by evil means.

[11] Unlike the condottieri, who did not wish to end their profitable employment, Machiavelli believed in a quick, decisive battle and in the "unconditional" defeat of the enemy. For treatments of Machiavelli's position regarding reason of state, see Friedrich Meinecke, *Machiavellism: The Doctrine of Raison d'Etat and Its Place in Modern History*, tr. from the German by Douglas Scott (London, 1957), and C. J. Friedrich, *Constitutional Reason of State: The Survival of the Constitutional Order* (Providence, R.I., 1957).

[12] Cf. Felix Gilbert, "Machiavelli: The Renaissance of the Art of War," in *Makers of Modern Strategy*, ed. by Edward Meade Earle (Princeton, 1944), p. 14. This article is concerned with the relationship of Machiavelli's military interests with his political ones; according to Gilbert, Machiavelli "became the first military thinker of modern Europe." Gilbert's article is both brilliant and fascinating.

It is a constant surprise to read assertions that Machiavelli was amoral when he so clearly recognized "wicked means" for what they are. He was willing to employ them, we believe, because he was trying to be scientific and "realistic."

There is another explanation for Machiavelli's attitude to political morality. It is offered by Charles Singleton, in his stimulating article "The Perspective of Art." According to Singleton, Machiavelli took political activity out of the medieval category of "doing" into the category of "making." This shift is important because, even according to Aquinas (who, of course, left politics in the traditional category of doing), the moral intention of the craftsman or artist is not important in relation to things that are made; only his artistic ability is involved. If we relate this to what we said earlier about the Renaissance state becoming a "work of art," we can see that Machiavelli could treat the prince as an artist shaping the state and thus as an amoral character.[13]

[13] Allan H. Gilbert, in *Machiavelli's "Prince" and Its Forerunners,* suggests of Machiavelli that "an artist he has been called, and properly enough. His mystery may be explained by asserting that he is not primarily a *politico*—whether political theorist or crafty politician—but rather an artist in politics. . . . If the *Life of Castruccio Castracani* and even *The Prince* itself are imaginative creations, if the structure of *The Prince* is determined less by considerations of political science than by those of art, the clue of the labyrinth is perhaps in our hands" (p. 45). This is a penetrating remark. It goes too far, however, in largely subordinating Machiavelli's science to art; we should prefer to subordinate the artistic element in Machiavelli to his science.

John Morley, *Machiavelli* (London, 1897), a short lecture which indicates and weighs opinions about Machiavelli through the centuries, concludes that even Machiavelli's "science" is of no avail in justifying his immoral position. H. Butterfield, *The Statecraft of Machiavelli* (London, 1940), while attempting to restore him to the context of his time and circle, concludes that the judgment of Machiavelli's intentional wickedness by the Anti-Machiavels and even the Elizabethan dramatists was "not so wilfully wide of the mark as some writers have assumed." Gerhard Ritter, *The Corrupting Influence of Power,* tr. from the German by F. W. Pick (Hadleigh, Essex, England, 1952), compares the teachings of Machiavelli and More on power politics and welfare politics. Sympathetic to Machiavelli is the famous essay by Thomas Babington Macaulay, conveniently found in the Everyman's Library ed. of *Critical and Historical Essays* (London, 1946), Vol. II. Originally written in 1827, it opened new ground, especially in England, in the critical judgment of Machiavelli. Macaulay insisted that Machiavelli could not and should not be judged out of the context of his times. This essay sheds light equally on Macaulay and on Machiavelli, and is well worth reading. Louis Dyer, *Machiavelli and the Modern State* (Boston, 1904), agrees with Macaulay that circumstances excuse *The Prince.* Having defended Machiavelli, Dyer then

Machiavelli's profound error, we believe, was not that he was amoral or immoral, but that he did not sufficiently reckon with the moral factor in politics. In short, his "science" was faulty. He should have seen, for example, that his hero Borgia failed to build a princedom because the only bond that connected him with his condottieri and the people whom he ruled was that of naked self-interest. Inevitably, a conflict developed between Borgia's desire for personal gain and a similar desire by his mercenary captains. Machiavelli cited Cesare Borgia's cruelty in the Romagna as one which brought order and peace. In 1502, Borgia met four of his own captains who, having previously rebelled, were extremely suspicious of him. He surrounded them with kindness and overwhelmed them with good humor; then, as soon as they left the body of their own troops, he had them strangled. Machiavelli's admiration for this incident is of a piece with what we have said about many of the great thinkers of his period, Leonardo among them: that admiration of the brutal, successful man was part of the tough, secular, condottiere tradition. Yet at bottom, Borgia's solution was short-lived and useless. It reflected no moral impulse animating his army and merely increased the suspicion and distrust existing between him and the remaining soldiery.

Because it underestimates the spiritual factor in political actions, Machiavelli's analysis of statecraft in *The Prince* could be only partial. It is a weakness in his theory very similar to his erroneous assumption that there is no problem in the state involving the relations of the ruled and the ruler. It is the typical omission of the hard-boiled "realist" who thinks that only material factors count. However, Machiavelli's attempt to organize a Florentine citizen army in place of the mercenary bands showed that he was capable, in practice, of transcending his own limited theory as expressed in *The Prince*.

---

presses his own charges: that Machiavelli overlooked the idea of progress, that he slighted the importance of Florentine and Italian commerce, and that he dealt too much with the given and did not, like Plato, see the value of "fictitious cases." Charles Benoist, *Le Machiavélisme*, 3 vols. (Paris, 1907, 1934, 1936), treats of Machiavellism as a perpetual "method"—a species of positivism devoted to politics—which should be studied and followed. On all questions related to Machiavelli's *Prince*, Federico Chabod's *Machiavelli and the Renaissance*, which was referred to earlier in Chap. 1, should be consulted.

VIII

Why did Machiavelli turn to a prince when he so obviously favored republican institutions? It is evident that he would have preferred a republic if it were possible, and he made no attempt to hide his admiration for the then-existing Swiss republican institutions. Further, he anticipated Montesquieu in his enthusiasm for a mixed government. He thought that a mixed government would be the most likely to maintain itself in a balance, and so prevent the inherent defects of any of the pure forms of government from bringing down the state. In the *Discourses,* he commented:

> I say then that all kinds of government are defective; those three [monarchical, aristocratic, and democratic] which we have qualified as good, because they are too short-lived, and the three [tyranny, oligarchy, and licentiousness] bad ones because of their inherent viciousness. Thus sagacious legislators, knowing the vices of each of these systems of government by themselves, have chosen one that should partake of all of them, judging that to be the most stable and solid. In fact, when there is combined under the same constitution a prince, nobility, and the power of the people, then these three powers will watch and keep each other reciprocally in check.

This was Machiavelli's ideal. Nevertheless, he felt that man's control over his political destiny was limited. Two elements, Fortune and Virtue, dominated the political destiny of man. By means of Virtue, or right knowledge (actually, the Virtue of a statesman is his "excellence," and that "excellence" is to understand politics), man could control certain elements of the situation. He could accelerate the rise of a state or brake its downward revolution. Fortune, however, set the conditions under which this modifying action could take place. Thus, Virtue consisted in a perpetual adaptation of life to Fortune, in a sort of "opportunism" (a charge often made against Machiavelli).

The opportunities of his time in Italy favored the existence of a prince. In Machiavelli's analysis, a republic was "out of season." Only a prince could save Italy from the barbarians. Machiavelli's discussion of dictators in the Roman Republic showed his willingness to accept this solution during difficult times.

Accepting, therefore, the condition posed by Fortune, Machiavelli set himself to discover the Virtue required of a prince. He conceded that Fortune "is the ruler of half our actions" and advised conforming to the temper of the times. He insisted that "fortune varying and men remaining fixed in their ways, they are successful so long as these conform to circumstance, but when they are opposed then they are unsuccessful." Machiavelli made no attempt to oppose his times or Fortune; in this, he thought he was being scientific.

IX

Let us close this brief description of the essential content of *The Prince* with two further points. One is that the prince does not hesitate to fool his people and to deceive them: he is, in modern terms, what we call a good propagandist. Machiavelli believed that people are easily fooled, and he thought it to the prince's advantage to spread false doctrines among the people. But, as with reason of state, the prince's actions are not intended for his own benefit. These false doctrines, Machiavelli implied, eventually redound to the advantage of the people themselves. The prince's lies—his propaganda—preserve the state from upheaval and insure tranquillity and stability.

The second point is that the prince does not employ useless cruelty. He does not, like an Oriental tyrant, cut off people's heads if they happen to displease him. Machiavelli admitted that, if a choice must be made, it is better for the prince to be feared than loved. But he drew a distinction between being feared and being hated, and pointed out that the prince ought to acquire the favor of his people if for no other reason than to avoid conspiracies. Along these same lines, Machiavelli cautioned the prince to respect the women and property of his citizens, as despotic attacks on these lose him all support from the people.[14]

Nevertheless, Machiavelli was not at all squeamish in calling for repression—even cruel repression—when it was necessary. He intended it, however, to be used primarily against the few competitors who threatened the power of the prince (literally, his top *status*). And he intended it primarily to avoid further cruelty. Thus, Machiavelli insisted that the prince "will be more merciful than those who, from excess of tenderness, allow disorders to arise, from whence spring bloodshed and rapine; for these as a rule injure the whole com-

[14] Compare the maxims of Jacopo Sforza, p. 11.

munity, while the executions carried out by the prince injure only individuals."

X

These two last points, the advocacy of propaganda and of "surgical" cruelty, are alike in that Machiavelli advised their use for the benefit of the people. Thus they are of a piece with all his advice about statecraft for the benefit, not only of the prince, but of the entire community.

Machiavelli was a practicing politician; he knew the twists and turns necessary to political existence. He was also a humanist; he knew that conditions had been similar in Roman times. His humanism further made him think of politics as a secular and not a religious affair. And, finally, he was a scientist; he wished to look at his materials empirically and analytically. His science told him that "all human things are kept in a perpetual movement and can never remain stable . . . states naturally either rise or decline."

As a "Virtuous" politician, a wise man, and a scientist, Machiavelli recognized that he could do little to affect the rolling tides which inundated man's small political world. Nevertheless, that little he set himself to achieve by maintaining the required coolness and detachment. The result was that, rightly or wrongly, he acquired the reputation of the devil's advocate. This was the price he paid for making the first attempt to develop a modern political science.

CHAPTER 4

Thomas More

I

THE BOOK with which we shall now compare *The Prince* is the *Utopia* of Sir Thomas More. If Machiavelli gave his name to a way of thinking, More's book has given its name to a whole genre of literature—Utopian literature—as well as to a way of thinking.

In general, Machiavellism and Utopianism can be taken to be sharply opposed; the one realistic and the other idealistic and dreamlike. Yet More's *Utopia* is an extraordinarily realistic book. It is, indeed, much closer in attitude to *The Prince* than is generally conceded. More, like Machiavelli, was a statesman-writer who clearly perceived political reality and dealt with the actual problems of his time. He was also, like Machiavelli, a humanist who used classical models—in his case, Plato—as a means of going beyond the mirror-of-princes literature. He, too, tried to penetrate the causes of the political evils of his time and to offer concrete and carefully thought-out solutions in place of the conventional sentiments of the time.

More's solutions, however, were vastly different from those of Machiavelli. They reflect the fact that he belonged to a different tradition from that of power politics followed by Machiavelli. More's tradition was one which, with its roots deep in English literature, went back to Chaucer and Langland. It is characterized by two traits: an intimate concern with the suffering of the common people, and a feeling that the state exists for its members.

Both these traits were lacking in Machiavelli. He looked upon the state as a work of art, and valued its people only as pieces in this greater mosaic. And, although he concerned himself about the welfare of the community, his concern for the common people was impersonal. We see the difference between More and Machiavelli clearly in the issue of magnificence. Machiavelli extolled it and thought pomp and ceremony part of a prince's majesty. More hated the idea of men drudging out their lives making "vain and useless things," and commented bitterly that of all God's creatures only men are greedy out of "pride alone, which counts it a glorious thing to pass and excel others in the superfluous and vain ostentation of things." He lashed out at those who "count themselves nobler for the smaller or finer thread of wool" in their coats.

More was a man of profound Christian belief. His Utopia is at bottom inspired by the Sermon on the Mount. All goods are held in common. There is no money. People spend their day joyfully in doing good deeds one to another. They live in a fine fellowship which, looked at cynically, now makes us think of a glorified Boy Scout troop. But this good society is not built on misty-eyed emotion and wishful thinking. More firmly erected his Utopia on an acute and realistic analysis of the cause of human misery. He found the cause of social evil, not in the whims of God or Fate, nor in the curse of original sin, but in the social structure built by man. More, who lectured on Augustine's City of God in his youth, wanted to build a City of Man, which God might not be ashamed to inhabit for a time.

II

Thomas More, son of a prominent judge, was born in England in 1478. After serving in his youth as a page to Cardinal Morton, he pursued his education at Oxford, studying Greek and Latin, and then he studied law at London.

Rapidly, he became a very successful lawyer, earning huge fees and becoming involved in public service. In 1504, he served as member of Parliament. In 1510, he was under-sheriff of the city of London, earning about 400 pounds sterling, equal to perhaps 10,000 pounds today (although all such equations are in part guesswork). He rose to be speaker of Commons in 1523 and, finally, lord chancellor of England from 1529 to 1532.

We cite these posts and salaries for two reasons. First, they show that More, like Machiavelli, had had political experience—we have still to mention his various diplomatic posts—and experience of a higher order than that of his Italian contemporary. Second, they illustrate the fact that More was a substantial member of the middle class, with an excellent chance of entering the nobility if he had wished. As his friend Erasmus wrote: "The study of law has little in common with true learning, but in England those that succeed have a great position; and with good reason, for it is from their ranks that the nobility is for the most part filled."[1]

More had a great position. But he chose to sacrifice it for something which he considered greater and more profound. We shall see what this was when we deal with the circumstances of his death, which occurred in 1535.

III

The *Utopia* was written almost at the same time as the *Prince*. It was written between the summer of 1515 and the fall of 1516 (although not translated from the Latin into English until 1551). The two dates are important because they mark the fact that the book was written, not as a cohesive whole, but as two parts which "represent two different and separate sets of intentions on the part of the author."[2]

The part which is the discourse on the Utopia proper is the second part, but it was written first. It was written by More in 1515 when he was ambassador for Henry VIII in the Netherlands, after a supposed encounter with a Portuguese philosopher named Raphael Hythloday, who had sailed with Amerigo Vespucci on the last three

[1] Erasmus to Hutten, July 23, 1519, quoted in *Sir Thomas More: Selections from his English Works and from the Lives by Erasmus and Roper,* ed. by P. S. and H. M. Allen (Oxford, 1924), p. 5. This is a very convenient short introduction to More's works other than the *Utopia.* For More's letters, see *The Correspondence of Sir Thomas More,* ed. by Elizabeth Frances Rogers (Princeton, 1947), where most of the letters are printed as More wrote them, in the Latin. Actually, Erasmus was exaggerating the ease with which lawyers entered the nobility; in general, they were merely knighted.

[2] J. H. Hexter, *More's Utopia: The Biography of an Idea* (Princeton, 1952), p. 28. Hexter's book is a brilliant piece of textual criticism, in which he investigates the relation between the structure of *Utopia* and the intent of the author; it is certainly the best discussion available of the circumstances attending the twofold composition of the *Utopia.* In the Everyman's Library edition of *Utopia* (1941, tr. by Ralph Robinson; the 1955 Everyman's edition has been revised and the spelling modernized), the break is on p. 18; the first 18 pages were written in 1515, and what follows of Book I, in 1516.

of his four voyages. It is a description of an island, called Utopia, which exists "nowhere," and it tells how people in this ideal state lived.[3]

The first part of the book, largely dialogue, which deals with advising a king, was written by More after his return to London. It was written rapidly, which accounts, as Erasmus said, "for some unevenness in the style." It reflected a new set of problems More was facing: the New Statesmanship, "Machiavellian" in character.

Let us defer treatment of the political part of More's *Utopia* and deal now with the part of the book which, though placed second, was written first: the Utopian commonwealth. This part of the book is concerned primarily with the problem of economics. With a close eye on the economic evils of his own time in England, More, a highly paid lawyer, decreed that there should be no money in Utopia, and said of the Utopian people: "Their common life and community of living without any traffic in money, this alone, which is the ultimate foundation of all their institutions, overthrows all excellence, magnificence, splendor, and majesty—the true properties and distinctions of a commonwealth according to the common opinion."[4]

In itself, this is pleasant Christian or medieval doctrine. But when one reads the *Utopia*, one feels it imbued with More's urgent sense that the medieval world was, in fact, falling to pieces and that a new kind of economy was conquering England. It was an economy of which he was afraid.

The most striking passages on this theme occur when More wrote against the enclosure of common land and peasant tillages. England has had two major enclosure movements, both with tremendous social consequences. One, which we shall have occasion to analyze

[3] There is an amusing book by Arthur E. Morgan, called *Nowhere Was Somewhere* (Chapel Hill, 1946), which compares More's Utopia to the Inca state in Peru.

[4] It is well to point out that the abolition of money means, automatically, that a state like Florence or Venice could not exist. The absence of money and money exchange would obviously prevent the accumulation of the kind of treasure which Lorenzo the Magnificent or the Doge of Venice possessed. Only a power state like Sparta would still be feasible.

Karl Kautsky, *Thomas More and His Utopia,* tr. from the German by H. J. Stenning (New York, 1927), conceives of Thomas More as one of the two great figures (the other is Thomas Münzer) on the threshold of socialism. Kautsky is perceptive and original, but all his ideas are set within the rigid frame of Marxist interpretation; thus, there is a "simplistic," uncritical attitude in his book.

later, was at the end of the eighteenth century. The other, about which we are now speaking, took place during More's lifetime and was an attempt to enclose land so that sheep might graze on it. It was part of a movement to turn sheep farming into a large-scale industry, and involved the end of small-scale farming.

More wrote against enclosure in a white heat of protest. He said, quite literally, that the sheep, formerly "wont to be so meek and tame" now "eat up and swallow down the very men themselves." Obviously, this was not written by a man who was merely laying down general dicta about how a state should be nicely run. It was written by a man who felt passionately that a change was coming over the landscape of his country—a change which terrified him. England, in fact, was becoming increasingly concerned about raising sheep on a large scale in order to make woolen cloth, its main staple of export from More's time to the eighteenth century.

There are other examples of how realistically More saw the changes in the social landscape of his time. He was keenly aware, for example, that owners of large estates were living in too great luxury, with too many retainers in livery. He was disturbed at the number of men being turned into vagabonds, largely because they were unemployed workers or discharged soldiers. Also, More wrote constantly against the delays of the law, its iniquities and "class" bias. In a characteristic outburst, Hythloday is made to say:

> Therefore when I consider and weigh in my mind all these commonwealths, which nowadays anywhere do flourish, so god help me, I can perceive nothing but a certain conspiracy of rich men procuring their own commodities under the name and title of the commonwealth. They invent and devise all means and crafts . . . to hire and abuse the work and labor of the poor for as little money as may be. These devices, when the rich men have decreed . . . be made laws.

More was equally violent against commercial iniquities. He sympathized with the difficulties of the small cloth weavers, and attacked engrossing and forestalling. These were the two great catchwords in all medieval discussions of economics. They are words to describe the control of the market by manipulating the flow of goods to it. To engross is to buy all that is coming to the market and keep it from the market until the price rises; to forestall is to buy futures. More could have been looking, from a remove of 400 years, at the pre-1929 manipulation of the New York stock market. He was saying that this is no way to order man's economic life.

In saying all this, More was setting himself against the New Economics (as later he set himself against the New Statesmanship). His position was that of the medieval church, with its doctrine of the just price: the doctrine that price should be determined by the needs of the two parties in the exchange to maintain themselves in their given status. The idea that price should be set by supply and demand in the market was repugnant to More. He was dismayed at the new economy, based on the free use of capital, which was about to conquer Europe.

Against the New Economics he proposed his Utopia. At one stroke, he cut away the ground from under the rising new capitalism. He eliminated the very element—gold—which inspired the discoverers and *conquistadores* of that part of the globe in which he set his island. Taking literally the maxim that "money is the root of all evil," More eradicated it from his ideal community. In his own mind, he was not being naïve or foolish; he was being realistic. He was going to the core of the matter.[5]

IV

Let us now look at the inspiration for More's doctrines. We have here a sixteenth-century man who served his king as lord chancellor

[5] We may compare More's solution to England's economic problems, as developed in the *Utopia*, with that proposed by another statesman, William Cecil, toward the end of the sixteenth century. Cecil's idea for relieving the farmer's plight in the face of enclosure was to make farming profitable. He helped exports and provided a domestic market for crops by encouraging industry. The result, according to the English economic historian William Cunningham, was that by the time of Elizabeth's reign the people were prosperous. (Cf. *Cambridge Modern History*, Vol. I, p. 524.)

Cecil's work also had political consequences. The result of the new economic measures was a wealthy industrial population under a relatively poor Crown. This was an important factor in the people's ability to maintain their liberties. Now, if we take it as true that More's successor, the arch-Machiavellian Thomas Cromwell, instituted an administrative revolution in Tudor government which was passed on to William Cecil (who was the disciple of Cromwell's friend and disciple, Sir William Paulet), we have an interesting contrast at hand. (For the administrative revolution, see G. R. Elton, *The Tudor Revolution in Government* (Cambridge, 1953); for Thomas Cromwell, see H. R. Trevor-Roper's review of Elton's book in *The New Statesman and Nation*, March 6, 1954.) Which solution, the Machiavellian or the Utopian, was or would have been the best? Or are they both extremes, awaiting a more balanced political philosophy to take their place?

and was associated with the upper classes. Yet he wrote with an edge of fierce passion about the changes in the life, not of princes and of courts, but of the simple, ordinary individuals in his country. Why?

One part of the answer is to be found in a consideration of More's character. He was, essentially, a religious man, with a strong ascetic bent. His real mission was a spiritual one, and we know that he seriously thought of entering a monastery. In fact, he lived for a few years with the Charterhouse monks while he was studying law. Even when a successful lawyer, statesman, and man of society, More kept the medieval close to him by wearing a hair shirt and by sleeping on a plank with a log for a pillow. Indeed, he was a strange sort of courtier!

More, however, decided not to embrace the monastic order. He married and had children; he entertained and made a merry home. Nevertheless, at the end of his life, while in prison, he still thought yearningly of the monastic life. He declared to his daughter that if it had not been for his family, "I would not have failed long ere this to have closed myself in as straight a room, and straighter too."

The important thing, however, is that More did decide not to enter a monastery or to become a churchman. He decided to live the religious life in society and to take the world as his monastery. In this, he appears very much a precursor of the Puritan, as Max Weber has painted him, who seeks to fulfill God's purposes by activity in this life. Further, the fact that, in spite of his intense longing, More became a lawyer and a government official illustrates the increasing dominance of secular interests in sixteenth-century Europe. It shows the tendency of scholars to serve the state rather than the church.[6]

Nevertheless, More's monastic longings, his hair shirt, give us a glimpse into the psychology of the man who wrote *Utopia*. It informs us that More was "looking backwards," toward the medieval ideal. It hints to us that Utopia is really an enlarged monastery, with a revised and expanded Rule of St. Benedict. The same "rational method" lies behind Utopia as behind the medieval monastery. But

[6] This growing secularism is made even clearer when we add that More's decision to "serve in the world" was partially the result of persuasion by his confessor, Dean Colet, and of the advice of Cardinal John Morton, in whose house More had served as a page.

Utopia is a society for all the people rather than for the chosen few.[7] And this last follows from More's original decision to live the good life in society.

Utopia is a society under rules. Thus, we have no such time-wasting and corrupting activities as games or dicing (one of More's pet dislikes).[8] Instead, "It is a solemn custom there, to have lectures daily early in the morning. . . ." Further, More goes into a detailed description of the clothes to be worn, which are extremely simple and the same for everyone. They resemble nothing so much as the garb of the Franciscans. It is illuminating to contrast this society with Rabelais' abbey of Thélème; one notes the medieval quality of More's "monastery."[9]

V

More was not, however, a simple man. He was not solely medieval in viewpoint, nor does the *Utopia* embrace merely the medieval ele-

[7] Hexter makes the point that Calvin's Geneva came closest to realizing More's ideal in actual life. Cf. Hexter, *op. cit.*, pp. 91–92.

[8] As Erasmus commented, "Games of ball and hazard and cards and the rest, with which many men of position wile away the hours, he cannot endure" (*Selections,* ed. by Allen, p. 4).

[9] These qualities—the monastic and the medieval—in More ran strongly counter to the currents of sixteenth-century England. We can see this clearly in two examples. The first is that the desire of the king and the rising middle class to confiscate the monastic property was opposed by More's support of the monastic, common life. The other is that he favored clerical immunity from the law; this ran counter, for example, to the feeling of Parliament, which in 1515 passed a measure against clerical immunity. (cf. A. F. Pollard, *Henry VIII,* London, 1905, p. 234). Only on one point, indeed, was More supported by the bulk of his contemporaries. That was on his assertion of the immortality of the soul. More felt so strongly about this that his Utopia decrees the death penalty for anyone who publicly denies it. (Those who deny the immortality of the soul but keep their beliefs to themselves are merely deprived of the rights of citizenship.) Nevertheless, even on this point, More was medieval rather than "modern"; voices here and there had begun to question. For example, in the very same year as the *Utopia,* 1516, the Italian philosopher Pomponazzi published a work denying the soul's immortality. Unbelief had gone so far that in 1513 Pope Leo X had considered it necessary for a Lateran Council to confirm by statute the doctrine of the immortality and individuality of the soul (cf. H. W. Donner, *Introduction to Utopia,* Sidgwick and Jackson, Ltd., printed in Upsala, Sweden, 1945, n. 3, p. 50). More, however, shunned all these new, secular thoughts. Whether in conformity with the majority of his time or against them, whether dealing with monasticism, clerical immunity, or the immortality of the soul, More was consistently medieval in viewpoint.

ment. In fact, in one sense, the book is a criticism of the medieval as well as of the sixteenth-century world, for More went beyond the medieval society to the Christian ideal which supposedly animated it. Thus, in opposition to medieval hierarchy, his society is equalitarian. More logically continued this by forbidding specialization; everyone works, fights, and studies in More's Utopia. There are no mere artisans, warriors, and scholars, *per se;* More was resisting the division of labor.

This classless and unspecialized nature of Utopia distinguishes More from Plato in the *Republic.* Yet, Plato was a main source of More's inspiration. Indeed, Erasmus tells us that "while still a young man, he worked upon a dialogue, in which he maintained Plato's principle of community in all things, even in wives." More abandoned the community of wives in the *Utopia,* but one has the suspicion that the Utopian practice of premarital physical inspection of mates in the nude may have been influenced by Plato's attitude to physical nudity.[10] In any case, More himself acknowledged his debt to Plato. In one place, he put it subtly when he declared: "Therefore when I reflect on the constitution of the Utopians, and compare with them so many other nations . . . I grow more favorable to Plato, and do not wonder that he resolved not to make any laws for such as would not submit to a community of all things."

Plato accorded with the early Christian belief in the "community of all things"; it was, undoubtedly, this which attracted More to him. But More, we must remember, was a humanist scholar as well as a statesman of the sixteenth century. In his younger days, he had written poetry and tried his hand at a reply to Lucian's *Tyrannicide.* He also wrote a *Life of John Picus, Earl of Mirandola,* the great wonder of Italian learning, and a *History of Richard III.* He was familiar with much of ancient literature, had studied Greek with Grocyn and Linacre, two of the leading scholars, and counted as perhaps his closest friend the foremost "saint" of the New Learning, Erasmus.

It is clear, then, that More was part of the humanist movement of the sixteenth century, and that the *Utopia* strongly reflects this movement. In fact, R. W. Chambers, one of More's most acute biographers, has attempted to show that the *Utopia* is based solely on the four

[10] On the other hand, it may have been a device to detect whether the prospective partner had syphilis, the new and spreading disease of More's time.

cardinal (i.e., pagan) virtues and not on the three Christian ones. Chambers draws the conclusion that Utopia is, thus, not even More's ideal state. It is merely an indication of what European states might be if based simply on human reason. In this interpretation, Utopia is not a Christian ideal; it is a secular, humanist one.[11]

VI

We have still not exhausted the different and converging sources of More's inspiration. So far, we have mentioned the medieval, the Christian, and the Greek-humanist influence. Now we must add that the New World discoveries also had a strong effect upon More's writing. This is most obviously shown in the first book of the *Utopia,* which introduces us to the Portuguese wanderer Raphael Hythloday, who has just returned to Antwerp (which is the growing, new port for the trade with America and the Far East) after voyaging with Amerigo Vespucci. We mentioned this earlier; but we did not mention that the *Utopia* was based to some extent on a factual account by Vespucci of his travels, first published as an appendix to the *Cosmographiae Introductio* by Waldseemüller in 1507. More himself refers to the *Voyages* as "now in print and abroad in every man's hands."[12]

In addition to this general interest in the New World, More had a more personal interest. We know that More's brother-in-law set out for the New Found Land six months after *Utopia* was published and that More was financially interested in the project.[13] Nothing important came of this affair, but it offers us additional evidence of More's interest in the lands to the West.

The New World influenced the writing of *Utopia;* but *Utopia* also

[11] More, of course, could not have kept up the fiction of Utopia as an undiscovered island if it had previously been visited by Christians. However, he did have the Utopians immediately assent to Christian doctrine as soon as they first heard it propounded. The notion that Utopia is a picture of the state of society to which man can attain without revelation, i.e., without Christian dogma, is developed by R. W. Chambers in *Thomas More* (London, 1938) and supported by H. W. Donner in *Introduction to Utopia,* previously cited. Mr. Donner has used German and Swedish works on the *Utopia,* in writing his book, and short glimpses into these, in English, are probably to be found only in his book.

[12] It is in Waldseemüller's volume of 1507 that the New World, owing to a misunderstanding, was first named America after Amerigo Vespucci.

[13] Chambers, *op. cit.,* p. 142.

had a reciprocal effect on the New World. There is a story that the description of Utopia was so realistic that a pious divine, thought to have been Rowland Phillips, Vicar of Croydon, wished to voyage to Utopia as a missionary.[14] More seriously, there is the story of Vasco de Quiroga who went to Mexico as *oviedor* (judge and government inspector) in 1530. He had read More and thought that the Indians might be able to live as More said the Utopians did. Ordained as Bishop of Michoacán, Quiroga built hospitals and schools, where the Indians were taught in More's ways. He even had books printed in Indian dialects to accompany Greek and Latin texts, instead of Spanish ones. Gradually, the Utopian customs—such as support of the aged and infirm, communal workshops, periodic redistribution of land—became indigenous to the natives. In short, the "ideal" Utopia took on geographical existence.

Alas, it came about that the very success of Quiroga's efforts led to outside jealousy and eventually the overthrow of his work.[15] More had borrowed some of the ideas in his book from stories about the Arawaks—their contempt for gold and money, their feelings toward nakedness, and their communal meals at long tables. He had incorporated these, and other, notions in the *Utopia*. But when Utopia crossed back to the New World whence it had taken its imaginary impetus, it soon failed. The silver mines of Potosí and not the Utopia of the Bishop of Michoacán dominated the minds of European men.

VII

One other point should be mentioned in this enumeration of the intellectual sources for More's *Utopia*. We have talked about medievalism, Christianity, Greek learning, and the New World voyages as having influenced More; there was also that vague entity known as the "climate of opinion." Our point is that the climate of opinion in the sixteenth century prepared the way for More's imaginary commonwealth. People were ready for new extensions of their experience. We can see this, oddly, even in mathematics, where the development of negative, irrational, and imaginary numbers was taking place. As Ernst Cassirer remarks, "Negative numbers first appear in the

[14] Donner, *op. cit.*, p. 25.

[15] See Silvio Arturo Zavala, *La 'Utopia' de Tomás Moro en la Nueva España* (Mexico, 1937); also, his *Ideario de Vasco de Quiroga* (Mexico, 1941).

sixteenth century in Michel Stifel's *Arithmetica Integra*—and here they are called 'fictitious numbers' (numeri ficti)."[16]

The ability to deal with the imaginary and nonexistent in an attempt to solve real problems was an innovation of More's period. It was transferred by More from the realm of abstraction to the world of politics and society. More's "vision," his view of the imaginary, instead of being induced by monastic asceticism, was given to him by involvement in the secular life of his time. More expressed the new view of the world, obtained by his age, in the very title of his work *Utopia,* which means "nowhere."

VIII

Now we should like to turn back to the first part of Sir Thomas More's *Utopia,* the part which was written after his return to England in 1516. This was also the year in which the entire book was published at Louvain, Belgium, where Erasmus saw it through the press. More, in 1516, was writing what turned out to be the first part of his book because he had, apparently, received an offer to enter the government service as an adviser to Henry VIII. Therefore, he had to give a good reason—not only to the king but to himself—for whatever decision he took.

For the king, More had several good reasons for refusing. One was that he could not earn nearly as much in the king's service as he was earning as a lawyer in private life. In fact, he had just returned from the embassy to the Netherlands of 1515, in which he had spent, during six months, twice his ordinary expense at about half his usual income. And, as he had a large family and lived up to his income, this was a telling reason.[17]

But this was certainly not the reason More gave to himself. In the first part of the *Utopia,* he was obviously casting about and putting to himself this question: "All right, you educated people, like Erasmus, like More, people who are outstanding in the intellectual life of your times. Is it not very natural that the government should come and say, 'You have given great signs of ability in your own art or profession. Wouldn't you come and advise the government?' " And More, who believed that "from the prince as from a perpetual

[16] Ernst Cassirer, *An Essay on Man* (New York, 1953), p. 84.

[17] The king, of course, agreed to match More's salary; but More knew that kings were notoriously untrustworthy in financial matters.

wellspring comes among the people the flood of all that is good or evil," realized that to be an advisor to the prince might mean great influence for good or for evil.

Nevertheless, More did not accept this argument. He was aware instinctively, rather than rationally, of the dangers involved for him in such a post. He sensed that it would be quite difficult for a man of his temperament to act as advisor to a king and to keep either his beliefs or his head. More put his doubts into the form of a dialogue called "Of Counsel." Let us sketch it roughly, in a paraphrase of More.

The first person in the dialogue says: "My views about government are well known; my entire book shows clearly what is my position. If I go into the king's presence and give him the kind of advice which I think he should follow, he will obviously pay no heed to me."

To this the second party in the debate replies: "You don't have to charge in like a bull in a china shop and say to the king, 'I don't think you ought to do so-and-so.' One doesn't have to be as blunt as all that. If you wish to persuade the king to act as you think right, you suggest it to him subtly, leading him to believe that it is his idea."

However, the first person to the dialogue rejects this approach. He suggests that if one is opposed, lock, stock, and barrel, to everything being done in government, the position is hopeless. In More's own words: "For either I must say otherways than they say, and then I were as good to say nothing; or else I must say the same that they say, and . . . help to further their madness."[18]

More was saying, implicitly, that the prince is surrounded by people who are exactly like Machiavelli. They are always proposing to the prince the identity of the state with his own personal wishes. And More could not accept this. He rejected such a solution out of hand. As late as 1532, he advised Thomas Cromwell, the new chancellor, "In your counsel-giving unto his Grace, ever tell him what he ought to do, but never what he is able to do. . . . For if a lion knew his own strength, hard were it for any man to rule him." Was More thinking of Machiavelli's metaphor of the lion and the fox? The irony of it was that Cromwell had returned from Italy, bringing in his luggage the first copy of *The Prince* to enter England.

[18] However, More's friend, John Colet, did oppose the government and escape with only a light censure; when England was preparing for war against France, Colet declared publicly that an unjust peace was preferable to the most just war (cf. Pollard, *op. cit.*, p. 134).

In any case, More knew that the philosopher and the king could never join hands. He satirically commented, "This school philosophy is not unpleasant, among friends in familiar communication, but in the council of kings, where great matters be debated and reasoned with great authority, these things have no place." From the things which "have no place" in the councils of kings, More turned to Utopia. In the person of Raphael Hythloday, he voiced his fundamental objection: "Howbeit, doubtless, Master More (to speak truly as my mind giveth me), where possessions be private, where money beareth all the stroke, it is hard and almost impossible that there the weal-public may justly be governed and prosperously flourish."

To the doubts of More and Peter Giles, another listener, as to the feasibility of a true "common wealth," Hythloday gives his answer: a description of Utopia. "You conceive in your mind," he says, "either none at all, or else a very false image and similitude of this thing. But if you had been with me in Utopia and had presently seen their fashions and laws, as I did . . ." We are on our way, outward from the courts of kings . . . on a voyage to Utopia.

IX

While More was carrying on this scrupulous debate with himself, he was being pressed to join the government of Henry VIII by Wolsey, who was cardinal as well as chancellor of England. Wolsey was the last of the great ecclesiastical statesmen of England. In this, he serves as a fitting contrast to More, the layman; and especially as, in effect, Wolsey was Machiavellian while More was Christian in emphasis. Nevertheless, although Wolsey was a much more worldly man than More, he did feel strongly that such evils as the delays of justice and the tendency of the courts to become instruments of the upper classes had to be corrected. Therefore Wolsey encouraged courts, like that of Star Chamber, which served to keep down the powerful subjects and to give the poor man protection from them. He also had measures against enclosure passed in 1518.

It may be that More, who cherished the reputation of being a "just judge," ignored his own beliefs, expressed in *Utopia,* and succumbed to the blandishments of government on the ground that he might thereby further the carrying out of justice. In any case, about twenty months after the dialogue "Of Counsel" he entered the service of Henry VIII. In fact, he served as speaker of the House of Commons in

1523 and acted at that time rather as a tool for the king. Increasingly, More found himself involved in government service. When Chancellor Wolsey finally fell from power in 1529, Sir Thomas More took his place; the only safeguard against his having to become a Machiavellian advisor or losing his head was that Henry VIII promised him "freedom of conscience." We shall soon see how much this promise was worth.

Wolsey's fall was directly occasioned by the private difficulties that Henry VIII was having. This is really all a part of the history of the Reformation, which we shall begin to deal with in the next chapter. Henry VIII was at that time married to a queen, Catherine of Aragon, who was somewhat older than he (40 to his 34), and who, in any case, was not producing male heirs. Henry's father, Henry VII, had just managed to unite England at the end of the fifteenth century, after the exhausting Wars of the Roses between the Lancaster and York successions; and he had managed to unite it by marriage. It was, therefore, of great importance that the royal line should be continued. Henry VIII intensely desired a son.

To complicate matters, Catherine had been the wife of Henry's older brother, Arthur, and a papal ruling had been needed to set aside the canon law rule that Henry could not marry his deceased brother's wife. When Catherine could produce no sons for him, Henry took this as a sign of the divine displeasure at his illicit union. He petitioned the papacy to annul his marriage. This, however, the papacy, although at first favorable, could not do because Catherine's nephew was Charles V, whose troops at that moment (1529) controlled Rome and the pope. The cause of Wolsey's fall was his inability to obtain the annulment for Henry.

A contributory factor in Henry's resentment toward the papacy was his feeling that since 1511, and earlier, England had fought Continental wars for the pope without reward. Now, he thought, in its first request, England's interest (for so Henry saw his need of a son) was being sacrificed to Spain. Henry's answer to an "international" church, in reality controlled by Spain, was the determination to make a national Church of England.

At this particular stage, Henry cast his eyes on Anne Boleyn, who remains the best-known of his six wives. He wanted to make her queen. In a sense, Anne Boleyn was the "candidate" of the Lutheran faction at Henry's court and thus an important tool in the conflicts then taking place. The break with Rome was now hastened along. By various acts of Parliament, Henry was given power to stop, at his discretion, payment of annates to Rome. His marriage to Anne Boleyn was legalized, and the succession to the crown vested in his heirs by her. Finally, the Church of England was made directly responsible to the king instead of to the pope in Rome. Thus was made in England, on a political level, that breach with the Church of Rome which was another part of the religious Reformation then taking place elsewhere under the leadership of Luther and, later, Calvin.

More did not support these actions, although he did not actively oppose them, and in 1532 he resigned. He was tried shortly thereafter, but the charges made against him at that time were not sustained. However, when it became necessary for all subjects to swear to Henry's marriage with Anne Boleyn and to any issue from it, More demurred. He took the position that he was willing to swear to the royal line, i.e., to any sons who would be born, but that he was not willing to swear that Anne Boleyn was the rightful Queen of England until she was so designated by the pope. As a result of this refusal, in spite of Henry's promise of "freedom of conscience," More was beheaded in 1535. His end was proof of his own advice: it was not wise for men like him to become embroiled in royal politics.

X

The circumstances of his death made Sir Thomas More into a martyr and, thus, like Socrates, more important in his death than in life. The Catholic Church has canonized him. Laymen have remembered him for his boldness in upholding individual conscience against the omnicompetent authority of government. Indeed, he has come to be the symbol of the modern intellectual who holds to his beliefs against the power of the state.

If we examine the matter closely, we shall in fact come to see that what was "liberal" about More was not his beliefs but his manner of dying for them. Both in his own life and in his rules for Utopia, More held that the individual might not publicly preach against the accepted dogma. Such action might lead to tumult and sedition.

Against Henry VIII, More's only claim was that he be allowed peacefully to hold his belief; he agreed that he had no right to convert anyone else to his view. More thus ignored the danger that this could be a damaging and illiberal doctrine; that it could lead to the notion that the realm of the spirit was free and the political servitude of the body of no consequence.

Yet even if we think this a shortcoming in More's approach, we cannot deny his importance as a shining figure of martyrdom. Nor can we deny his importance as the writer of *Utopia,* and as a synthesis, in his person, of many of the trends of the sixteenth century. A Renaissance scholar devoted to the New Learning, a successful lawyer emerging from the rising middle classes; these were both traits of the modern sixteenth century. An opponent of the New Economics and the New Statesmanship, an upholder of the Christian monastic ideal: these were both elements of the medieval sixteenth century. Caught between the currents of his time, More entered the service of the New State while retaining his old loyalties. The result was that he perished on the scaffold, symbol of the triumph in the sixteenth century of stronger and more brutal ideas and values than his own.[19]

[19] The following books on More may be found helpful: Frank and Majie Padberg Sullivan, *Moreana, 1478–1945* (Kansas City, Mo., 1946) is a bibliography on everything to do with More. It should be supplemented by the useful review article, "Sir Thomas More," by Frederick L. Baumer, in the *Journal of Modern History* (Dec., 1932). The standard edition of More's great work is *The Utopia of Sir Thomas More,* in Latin from the edition of March, 1518, and in English from the first edition of Ralph Robynson's translation in 1551, with additional translations, introductions, and notes by J. H. Lupton (Oxford, 1895). Another useful edition of Robynson's translation is that by J. Churton Collins (Oxford, 1904). Roper's life of More (any edition) is basic. A more modern book is E. M. G. Routh's *Sir Thomas More and his Friends, 1477–1535* (Oxford, 1934). Russell Ames, *Citizen Thomas More and His Utopia* (Princeton, 1949), is, as its title suggests, concerned not with More the saint but with More the London citizen. According to Ames, More "criticized decadent feudalism in the interests of the 'best' aspects of rising capitalism"; thus, the *Utopia* is a "product of capitalism's attack on feudalism, a part of middle-class and humanist criticism of a decaying social order." While provocative and stimulating in parts, Ames's book must be used with great caution; he makes important assumptions which are often unsupported by any substantial evidence and he tends to fit his materials too readily into a preconceived framework of social history. For a comparison of More's Utopian and social ideas with those of Erasmus, Bacon, William Postel, and Campanella, see Emile Dermenghem, *Thomas Morus et les Utopistes de la Renaissance* (Paris, 1927).

CHAPTER 5

Erasmus and the Humanists

I

DESIDERIUS ERASMUS is more than any other man the symbol of humanism. This powerful movement which had begun in the Renaissance culminated in Erasmus: his personality was formed by it, and for a lifetime he expressed humanism for all men. He was born about 1466 in Holland, but his mind was cosmopolitan, and it dominated intellectual Europe in his age as the mind of Voltaire dominated a later age. One of his friends confessed: "I am pointed out in public as the man who has received a letter from Erasmus."

The movement of humanism which Erasmus personified was (if one must find a single phrase for it) a liberal movement. Its history and its defeat, therefore, have a special interest for our time. The life of Erasmus has a modern moral and, indeed, a very modern ring. He had the respect of thinking men and, like his friend Sir Thomas More, for a time he had the ear of princes. Then, in 1517, the Reformation divided Europe into two religious camps, and soon each side outdid the other in dogmatic bitterness. Erasmus was helpless between the two forms of intolerance, and the last years of his life (he died at the age of 70 in 1536) are marked with his own sense of failure.

Thomas More had lived the tragedy of an individual martyr. Erasmus lived the tragedy of a whole generation of intellectuals—and of later generations too. His rise showed that a movement of tolerance, such as humanism was, can inspire men so long as it confronts a single intolerance. And his decline showed that tolerance as an ideal no longer moves men when two opposing intolerances clamor for their loyalty. This has been the dilemma of the liberal spirit in every age since Erasmus.

II

Humanism was a movement in which many strands were woven together: the strand which leads directly to Erasmus was the new interest in the classical writers of Greece and Rome. This interest, which was strong throughout the Renaissance, goes back in its beginnings at least to the fourteenth century in Italy. It is first clearly expressed at that time in the poems of Petrarch. And these poems already show the characteristic coupling of ideas in humanism: classical literature is thought of not as an end in itself, but as the expression of a wider love for man and nature.

In one sense, humanism was a pagan movement. It was impatient of the narrow asceticism which the church laid down; it was not willing to abhor nature as a beautiful snare, to think the flesh evil, and to find virtue only in a monastic renunciation of life. The doctrine of the medieval church was original sin—the belief that the soul and the body are sharply divided and that, because man cannot express his soul except through his body, he carries an unavoidable sin. The doctrine of humanism was original goodness—the Greek belief that the soul and the body are one, and that the actions of the body naturally and fittingly express the humanity of the soul.

Just as the churchmen leaned on the Bible and the Church Fathers, so humanists turned for support to the pagan classics. Therefore, the literature of Greece and Rome came to be regarded as a golden ideal in all things. Aeneas Silvio Piccolomini, an early humanist who later became pope, wrote, "Literature is our guide to the true meaning of the past, to a right estimate of the present, to a sound forecast of the future. Where letters cease, darkness covers the land; and a Prince who cannot read the lessons of history is a hopeless prey of flattery and intrigue." In the same spirit, Machiavelli found it natural to support his realistic advice on the conduct of politics by references to Livy's *History of Rome*.

But the appeal to classical literature was, at its best, an appeal to its spirit. Humanism was not a literary but an intellectual movement, a shifting of values and a new self-consciousness of the human spirit: it was, in the words of the scholar Ferdinand Schevill, "a movement of the human mind which began when, following the rise of the towns, the urban intelligensia slowly turned away from the transcendental values imposed by religion to the more immediately perceptible values of Nature and of man."[1]

In the setting of those times, it was, of course, impossible for humanists to think themselves antireligious. Like all reformers, they felt their protest to be a protest only against abuses of religion. They criticized churchmen and scholastic philosophers; but in this, they felt, they were not opposing Christianity, they were merely correcting the errors which the medieval church had put on it. When Lorenzo Valla, a papal secretary, wrote a book which he called *Pleasure as the True Good,* he insisted that its moral was to show that elegant living was an expression of Christian virtue.

III

The theme of Christian virtue ran through Renaissance humanism, all the way from Petrarch to Erasmus. The splendid flesh tones of Raphael and the lyrical treatment of naked muscle in Michelangelo were, to these humanists, elements in a devotional art. The greatest architectural monuments of the Renaissance are churches; its greatest books take moral virtue as their theme. The clearest note in Renaissance literature is a constant wish to show that virtue in the Greek sense and in the Christian sense are one.

There were, indeed, elements of Greek Stoicism in the model of Christian virtue which the medieval church had set up. But, at bottom, the link which the humanists tried to find between the medieval Christian vision and the vision of the Greeks was false. The church idealized the ascetic and monastic virtues, and allowed man the pleasures of the flesh only because man was by nature weak. By contrast, the pagan vision glorified the flesh; and for a time the humanists converted at least some leaders of the church to accepting this vision. For a time, humanism persuaded the church to take as its ideal the complete, the universal man.

In doing so, humanism had to attack the monastic virtues, and therefore had to represent these as false doctrines imposed on the

[1] Ferdinand Schevill, *History of Florence* (New York, 1936), pp. 316–317. For a bibliography on humanism, consult Federico Chabod, *Machiavelli and the Renaissance* (London, 1958).

true structure of Christianity. The work which made the reputation of Erasmus was a bitter satire on this theme, *The Praise of Folly,* in which he mocked both the monastic life (he had spent six unhappy years in a monastery) and the indulgences and abuses of the church.

The attack on abuses is always an attractive refuge for those who do not want to be deeply involved in principle. By making fun of superstition, by showing the bigot at his most absurd, the critic can keep aloof from the deeper issues which drive men to commit themselves. But the critic deceives himself if he thinks that the attack on an established way of life can stop at what seem to be its accidental faults. What Erasmus said about the corruptions of the church in fun, Luther soon said in earnest. And for Luther these corruptions became not accidents but essentials—evils which grew out of the structure of the Catholic Church itself. Humanism undermined the belief in medieval tradition and practices, and inevitably Luther turned its attack into a new theology.

Even the scholarship of the humanists had the effect of destroying the respect for the medieval church. When Lorenzo Valla studied the *Donation of Constantine,* he proved it to be a papal forgery; and other critics uncovered the spurious history of other Christian texts. Research in history and in languages, which flowed naturally from the interest in classical literature, turned out unexpectedly to throw doubt on the honesty of much that was revered in the church. As a result, the authority of the church came to be doubted in other fields, and Aristotle and the Christian Fathers were no longer accepted as infallible. Luther took advantage of these infectious doubts, though ironically the new dogmatism that he created soon sustained itself by means no more scrupulous than the old.

IV

Erasmus was an illegitimate child, as Leonardo da Vinci had been; and like Leonardo, he seems to have felt the slur. As a young monk, he believed that for this reason he could hope for no great career in the church. And when he was at the apex of his fame, in 1516, he wrote to the pope in some embarrassment to ask him to lift the bar by which, as an illegitimate child, he could not legally hold church office.

Erasmus' childhood, however, was not unhappy or isolated. His parents lived together and had another son, and Erasmus went to a school run by the lay society of the Brethren of the Common Life. Here the stress was on the spiritual teachings of Christ, on the Bible, and on the good life.

These years of simple piety ended when Erasmus was 14; his mother died of the plague, and his father died soon after. His guardian was anxious to be rid of responsibility for the two boys, and had them prepared for the monastery. There was no escape; reluctantly, Erasmus became an Augustinian monk at the age of 21. Even in the monastery, the writers he cared for were Aeneas Silvio Piccolomini and Lorenzo Valla. His first work, which he called *The Book Against the Barbarians,* was modeled on Lorenzo Valla, and argued that the new learning of the pagan writers was not opposed to Christian virtue.

In 1492, Erasmus became a priest and was able to move from the monastery to the court of the Bishop of Cambrai; and at last, in 1495, he was able to go to the University of Paris, the most famous school in Europe. But here, where he had hoped for a new spirit, he found that the theology again shocked and disappointed him: the scholastic arguments were empty. Paris was under the influence of the followers of Duns Scotus, who derived from the idealistic philosophy of Plato and St. Augustine. But the same preoccupation with formal detail and with empty propositions also filled the followers elsewhere of Thomas Aquinas, and the followers of William of Ockham, who each in different ways derived from the more empirical writings of Aristotle. All three schools of the Schoolmen were remote from the content either of religion or of daily life. As Erasmus wrote privately: "Those studies can make a man opinionated and contentious; can they make him wise? . . . By their stammering and by the stains of their impure style they disfigure theology which had been enriched and adorned by the eloquence of the ancients. They involve everything whilst trying to solve everything."

The Schoolmen, who repeated the traditional philosophies either of Plato or of Aristotle, were bitter opponents of the New Learning. Erasmus describes their bigotry in his *Letters:*

It may happen, it often does happen, that an abbot is a fool or a drunkard. He issues an order to the brotherhood in the name of holy obedience. And what will such an order be? An order to observe chastity? An order to be sober? An order to tell no lies? Not one of these things. It will be that a brother is not to learn Greek; he is not to seek to instruct himself. He may be a sot. He may go with prostitutes. He may be full of hatred and malice. He may never look inside the scriptures. No matter. He has not broken any oath. He is an excellent member of the community. While if he disobeys such a command as this from an insolent superior there is stake or dungeon for him instantly.[2]

And Erasmus saw that the formalism which withered the minds of these men also withered their lives. If thinking was merely an arrangement of traditional arguments, then living was merely an arrangement of traditional observances. In 1501, Erasmus wrote a *Handbook of a Christian Warrior* in which he contrasted this mechanical worship with the warm piety of the Brethren of the Common Life.

Many are wont to count how many masses they have heard every day, and referring to them as to something very important, as though they owed Christ nothing else, they return to their former habits after leaving church. . . . You worship the saints, you like to touch their relics; do you want to earn Peter and Paul? Then copy the faith of the one and the charity of the other and you will have done more than if you had walked to Rome ten times.

Erasmus put this contrast into many forms in his book. Historically the most striking form is this.

Thou believest perchance all thy sins and offenses to be washed away at once with a little paper or parchment sealed with wax, with a little money or images of wax offered, with a little pilgrimage going. Thou art utterly deceived and clean out of the way!

It was the abuse of indulgences which tipped over Luther's patience in 1517; and the sentences of Erasmus are, therefore, the prophetic rumblings, sixteen years before the thunderclap, of the storm which was drawing together over Rome.

[2] J. A. Froude, *Life and Letters of Erasmus* (New York, 1894), p. 68. Froude's book is still valuable and interesting. Another book containing selections from Erasmus' letters is the splendid work by J. Huizinga, *Erasmus of Rotterdam*, which, written in 1924, has been translated by F. Hopman and is published in an excellent edition by the Phaidon Press (London, 1952) with 32 accompanying plates. Huizinga's book gives a fine portrait of Erasmus and his spirit. Margaret Phillips, *Erasmus and the Northern Reformation* (London, 1949), is a very satisfying treatment of Erasmus seen against the background of Reformation events. For specialized information on the educational background of Erasmus' life, see Albert Hyma, *The Youth of Erasmus* (Ann Arbor, 1930).

V

When the *Handbook of a Christian Warrior* was written, Erasmus had already made a visit to England in 1499 which deeply changed his life. There he had met Thomas More and other English humanist scholars, among them Grocyn, Linacre, and Colet. They were devout and even ascetic men, but their virtues seemed to grow naturally out of their personalities, and their lives and their minds were of a piece. Among these English idealists, Erasmus felt, Christianity was truly an expression of the spirit, and of the classical spirit. Argument and worship were not brittle forms here; the search for truth was generous; and faith was not, as he had felt it to be in Paris, a dead superstition.

Erasmus had always longed for the liberal and humane vision of the classics, and had always believed that it expressed the best in Christianity. Now he felt that he saw that best in action, and that Christianity could be an expression of broad and tolerant virtues, of the whole man. In the houses of Sir Thomas More and his friends, Erasmus could feel that his longing was realistic, and that he in his own person could bring this vision to Europe. This, he saw, should be his life's work: the reconciliation of the classics with Christianity. To a later age, the noble savage became the model for a natural morality; to Erasmus, the simplicity of the classics spoke with the same inspiration. The classics were a natural gospel; reading Cicero and other moralists, he was carried away: "A heathen wrote this to a heathen, and yet his moral principles have justice, sanctity, truth, fidelity to nature, nothing false or careless in them. . . . When I read certain passages of these great men I can hardly refrain from saying, St. Socrates, pray for me."

All this Erasmus believed, but believed in part on hearsay, from his English friends. For, in fact, Erasmus, like others trained in the monastery, did not at this time read Greek. Yet his belief was so strong that he at once began to learn Greek when he went back to Paris, though he was already 34, was in need of money, and was often ailing. He wrote, "I am determined, that it is better to learn late than to be without that knowledge which it is of the utmost importance to possess. . . . We see, what we have often read in the most weighty authors, that Latin erudition, however ample, is crippled and imperfect without Greek." And he mastered Greek in three years.

He now began to translate, to edit, and to popularize the works of antiquity. He had already published, in 1500, a collection of about 800 *Adages* or tags from the Latin classics which, like the collection of wise saws which Benjamin Franklin made later, went through countless popular editions. He enlarged this in time to more than 3000 sayings, many of them now drawn from Greek authors. He translated Aristotle and Euripides, Plutarch and Lucian and Seneca.

At the same time, it was part of Erasmus' sense of his own mission that he should also translate and edit Christian documents. His work here has been called "the foundation of modern critical study of the Bible and the Fathers."[3] He published editions of a number of Church Fathers, among them St. Jerome and St. Augustine. The great edition of St. Jerome was printed by the famous Swiss printer Froben in nine volumes in 1516.

St. Jerome had translated the Greek Bible into Latin, and this translation was the accepted Vulgate. This was the center of Erasmus' interest in St. Jerome, and in the same year, 1516, he printed his own translation of the Bible, in Greek and Latin together. On one page stood the Greek text as Erasmus had revised and edited it, and on the opposite page his translation into Latin, which differed markedly from the Vulgate of St. Jerome. Erasmus felt that he was giving the Bible freshly to common men, as the Brethren of the Common Life had given it to him. He wrote in his preface: "I wish that all women might read the Gospel and the Epistles of Paul. I wish that they might be translated into all tongues of all people, so that not only the Scots and the Irish, but also the Turk and the Saracen might read and understand. I wish the countryman might sing them at his plough, the weaver chant them at his loom, the traveler beguile with them the weariness of his journey." In a few years, Luther broke tradition still more abruptly by translating the Bible into the everyday language of his country, German.

VI

For some years from 1504 on, Erasmus had traveled through Europe, and in particular had spent time in Italy. In those years some of the greatest Renaissance painters were pouring out their work:

[3] P. S. Allen, *The Age of Erasmus* (Oxford, 1914), p. 4. A more popular and superficial treatment of Erasmus and his times is Stefan Zweig, *Erasmus of Rotterdam* (New York, 1934).

Raphael and Michelangelo in Rome, Giorgione and Titian in Venice, and many others. It is an odd quirk of character that the humanist Erasmus took no interest in their art. With so subtle a gift of thought, with so rich a gift of words, he plainly had no gift of visual imagination; like another great humanist and satirist of a later age, George Bernard Shaw, he had no sensuous appreciation of the color, the texture, the shape of things. He may also have been lacking in sensuality in his private life.

When Henry VII died in 1509, Erasmus' English friends urged him to come there in the hope that he might find advancement under the new king, Henry VIII. He left Italy at once; and it was while he was crossing the Alps on his way to England that he conceived the idea of writing his famous satire on monkish life. He wrote the satire in a week in the house of Sir Thomas More, with whom he again stayed in England; and by way of acknowledgment, he gave it a title which was meant as a pun on the name of More: *Moriae Ecomium,* or in English, *The Praise of Folly.*

The Praise of Folly was published in 1511, and was at once read with delight everywhere. It was printed in many languages and editions, and in 1517 Hans Holbein the younger, who was then 20, added a set of marginal drawings to it. It inspired many other satiric books, among them those of Rabelais.

The satire in *The Praise of Folly* seems oddly lacking in humor to us now. The attack on the formalism of churchmen and the greed and stupidity of monks is not noticeably gayer than it had been in Erasmus' serious books. For example, Erasmus writes in *The Praise of Folly:*

> Perhaps it were better to pass over the theologians in silence, and not to move such a Lake Camarina, or to handle such an herb *Anagyris foetida,* as that marvellously supercilious and irascible race. For they may attack me with six hundred arguments, in squadrons, and drive me to make a recantation; which if I refuse, they will straightway proclaim me an heretic. . . . They are protected by a wall of scholastic definitions, arguments, corollaries, implicit and explicit propositions. . . . The methods our scholastics pursue only render more subtle these subtlest of subtleties; for you will escape from a labyrinth more quickly than from the tangle of Realists, Nominalists, Thomists, Albertists, Occamists, Scotists.[4]

[4] We use here the translation of *The Praise of Folly* by Hoyt H. Hudson (Princeton, 1941).

There is little to distinguish this from the earlier text by Erasmus in his own person on the dreary disputation of the Schoolmen. Yet, to his own generation, Erasmus in *The Praise of Folly* seemed somehow nimbler and more carefree; it was possible to side with him in laughter without being committed to a more profound criticism. The fool was a familiar device in the tales of the times; and by speaking in the universal person of the fool, Erasmus made himself one with all his readers.

Erasmus was speaking the discontent of his age, in his satire as much as in his serious translations. The monks and the Schoolmen had ceased to be a vital intellectual force; they no longer reached the minds of their hearers, nor did their own minds give anything fresh to their doctrines. Thus the churchmen no longer commanded intellectual respect. But since they claimed that their doctrine spoke to men's minds, there was no other form of respect that could be given to them. They were, therefore, seen simply as figures of pomp, offering empty words of superstition.

The age had had enough of clerical pomp and of obedience without respect. In fun and in earnest, Erasmus voiced the discontent of the powerful minds of the age; and princes and popes heard him with pleasure and were his friends. The simple minds of the age felt the same discontent; but for them it was voiced more dramatically by Martin Luther.

VII

Martin Luther nailed his Ninety-Five Theses on indulgences to the door of the church at Wittenberg on October 31, 1517—the eve of All Saints' Day. With that gesture, he turned discontent into action. The church could no longer smile at its own weaknesses, as the popes who befriended Erasmus had done.

Luther had studied the works of Erasmus and had been guided by them—by the *Adages*, by *The Praise of Folly*, and above all by Erasmus' edition of the Greek New Testament, which Luther used as the basis of his own lectures. In 1516, he had prompted a friend to write to Erasmus to criticize his interpretation of St. Paul's Epistle to the Romans—characteristically a text on which the liberal and the zealot would fall out. Luther sensed from the outset that Erasmus was not, either in temperament or in opinion, a man to go far enough for him. Six months before he nailed up the Theses, Luther already wrote about Erasmus that "human considerations prevail with him much more than divine."

Erasmus was a supporter of the Ninety-Five Theses, in principle; he sent copies of them to Thomas More and to Colet in England, with a letter of approval. But Erasmus was not—and this again both by temperament and by opinion—a man to push the criticism of the church so far that both sides would find themselves committed to positions which allowed no movement. A year after the Theses, in October, 1518, Erasmus wrote to a supporter of Luther, John Lang, approving them but pointing out that their result was likely to be just this: that those who were allied to the church would be forced to take up an inflexible position. "I see that the monarchy of the Pope at Rome, as it is now, is a pestilence to Christendom, but I do not know if it is expedient to touch that sore openly. That would be a matter for princes, but I fear that these will act in concert with the Pope to secure part of the spoils."

Luther had now been commanded by the church to recant, and had refused. In general, the humanists supported him. Erasmus' Swiss publisher, Froben, printed a book of Luther's pamphlets. Their violence alarmed Erasmus; he was both more timid and more farsighted than others; above all, he saw that humanism itself, the revival of learning, the cause of "good letters" which he had nursed so long, would be threatened. He wrote privately to Froben to advise him not to publish Luther's writings, "that they may not fan the hatred of the *bonae literae* still more."

Meanwhile, Luther in his first struggles needed what support he could get, and he particularly needed the open support of Erasmus. He therefore wrote to him in March 1519:

> Greeting. Often as I converse with you and you with me, Erasmus, our glory and our hope, we do not yet know one another. Is that not extraordinary? . . . For who is there whose innermost parts Erasmus has not penetrated, whom Erasmus does not teach, in whom Erasmus does not reign? . . . Wherefore, dear Erasmus, learn, if it please you, to know this little brother in Christ also; he is assuredly your very zealous friend, though he otherwise deserves, on account of his ignorance, only to be buried in a corner, unknown even to your sun and climate.

But Erasmus was not to be drawn. In his reply, he carefully dissociated himself from Luther's writings:

> Dearest brother in Christ, your epistle showing the keenness of your mind and breathing a Christian spirit, was most pleasant to me. I cannot tell you what a commotion your books are raising here [at Louvain]. These men cannot be by any means disabused of the suspicion that your works are written by my aid and that I am, as they call it, the standard-bearer of your party. . . . I have testified to them that you are entirely unknown to me, that I have not read your books and neither approve nor disapprove anything. . . . I try to keep neutral, so as to help the revival of learning as much as I can. And it seems to me that more is accomplished by civil modesty than by impetuosity.[5]

What Erasmus wanted from both sides was moderation. He did not want Luther to be wronged: on the contrary, he tried to guard him from persecution, and he even wrote to the Archbishop of Mainz to plead for Luther's safety—and this though the indulgences which Luther had attacked in his Theses had been preached precisely for the coffers of this Hohenzollern archbishop.

At the same time, Erasmus wanted Luther to be moderate. In encouraging John Lang, he wrote in a tone which is wishful to the point of being absurd: "All good men love the freedom of Luther who, I doubt not, will have sufficient prudence to take care not to allow the affair to arouse faction and discord." It was, in fact, absurd to believe that, in such a quarrel, either side could be reasonable. And Erasmus knew that Luther was a less moderate man, indeed less a humanist, than many church dignitaries. What made Erasmus helpless was that he believed Luther's criticisms of the church to be just, but that he knew that they would merely entrench in the church the uncompromising men, the monkish bigots whom the humanists had worked so hard to displace. If Luther was defeated, then the reactionaries would also sweep away all that the humanists had gained. "I am deeply disturbed about the wretched Luther. If they pull this off, no one will be able to bear their intolerance. They will not be quiet until they have utterly ruined the study of languages and 'good

[5] Preserved Smith, *Erasmus* (New York, 1923), p. 222. For Erasmus' constant opposition to conflict and his interest in peace, see José Chapiro, *Erasmus and Our Struggle for Peace* (Boston, 1950), which also offers a translation of Erasmus' "Peace Protests."

letters.' . . . This tragedy has sprung from the hatred of 'good letters' and the stupidity of the monks."

In the summer of 1520, a papal bull declared Luther a heretic, giving him sixty days to recant or be excommunicated. Luther's answer was to burn the papal bull, and the canon law with it, in public. After this, in spite of further searches for a compromise, there was in effect no going back. Erasmus was already under attack from the University of Louvain, where he had lived since 1517, and where the churchmen now accused him of double dealing. The church was making it clear that those who were not openly against Luther must be counted to be for him. Albrecht Dürer made a last appeal to Erasmus to take the side of Luther, at a time when Luther was thought to be dead or in hiding: "O Erasmus of Rotterdam, where will you be? Hear, you Knight of Christ, ride forth beside the Lord Christ, protect the truth, obtain the martyr's crown. . . . I have heard you say that you have allowed yourself two more years, in which you are still fit to do some work; spend them well, in behalf of the Gospel and the true Christian faith. . . . O Erasmus, be on this side, that God may be proud of you."

Luther had recently appeared before the Diet of Worms, in April, 1521, but had refused to retract anything of his doctrine. Duke Frederick of Saxony, with prompt political foresight, had had him seized on his return from Worms and hidden from the coming storm. It was this defensive stroke which had set off the rumor that Luther was dead. And indeed, as the duke had foreseen, the emperor almost at once gave in to the papal persuasion, and signed the edict which outlawed Luther and commanded his books to be burned.

Erasmus knew that he was not the man for such heroics: and in his view, the heroics had already done harm to his cause. With that unposturing simplicity which gives all his writings their modest personal air, he wrote sadly: "All men have not strength for martyrdom. I fear lest, if any tumult should arise, I should imitate Peter. I follow the just decrees of popes and emperors because it is right; I endure their evil laws because it is safe."

VIII

It was not only his temperament which made Erasmus retreat from the side of Luther. He found Luther's opinions more and more distasteful. He did not care for his German nationalism, for his fanaticism, his intolerance, and above all for his belief in the essential helplessness of man under the divine will. For Luther was now outspoken in beliefs which we should call Calvinist, and which left no room for the humanist belief in the goodness of man. To Erasmus, Luther's belief in predestination was no better than the medieval belief in original sin.

Therefore, when the church pressed Erasmus to speak out against Luther, he chose an issue, Free Will, on which he was indeed intellectually opposed to Luther and to the rising shadow of Calvinism. Luther replied by writing *The Bondage of the Will,* and left no doubt that there was no longer common ground between them. He sent a copy of *The Bondage of the Will* to Erasmus, with a letter which at last stung Erasmus to speak his mind about the destruction of his ideals:

> Your letter was delivered to me late and had it come on time it would not have moved me. . . . The whole world knows your nature, according to which you have guided your pen against no one more bitterly and, what is more detestable, more maliciously than against me. . . . The same admirable ferocity which you formerly used against Cochlaeus and against Fisher, who provoked you to it by reviling, you now use against my book in spite of its courtesy. How do your scurrilous charges that I am an atheist, an Epicurean, and a skeptic help the argument? . . . It terribly pains me, as it must all good men, that your arrogant, insolent, rebellious nature has set the world in arms. . . . You treat the Evangelic cause so as to confound together all things sacred and profane as if it were your chief aim to prevent the tempest from ever becoming calm, while it is my greatest desire that it should die down. . . . I should wish you a better disposition were you not so marvelously satisfied with the one you have. Wish me any curse you will except your temper, unless the Lord change it for you.[6]

Alas, Erasmus had not succeeded in mollifying the church either. He left the Catholic University of Louvain and went to Switzerland. Catholic hotheads insisted that he was the man who "laid the eggs which Luther and Zwingli hatched." Although Erasmus protested that "I laid a hen's egg; Luther hatched a bird of a different breed," the eggs were all broken together. *The Praise of Folly* was placed on the index of forbidden books; the work on the New Testament was

[6] Smith, *op. cit.*, pp. 354–355.

expurgated; and Erasmus himself was condemned by the Council of Trent as "an impious heretic." His cause had failed; he was at home in neither of the two camps now at war; and he had lived beyond his time.

What had failed when Erasmus failed was not a man but an outlook: the liberal view. He gave his life to the belief that virtue can be based on humanity, and that tolerance can be as positive an impulse as fanaticism. Above all, he believed in the life of the mind. He believed that thoughtful men would become good men, and that those who knew and loved the great writings of all ages must live more justly and more happily in their own age.

When Erasmus was appointed to the court of the future Emperor Charles V in 1516, he wrote for him *The Education of a Christian Prince.* The word "Christian" in the title points the contrast to *The Prince,* which Machiavelli had written three years before; and so do the opening words of Erasmus' dedication, that "no form of wisdom is greater than that which teaches a Prince how to rule *beneficently.*" But the sense in which Erasmus used the word "Christian," his longing for universal good, could not survive the violence of both sides in the coming struggle.

Part of that struggle was national; Luther was very German, and the Reformation of Henry VIII was very English. In this also Erasmus was out of place; he had hoped to make humanism a movement of universal peace from one end of Europe to another. And in his great years, he had traveled Europe as if this empire of the mind, this free Christian community, had already been created. For a time the courts of Italy and England, the universities of France and Spain, the houses of cardinals and reformers were open to him. But the time was short, and it has not returned.

CHAPTER 6

The Reformation

WE COME directly now to a moment in history which we have been approaching and retreating from in our previous chapters. In talking about Leonardo and Machiavelli, the name of Savonarola, the reformer of Florence, was mentioned. We have discussed the changes in the relation between king and church in England which cost Sir Thomas More his head. And, in the previous chapter, we dealt with More's friend, Erasmus. All these were outriders of the great moment when Martin Luther posted his Ninety-Five Theses on the church door at Wittenberg and set in motion the series of events which we call the Reformation.

The consequences of the Reformation were manifold. It divided Europe, religiously. It gave to Europe, as a result of the religious wars which stemmed from it, the political shape which, more or less, it has kept ever since. And it supplied the European mind with a new ethos, a whole new sensibility, and a stock of novel political, social, and economic ideas. At a further remove, the Reformation carried forward into the current which later took the Pilgrim fathers out of England and led to the colonization of North America.

We shall be concerned mainly with the Reformation itself and its consequences; but first let us sketch a background for it by briefly recalling some of its predecessors.

THE PRECURSORS

I

It would be erroneous to think that the first time that anybody had ever found fault with the church was when Martin Luther nailed his Theses to the door at Wittenberg. The Catholic Church, in one form or another, had survived for 1200 years or more, and no institution survives for so long a time wihout internal rebellions. A glance at such figures as Peter Waldo, St. Francis of Assisi, John Hus, or Joan of Arc brings home the realization that great challenges had always confronted the church.

Sometimes the church dealt with these challenges by absorbing them and reforming itself: thus, there were waves of reform all through the Middle Ages, such as the Cluniac (eleventh century), the Cistercian (twelfth century), and the Franciscan (thirteenth century). Sometimes she dealt with them by anathematizing them as heresies and launching crusades against them: thus we have the Waldensian heresy (eleventh century), the Albigensian Crusade (thirteenth century), and the burning of John Hus at the stake (fifteenth century).

The church knew very well the dangers involved not only in protest but in mysticism; for example, St. Teresa, St. Joan of Arc, and even Ignatius Loyola, founder of the Jesuits, all lay under heavy suspicion, and sometimes persecution, by the church. Indeed, the line between accepted reform and rejected heresy was thin at times and, often, more the result of the existing political situation than of doctrinal determinants. Thus the Franciscans came very close to being judged heretical, and the Waldensians to being tolerated within the church.

Luther, of course, was not tolerated. Nor, on the other hand, was he successfully suppressed. The reasons for this are deeply embedded in the history of the Reformation. Before we deal with that history, however, let us describe briefly a comparable case, but one which ended in failure and in flames at the stake—the story of Savonarola.

II

Savonarola, although not part of the Reformation, gives us a typical indication of how it worked. He was, at the beginning of his career, a priest of little distinction, whose early preaching made no great impression. Gradually, however, he developed an evangelical and zealous style of oratory. Then, from his post as prior of San Marco he was brought, because he had impressed Pico della Mirandola, to the attention of Lorenzo the Magnificent, around the year 1490.

Savonarola's new type of preaching—in which people fell into ecstasy and wept—took Florence by storm. Savonarola, like Luther after him, achieved much of his success by painting lurid word pictures of the devil and hell. Through the hold on public opinion created by his preaching, Savonarola virtually controlled Florence by 1492. When, in that same year, Lorenzo the Magnificent was dying, he called Savonarola to his deathbed and asked for absolution. No one knows exactly what happened, but a wonderful story attaches to this event which highlights the coming drama of the Reformation.

Savonarola supposedly imposed three conditions before he would grant Lorenzo absolution.

1. He was truly to repent. This, a religious condition, the dying Lorenzo accepted in spite of his proclivities for the life of a Renaissance ruler.

2. He should give up his "ill-gotten" wealth. After much debate, it is said Lorenzo also accepted this condition, which attacked splendor.

3. He should renounce, on behalf of the Medici, any claim to rule Florence and allow the city to become a democracy (which meant, in this case, to be run by Savonarola). This condition Lorenzo refused because it was essentially a political condition. As a result, according to the story, Savonarola withheld absolution from him.

This story is interesting because it contains, in a way, a time capsule of the three major characteristics of the Reformation: the purely religious aspect; the protest against wealth and splendor; and the political issue. These three elements were inextricably involved in the Reformation.

Savonarola ruled Florence until the year 1498. During his sway, a great many of the works of art in Florence were destroyed. Savonarola constantly called on the population to abjure its immoral ways, to burn its books (Petrarch's poems, for example, were given to the flames), pictures, vanities, and treasures, and to return to the simple ways of God. His exhortations were even effective in turning some of the prominent artists, like Botticelli and Michelangelo, toward more religiously inspired paintings.

Savonarola regarded himself as the instrument of God and held that nothing must stand in his way. To enforce his new morality, he instituted a form of thought control, in which boys spied on their elders (we shall see more of this in Calvin's Geneva; it is the new form of the Inquisition). He did not, however, attack church dogma, but restricted his work to reviving religious fervor and morality.

Nevertheless, after a time, his independence of action incurred the opposition of the Borgia papacy. Between the enmity of Rome and the inevitable Florentine reaction against his rigid restrictions, Savonarola's rule of Florence was doomed. He met his end in 1498 at the stake, condemned by the very mob whose emotions he had aroused so many times before. Unrooted in criticism of fundamental dogma, left without the backing of prince or people, Savonarola's attempt at reform of church and society had failed. It was really a revivalist movement rather than a true reformation.

Luther

I

The Reformation proper was dominated by the figure of Martin Luther. He was born in 1483 and lived until 1546. It used to be a commonplace that Luther was the son of very poor parents, who earned their scanty living as peasants. More recent scholarship, however, has shown that Luther's father was a miner in the Mansfeld district of Saxony, who gradually rose to possession of furnaces and mines of his own. Thus, rather than a peasant, Luther's father was a skilled workman who was rising in the world and wished his son to rise also by becoming a lawyer.[1] In furtherance of this aim, after preliminary schooling in his birthplace of Eisenach, Luther, at about the age of 17, entered his name at the University of Erfurt.

When Luther attended Erfurt, it was the most flourishing university in Germany. It was also the center of the conflict, peaceful enough

[1] See E. G. Schwiebert, *Luther and His Times* (St. Louis, 1950), pp. 106–109. This is an excellent book, with much new material on various aspects of Luther's life and work. Standard works on Luther in English, all from a Protestant viewpoint, are James Mackinnon, *Luther and the Reformation*, 4 vols. (New York, 1928); Roland H. Bainton, *Here I Stand: A Life of Martin Luther* (New York, 1950), which has a bibliography and some lovely illustrations; Robert H. Fife, *The Revolt of Martin Luther* (New York, 1957), with a bibliography; and Heinrich Boehmer, *Luther and the Reformation in the Light of Modern Research*, tr. by E. S. G. Potter (London, 1930), and *Road to Reformation, Martin Luther to the Year 1521*, tr. by John W. Doberstein and Theodore G. Tappert (Philadelphia, 1946). An excellent work by a Catholic Luther specialist is Hartmann Grisar's *Luther*, tr. by E. M. Lamond, 6 vols. (London, 1913–1917).

at first, between the scholastics and the humanists. Luther, who enrolled in the faculty of philosophy, which prepared both for law and theology, seems not to have taken a definite side. However, he did take time out from his scholastic studies to read the Latin classical authors, such as Cicero and Virgil. He obtained his master's degree, in spite of these literary excursions, in a short time and in excellent fashion, standing second among the seventeen successful candidates. In 1505, a flourishing legal career seemed to lie ahead of Luther.

II

But at this point, aged about 21, Luther rejected the world. It is not very clear how, but in 1505 a conversion of some kind came over him. His own story is that he was struck by the omnipresent hand of God—the feeling that God was in everything—when he took refuge from a thunderstorm in the countryside. In any case, his mind must have been occupied by doubts and burdened by an overwhelming sense of guilt and fear for some time before this, and the incident of the thunderstorm only served to bring matters to a head. Luther's soul was unquestionably in torment, but the only words of his we have on the matter are that "he doubted of himself."

To relieve his doubt and sense of anxiety, he joined the Augustinian order of hermit monks. To allay his fears of not being saved if he remained "in the world," he fasted, prayed, and scourged himself relentlessly. This tremendous spiritual struggle gave him his stanchness of character and a sense of inner power. Freed from many of his mental uncertainties, Luther emerged with an awareness of confidence and freedom.

Ordained a priest, Luther continued with his theological studies, reading especially Augustine and Bernard. He also showed himself capable of handling administrative matters, and his abilities attracted to him the attention of Staupitz, the head of the order. In 1508, Staupitz saw to it that Luther was transferred from the monastery at Erfurt to Wittenberg. He was transferred there in order to teach and help at the new university which, founded in 1502 by the Elector Frederick of Saxony, was under Staupitz's administration.

Luther was chosen by Staupitz to succeed him as professor of philosophy, and rapidly became the leader in the fight to make Wittenberg a center of humanism rather than of scholasticism.[2] There is no doubt that Luther's sympathies here were with the humanists, that he based his lectures on the new philological and critical investigations of men like Valla (for example, in 1537, Luther translated Valla's work on the *Donation of Constantine* into German) and that he considered himself at first to be a follower of Erasmus.[3]

Nevertheless, Luther personally was more interested in preaching a religion of piety than in the study of polite letters. Shunning both the aridness of the scholastics and the worldliness of the humanists, he devoted himself to discovering God. Thus when, in 1510, a trip to Rome on business of his order interrupted his studies and his preaching, Luther's attitude toward the imperial city was that of a pilgrim rather than a humanist scholar. He climbed on his knees the steps of St. Peter's, which gave remission of penance; he knelt before altars and stood in awe before pious relics. Only one thing shocked him—the immoral life of the priests and cardinals; he was dismayed at the cynical and scoffing attitude they held to the church rites which they performed.

Returning to Wittenberg in 1512, Luther resumed his lectures and his preaching. He ignored completely the great scholastic systems of the Middle Ages and concentrated on expounding such documents as the Psalms, using Augustine and Bernard as aids, and the Epistles of St. Paul. But, in general, up until 1517, there was no particular reason to think that Luther was a dissatisfied member of the Church of Rome. That year, however, was to mark the crisis of the Reformation, when Luther made his famous protest against indulgences.

III

Previously, we have seen many strands running separately: the story of Savonarola, the writings of Erasmus, the business of Henry VIII

2 Cf. Schwiebert, *op. cit.*, pp. 194–195 and 268–302. José Luis Aranguren, in his very interesting book, *Catolicismo y Protestantismo como Formas de Existencia* (Madrid, 1952), makes the following comment: "Luther's antirationalism was favored by his cultural formation. Educated at Wittenberg [this, of course, is not quite true; Luther was educated mainly at Erfurt], a university without tradition, founded six years before his entrance, Luther there learned, above all, the 'modern' theology of Occanism. The traditional doctrine, the 'via antiqua,' he scarcely learned or practiced" (n. 21, p. 44).

3 For Luther's humanist background, cf. Preserved Smith, *The Age of the Reformation* (New York, 1920), p. 49, and Schwiebert, *op. cit.*, pp. 215–282.

and More brewing in England. These were all gusts and eddies of thought and action anticipating the Reformation storm. The discharge came on October 31, the eve of All Saints' Day, in 1517. On that day, Luther nailed to the door of the Wittenberg Church the Ninety-Five Theses relating to indulgences, offering, in typical academic fashion, to dispute them against all comers. The protestation at the end was bold:

> I, Martin Luther, Doctor, of the Order of Monks at Wittenberg, desire to testify publicly that certain propositions against pontifical indulgences, as they call them, have been put forth by me. . . . I implore all men, by the faith of Christ, either to point out to me a better way, if such a way has been divinely revealed to any, or at least to submit their opinion to the judgment of God and the Church. For I am neither so rash as to wish that my sole opinion should be preferred to that of all other men, nor so senseless as to be willing that the word of God should be made to give place to fables, devised by human reason.

The day chosen by Luther—All Saints' Day—showed his powers as a propagandist, for the town of Wittenberg was crowded with peasants and pilgrims come to honor the consecration of the church. Rapidly, word of the Theses spread through the crowd, and, spurred on by Luther's university friends, many people called for their translation into German. As a result, Latin and German versions were sent to the university press, whence they soon spread throughout Germany. It was, therefore, the new invention—the printing press—which permitted Luther to obtain wide support so quickly and so fully.

What was Luther actually attacking, why, and who supported him? These are the pertinent questions. They are, of course, linked, but we must try to answer them one at a time.

The particular indulgence which raised Luther's wrath was being sold in Germany by an agent named Tetzel. It was issued to raise money for two causes: for the pope's own construction of the basilica of St. Peter's; and for the new Archbishop of Mainz, who had borrowed money from the Fugger bankers in order to pay the papacy for his confirmation and was now being pressed by the Fuggers to pay back the loan.[4]

[4] For the unsavory financial details of this transaction, cf. Lucien Febvre, *Martin Luther: A Destiny,* tr. by Roberts Tapley (New York, 1929), pp. 73–96.

Luther, however, was also attacking indulgences in general. According to the church, indulgences, which had behind them a long tradition, took their efficacy from the surplus grace that had accumulated through the lives of Christ and the saints. The purchase of an indulgence put the purchaser in touch with this surplus grace and freed him from the earthly penance attached to a particular sin, *but not from the sin itself.*

This, however, was not the story Tetzel was passing out in his sales talk. His sales speech implied that the purchaser was freed from the sin as well as the penance attached to it. He also led his listeners to believe that an indulgence purchased for a relative in purgatory released that relative's soul so that it might forthwith fly to heaven. Tetzel's little ditty was as follows:

> As soon as pennies in the money chest ring,
> The souls out of their Purgatory spring.

Luther's comment on this was short and to the point: "It is certain that when the money rattles in the chest, avarice and gain may be increased, but the Suffrage of the Church depends on the will of God alone" (Thesis 28).

Luther claimed that it was not only Tetzel but the papacy itself which spread the false doctrine of the indulgence. The popes claimed plenary remission of *all* penalties. Luther retorted: "The Pope has neither the will nor the power to remit any penalties, except those which he has imposed by his own authority, or by that of the canons" (Thesis 5). This challenge, however, was not enough for Luther. He insisted that without "true inward penitence" even the pope's remission of penalty was not valid. In Thesis 35, Luther declared: "They preach no Christian doctrine, who teach that contrition is not necessary for those who buy souls out of purgatory or buy confessional licenses."

IV

It is not our purpose to involve ourselves in the intricacies of the indulgence issue. The important thing is to realize that Luther, by attacking indulgences, was attacking the entire theology and church structure which stood behind them. It was a short step from saying that without contrition the indulgence was invalid to saying that contrition alone, without any papal paraphernalia, was sufficient. Thus, by making salvation dependent on the individual's own faith and contrition, Luther abolished the need for sacraments and a hierarchy to administer them.

Luther claimed that he came to this position, and thus to an attack on indulgences, by following St. Augustine (and we would add the mystics). He put his trust in a pious reliance on God and arrived at the belief that faith alone, without the necessity of good works, would bring salvation. The germs of this belief had first stirred in him, apparently, when he had abandoned law and entered the monastery. At that time he had "saved" himself from doubts and anxieties about his salvation by putting his total faith in God. Only gradually, however, had he come to support his inner feeling by a theology, mainly based on the Epistles of St. Paul, and to realize the distance that separated him from Rome.

At first, Luther thought he might reform the church from within. His, however, was not to be one of those reform waves about which we talked earlier; it was to be a heresy which broke the unity of the Catholic Church in the West. Actually, it was not until around 1520, three years after the posting of the Ninety-Five Theses, that Luther's break with the church became evident and acknowledged. After that date, however, reconciliation was no longer really possible.

V

Support for Luther originally came from two sources: humanists and nationalists. The incompatibility of the two was not at first perceived. Erasmus and the other humanists, as we have seen, supported Luther initially in his attempt to reform the existing church. They agreed with his denunciation of the abuses of the church—its simony and peculation, its immorality, and its grubbing for money. They, too, wished to put the emphasis on piety and Christian virtue rather than on dogma and scholastic speculation. But when it became clear that Luther was attacking not only the abuses of the church but the church itself, most of the non-German humanists turned their backs on him.

The princes, knights, merchants, and peasants of Germany, however, did not. Their grievances against Rome were as much political and economic as religious. And these grievances had behind them a long history. As early as 1448, there had been German *gravamina*, or lists of grievances, attacking the "Italian" popes and priests. In 1508, the German Diet resolved not to let money raised by indulgences leave Germany. And in 1518, the Diet of Augsburg stated that the real enemy of Christendom was not the Turk but the "hound of hell" in Rome.[5]

United by these sentiments, much of the German nation looked to the Holy Roman Emperor, Charles V, for leadership. But he seemed to be without sympathy for Germany and was pursuing a universalist policy which he was unable to imagine without Rome as a pivot.[6] Therefore, the German nation turned away from Charles to Luther.

Luther, who had begun with a universal, Christian idea, ended as a German nationalist. Partly under the influence of the knight Ulrich von Hutten, Luther turned from reforming a world church to erecting a German church; his *Address to the Christian Nobility of the German Nation* (1520) marks in its title the mid-point of his evolution. We may, in this matter, consider Luther as the German equivalent of Machiavelli. Luther's words are similar to the latter's "Exhortation to Free Italy from the Barbarians," only in reverse.

Borne by the wave of national sentiment, Luther carried the fight against Rome. He exhorted Germany to seize the lands of recalcitrant churchmen, and his message fell on ready ears. Knights and princes joined in grasping at the lands of the church. The result was that for economic reasons, even if not for political reasons, many of the holders of military power were ready to support Luther's reformation. Thus what had begun as a religious reform became merged with a struggle for political and economic supremacy and identified with social and national aspirations.

VI

Earlier, we said that the Reformation rested on three levels. The first was the purely religious one. For Luther this meant that everyone had to decide in his own conscience how the Word of God should be read. In accord with this belief, he translated the Bible into the

[5] Cf. Smith, *op. cit.*, p. 46.

[6] For the policy of Charles V, see Karl Brandi, *The Emperor Charles V*, tr. from the German by C. V. Wedgwood (New York, 1939).

vernacular.[7] He struck the truly liberal note of his revolt from Rome when he declared: "I wish to be free. I do not wish to become the slave of any authority, whether that of a council or of any other power, or of the University or the Pope. For I shall proclaim with confidence what I believe to be true, whether it is advanced by a Catholic or a heretic, whether it is authorized or not by I care not what authority." In his declaration that it was neither safe nor honest to act against one's conscience, Luther sounded the really important note of Protestantism, one in accord with much of the intellectual movement of the sixteenth century, whether in the field of the humanities or of science.

The second level on which the Reformation rested was the revolt against the splendor with which the papacy had come to surround itself.[8] For example, it was Lorenzo's son, Leo X, who, as pope (1513–1521), enriched the Vatican with its extensive libraries and its priceless Raphaels and Michelangelos. Against this, people like Savonarola, Luther, and Calvin reacted strongly. In fact, one feels that they disliked modern art; that is, the modern art of their day—Leonardo, Michelangelo, and Raphael. They thought that it was all degenerate, and they disliked the notion that it could be approved and supported by dignitaries of the church. They wanted a bare religion, such as Puritanism established later on. Nietzsche recognized this when he asserted that the Reformation was a reaction of old-fashioned minds against the Italian Renaissance.

The third level on which the Reformation rested was in the development of political and social ideas. Luther would never have succeeded as a religious reformer without political support, and this he often purchased at the price of his original beliefs. Thus, believing originally in the efficacy of the word alone, he felt that what did not come from a man's conscience ought not to be forced by coercion. By 1525, however, he had come to justify the use of the

[7] Published in 1534, Luther's translation of the Bible had a tremendous influence on subsequent German literature. Luther, of course, believed that all consciences would conceive of the Bible in the same way; he was shortly to be disillusioned of this belief.

[8] In general, the popes of the time were very unsuitable persons. Alexander Borgia, pope from 1492 to 1503, got himself elected by lavish and unequaled bribery. Julius II, pope from 1503 to 1513, was a warlike general whose military prowess far exceeded his reputation as the Prince of Peace and earned a satire from the pen of Erasmus.

sword in support of the word. This shift paralleled his shift from a universal Christian idea to a German nationalist ideal, which we discussed earlier.

Luther was pushed toward his 1525 political position by three events, or forces: the Knights' War, the Peasants' Revolt, and the Anabaptists. The Knights' War broke out in 1522 as a response to Luther's exhortation to seize the lands of the church. Led by Ulrich von Hutten, the German knights, a class of feudal nobles, attempted to confiscate the lands of the Archbishop of Treves. The archbishop's military resistance, however, was unexpectedly strong, and when some of the territorial princes joined in his defense the knights were routed. The result was that many of the knights, with their feudal resistance to the princes' juridical and territorial claims, were eliminated; and this meant, of course, that one of Luther's independent sources of support was also eliminated. The defeat of the knights emphasized Luther's dependence on the princes.

This dependence was made even more thorough by the Peasants' Revolt. In 1525, the German peasants, goaded to despair by the increased extortions made upon them by the nobles (who were trying in this way to cope with rising prices) and animated by what they took to be Luther's doctrine that the Word of God had declared all men equal, rose in rebellion.

At this point, however, leadership of the revolt passed into the hands of what we may call "left-wing reformers": the Anabaptists. Luther had started the Reformation by rejecting the papal hierarchy and emphasizing the Word of God, to be found in Scripture. The Anabaptists carried this a step further and considered themselves directly in touch with the Holy Spirit and, thus, without need even of the Scriptures. Like the later Fifth Monarchy men of the Puritan Revolution in England, the Anabaptists believed the reign of Christ was imminent and the bloody purification of the church at hand.[9]

The Peasants' Revolt became permeated with Anabaptist doctrine. One recent author, Karl Mannheim, has considered this union of "chiliasm"—or belief in Christ's return to earth to reign during the millennium—with the demands of the oppressed strata of society as

9 The Anabaptists took their name from their opposition to infant baptism, which they opposed on the grounds that infants could not have faith, and without faith the sacrament was invalid.

marking the "decisive turning point in modern history." It introduced the era of social revolution and, according to Mannheim: "It is at this point that politics in the modern sense of the term begins, if we here understand by politics a more or less conscious participation of all strata of society in the achievement of some mundane purpose, as contrasted with a fatalistic acceptance of events as they are, or of control from 'above.' "[10]

Luther would have none of such "social revolution." The peasants, at first, naïvely had thought that Luther and the Anabaptists were preaching the same political and social doctrine. Luther shortly disillusioned them. He reiterated more forcefully his belief that the Gospel held no promise of worldly redemption. He declared that heaven was hereafter and not on earth. And, in advice that he was later to betray, he reminded the peasants that the word, and not the sword, was alone efficacious and to be resorted to.

When the peasants disregarded his moderating counsel, remained under their Anabaptist leaders, and increased their violent deeds, Luther wrote a virulent pamphlet, *Against the Thievish, Murderous Hordes of Peasants*. He advised the princes to "exterminate, slay; let whoever has the power use it." The princes hardly needed this injunction: by 1526 the revolt was crushed, with over 100,000 peasant dead.

The result for Lutheranism was a considerable loss of support by the peasants and the city masses. They turned away from Luther's doctrine that serfdom was necessary in a society of unequal classes. They spurned his view that feudal dues must be retained and that the people must be controlled by forceful authority. In reply, Luther invoked the secular arm to enforce his doctrine and, thus, the Word of God. He thereby tied Lutheranism closely to the cause of the princes.

In its final form, Luther's doctrine was well calculated to appeal to the existing political authorities. In essence, Luther advocated the doctrine of passive obedience. "Neither oppression nor injustice excuses revolt," he declared. He advised the serfs that "the only liberty for which you should care is spiritual liberty; the only rights you

[10] Karl Mannheim, *Ideology and Utopia*, tr. from the German by Louis Wirth and Edward Shils (New York, 1952), pp. 191–192. Although frowned upon by many scholars for its sociological "jargon" and for some Marxist ideas, this is a suggestive and valuable book.

can legitimately demand are those that pertain to your spiritual life."

Luther believed that the faith of a Christian had nothing to do with politics; the duty of a Christian was simply to obey constituted authority. Thus, he could say, in words somewhat like those Bismarck would use much later: "Therefore stern, hard civil rule is necessary in the world, lest the world become wild, peace vanish, and commerce and common interests be destroyed. . . . No one need think that the world can be ruled without blood. The civil sword shall and must be red and bloody."

Turning completely to the princes, Luther confirmed them in the righteousness of their power. The result was an alliance of church and state in which the former was subservient to the latter. In effect, therefore, Lutheranism made a total surrender of the practical life of the individual to state control. As Luther himself declared, "Our teachings have accorded to secular sovereignty the plenitude of its rights and powers, thus doing what the popes have never done nor wanted to do."[11] To secure the freedom and peace of the inner life, Luther had "rendered unto Caesar what was Caesar's."

For Germany, the result of Luther's thought was a division between the inner life of the spirit, which was free, and the outer life of the person, which was subjugated to unattackable authority. This dualism in German thought has lingered from Luther's day to this.

VII

Yet, in the overall picture, Lutheranism contributed to the breakdown of authority. Its anarchical origin is evident, for what Luther desired at the beginning was a return to primitive Christianity. His

[11] Georges de Lagarde, *Recherches sur l'esprit politique de la Réforme* (Paris, 1926) implies that the political ideas of the reformers prepared the way for the totalitarian state along two paths: by freeing the state from moral-religious restraints, and by making the state a moral entity with a mission. According to Lagarde, the reformers trod the two paths because they broke with the medieval tradition of natural law. John T. McNeill, in "Natural Law in the Teaching of the Reformers," *Journal of Religion* (July, 1946), denies Lagarde's assertion but aligns himself with M. E. Chenevière, *La Pensée politique de Calvin* (Geneva, 1937), and the view that Calvin did modify the tradition of natural law to give increased emphasis to conscience and to reduce the medieval emphasis on reason.

As will be evident when we come to treat the Huguenots, we believe that both increased state control and increased individual political freedom could and did emerge from the Reformation.

doctrine of justification by faith alone transformed most of the sacramental system into an unnecessary apparatus and broke the church's hold on the individual. His emphasis on simplicity eliminated the whole church paraphernalia of relics and saints. Monastic life was also abolished, and Luther himself married a nun, partly in order to set the example.

Later developments, it is true, came to work in the opposite direction, against individualism. When Luther was faced with the problem of constructing a church of his own, after he had razed the ancient church, he turned to authoritarianism. Nevertheless, in spite of the later authoritarianism of the Lutheran Church, the original mark of rebellion against authority clung to it and influenced men's thought. What had begun as an assertion of religious individualism continued into other fields.

We see this clearly in the growth of economic individualism. Once again, what Luther intended and what eventually emerged from his work were quite different and almost diametrically opposed. Luther himself hated the economic individualism of his age and attacked it fiercely in his pamphlets. Tawney has stated correctly that the element of individualism in Luther was "not the greed of the plutocrat, eager to snatch from the weakness of public authority an opportunity for personal gain. It was the ingenuous enthusiasm of an anarchist, who hungers for a society in which order and fraternity will reign without the 'tedious, stale, forbidding ways of custom, law and statute'."[12] But the result of Luther's divorce of man's inner life from civic activity was to free economics from ethical and religious restraints and to foster economic as well as religious individualism.

In intent a reactionary rather than a progressive movement, Lutheranism contributed largely, in practice, to the waves of change foaming through sixteenth-century Europe. Luther tried to erect a dam of absolute political authority to hold back the religious, economic, and political forces he had let loose. But, to use a religious image, "Man proposes. God disposes." Luther originated "Protestantism"[13];

12 R. H. Tawney, *Religion and the Rise of Capitalism,* Penguin Books ed. (New York, 1947), p. 81.

13 The name "Protestantism" derives from the written "Protest" made against the acts of the Diet of Speyer in 1529 by the princes supporting Luther.

the right of protest which he claimed for himself could not, later, be withheld from other men. Thus the Lutheran Protest opened the way for other protests.

Calvin

I

Calvin, born in 1509, was of a different generation from Luther. He was also of a very different character. Luther, the son of a miner, had been prepared for the law, but became a monk after a deep, personal experience during a thunderstorm. Calvin, on the other hand, came from petty-bourgeois townspeople, was destined for the church, and left theology for the law.[14] Unlike Luther, there was nothing mystical about Calvin; indeed, his character tended to the judicial and the narrow.

His early education was acquired in a family of local gentry near Paris. The family's most honored member was a bishop, and here Calvin imbibed good manners and a taste for the humanities. From 1523 to 1528, he studied theology at the University of Paris. In the latter year, he left to study law at Orléans, and then Greek at Bourges. His becoming a lawyer was in tune with the lay rather than clerical nature of the Calvinist reform. Luther drew most of his early reformers from the ranks of the Augustinian and Franciscan monks,

[14] The monumental work on Calvin's life is Emile Doumergue, *Jean Calvin, les hommes et les choses de son temps* (Lausanne, 1899–1927), 7 vols. Jean-Daniel Benôit, *Jean Calvin: La Vie, l'homme, la pensée* (Neuilly, 1933), presents what he calls "the quintessence of the seven big volumes of my old master, dear Emile Doumergue," in 275 pages. James MacKinnon, *Calvin and the Reformation* (London, 1936), as the author says, "is not a biography, but primarily a critical survey of the Reformer's work and influence, into which the biographical element only enters as far as it is relevant." In English, there is also R. N. C. Hunt, *Calvin* (London, 1933).

John T. McNeill, "Thirty Years of Calvin Study," *Church History* (Sept., 1948), is a basic and exhaustive bibliographical article. See, too, Roland H. Bainton, *Bibliography of the Continental Reformation; Materials Available in English* (Chicago, 1935). For a good treatment of Calvin by a Catholic, see Pierre Imbart de la Tour's work, which forms Vol. IV of that author's *Les Origines de la Réforme* with the title, *Jean Calvin: L'Institution chrétienne* (Paris, 1935). In his *Thought and Expression in the Sixteenth Century,* 2 vols. (New York, 1920), Vol. I, Chap. 17, Henry Osborn Taylor has 35 vivid pages on the thought, style, and personality of Calvin. André Fauve-Dorsaz, *Calvin et Loyola: Deux réformes* (Paris, 1951), is a very interesting work of comparative psychology and theology.

but Calvin's followers seem to have come mainly from the humanists.

Calvin himself received his initial impetus from the works of the humanists, and his first book, a commentary on Seneca's *De Clementia* (1532), relates to the classical Renaissance rather than to the Lutheran Reformation. From this study of Roman Stoicism may well have come Calvin's emphasis on stern ethical qualities.

By 1533, however, the influence upon Calvin of Erasmus' New Testament and certain writings of Luther can be perceived. Calvin had been caught up in the religious question. On All Saints' Day of that year his friend, Nicholas Cop, as new rector of the University of Paris, delivered an address clearly defending the doctrine of justification by faith; the address may have been largely written by Calvin. Both Cop and Calvin had to flee Paris.

Renouncing Catholicism, Calvin settled after a while at Basel and wrote the first sketch of his *Institutes of the Christian Religion*, in 1536. The *Institutes* is a remarkable work for a man of 26; at one blow it placed the young Calvin at the head of the reforming forces. Luther's writings had been passionate outbursts, expressive of his inner feelings but giving little in the way of definite, codified, external dogma. Calvin made up for this lack; in six chapters, slowly to grow by the last edition of 1559–1560 to eighty chapters, he set forth a tightly reasoned, logically arranged system of morals, polity, and dogma.[15]

The core of Calvin's dogma was that man was a helpless being before an omnipotent God. Calvin pushed Luther's arguments against Free Will to their absolute, logical conclusion, and emphasized that man could do nothing to alter his fate: he was predestined either for hell or to be saved. If he were saved, i.e., one of the "Elect," he would probably show this by his exemplary behavior on earth. This was a sign of God's favor; but it was only a sign, not a guarantee.

[15] John T. McNeill has edited Calvin's *Institutes of the Christian Religion* for the Library of Christian Classics. In this same Library is Calvin's *Theological Treatises*, tr. with introduction and notes by the Rev. J. K. S. Reid (Philadelphia, 1954), which presents aspects of Calvin's work as a teacher, administrator, and controversialist as well as theologian which might not be found in the *Institutes;* of especial interest is "The Genevan Confession of 1536," one of the best short statements of Calvinist beliefs. *Calvin: Textes choisis*, par Charles Gagnebin, préface de Karl Barth (Paris, 1948), offers a brief view of Calvin's writings in the original (mainly from his sermons and the *Institutes*) and permits us to see why his influence on French language and literature is considered to have been so strong.

Instead of Luther's discursive outpourings on these matters, Calvin offered fixed laws. By providing a dogmatic creed, he gave to the Swiss and French reformers of his time a rallying point. Thus, it was only natural that Guillaume Farel, who was attempting to convert Geneva to the reformed doctrine, should prevail upon Calvin to turn his hand to the work there.

II

Geneva was ripe for Calvin's experiment. It had just revolted against its bishop, who had ruled for the Duke of Savoy over the city; nevertheless, it had a tradition of being an ecclesiastical state. This fitted in perfectly with Calvin's belief that the state was subordinate to the church and that obedience to God came before that to the state.

At first, Calvin's power in Geneva was a moral power. Indeed, he was appointed on his arrival in Geneva a "Reader in Holy Scriptures," and never held a post higher than pastor. The Genevans, however, were proud to have as a minister the author of the *Institutes* and the leader of the Reformed Church. They groaned under his harsh standard (and at one point they turned him out of the city for about three years) but felt, like Americans under Prohibition, that it was good for them and their children.

When Calvin returned to Geneva, he did so only on the condition that the citizens accept his terms. These terms he embodied in the *Ordonnances ecclésiastiques* and the *Ordonnances sur le régime du peuple,* under which the people of Geneva henceforth lived; those who objected to Calvin's terms simply left, or were jailed or executed. This time, Calvin's power was not only moral but legal and political.

The Ordinances show Calvin's prowess as a legislator. He set up two main organs of government: the Ministry and the Consistory. The Ministry established, for the first time, instead of chance recruits, a disciplined army of Protestant preachers who adhered to a definite program and way of life. Candidates for the Calvinist ministry had to run the gauntlet of examination by the existing ministers (Calvin as one, of course) and approval by the city council. Then they were made to preach before the people, without whose approval they could not be called to a post. The latter condition, at least in theory (it became, in practice, a mere formality), introduced a democratic leavening into the theocratic government of Geneva.

This Ministry handled religious doctrine; questions of morality were dealt with by the Consistory, which consisted of six ministers and twelve elected elders. The Consistory examined charges and passed sentences and, eventually, even had the power to excommunicate. A decision by the Consistory was enforced by the civil officials. Thus questions of morality were turned into questions of law, and made subject to the power of the state.

Between the Consistory and the Ministry, which together met every three months in a Synod to review discipline, Calvin had at his disposal sufficient tools with which to create a "new man" as well as a Calvinist Church. Indeed, as A. M. Fairbairn phrases it, the Church was "not simply an institution for the worship of God, but an agency for the making of men fit to worship Him."[16] It is this idea of a civic church power creating its own citizens that was so strongly to attract and influence Rousseau later on.

The regimen Calvin imposed on Geneva was in many ways similar to that in More's *Utopia*—a regimen which included getting up very early, working very hard, and being always concerned with good morals and good reading. The virtues of thrift and abstinence were omnipresent.

Behind them, we must remember, stands what we have described earlier: the protest against the luxury of the church; the failure of the papal court to be close to the simple needs of the members of the church; and, above all, the feeling that, at bottom, religion is a matter of personal conviction and conscience. The result of these feelings is most familiar to us in the type of character produced by Calvin's regimen: the Puritan.

Calvin enforced his regimen with great vigor and, frequently, with outright ferocity.[17] One of his "citizens" was beheaded for writing a set of what Calvin called obscene verses. A card player was pilloried, and an adulterer whipped through the streets and then banished.

Among these, the persecution of Servetus was the gravest incident of Calvin's rule in Geneva. Servetus, who was a doctor and

[16] *Cambridge Modern History,* Vol. II, p. 364.

[17] Blame for these incidents is hardly mitigated by the fact that the citizens had the choice of accepting the Confession of Faith or departing, and that only those who elected to remain were expected to conform. Nor is it softened by the fact that this rigor was a reaction against the relaxed moral standards of the Renaissance Church of Rome.

scientist living in France, wrote a book attacking the orthodox doctrine of the Trinity. Thereupon, he and Calvin became engaged, by letter, in a violent theological polemic. Calvin's anger mounted to the point where, himself a heretic from the Catholic Church, he secretly accused Servetus of heresy to the Catholic Inquisition in France. Servetus was forced to flee; and, as bad luck would have it, his escape route took him through Geneva. Although his book had been neither written nor printed at Geneva, Calvin had Servetus seized and burned at the stake.[18]

Calvin's violent action in burning an early adherent of more or less Unitarian doctrines, who was also a scientist, suggests an important note about the Reformation. Both Luther and Calvin opposed not only the new art but the developing science of their time as well. In many ways, they were more fiercely antiscientific in their attitude than was the Church of Rome, and it has often been pointed out that Galileo, although he was badly treated by the Inquisition in Rome, would have suffered more severely if he had been unfortunate enough to live in the Geneva of Calvin's regime. Later, the twists and turns of history were to make the Puritans stanch supporters of the new science; but none of this was intended by Calvin's doctrine and discipline.

III

In all essentials, Calvin's state was a theocratic dictatorship.[19] Yet, as in the case of Luther, Calvin's movement, as it worked itself out in history, led to a greater independence of the individual. It

[18] This gave rise to a prolonged controversy about religious tolerance in which Calvin's defense of his action was challenged by Sébastien Castellion in 1554, and redefended by Calvin's disciple Théodore Beza. It might be added that Castellion's plea for freedom of religious opinion from persecution strongly influenced the development of Arminianism in Holland, whence the notion of tolerance spread to England. Cf. J. W. Allen, *Political Thought in the Sixteenth Century* (London, 1928). An interesting but somewhat novelized, emotional, and extremely anti-Calvin account of the burning of Servetus and of the resultant controversy between Calvin and Castellion is Stefan Zweig, *The Right to Heresy: Castellio Against Calvin,* tr. by Eden and Cedar Paul (New York, 1936).

[19] R. N. Carew Hunt, "Calvin's Theory of Church and State," *The Church Quarterly Review* (April, 1929), p. 71, denies this when he writes, "Did Calvin then establish a theocracy at Geneva? If by a theocracy we mean clerical domination he certainly did not." However, he finally admits, "But in strict definition Calvin not only aimed at a theocracy but by the end of his life he had come very near to creating one."

contributed, intentionally and unintentionally, to personal, economic, and political individualism.

For one thing, Calvin realized that an enlightened, trained ministry which controlled its flock by preaching instead of by sacraments implied an enlightened citizenry; and the Calvinist stress on Bible reading meant a literate populace. Thus, Calvin set up a system of education which had more to offer to the ordinary person than had earlier systems. In his schools all children had equal educational opportunities, regardless of birth or wealth.

Further, Calvin accepted (although he did not always approve of) the New Economics. He assumed the existence of a capitalist economic system for society and set up his ethics on that basis. According to R. H. Tawney, Calvin openly accepted the main features of a "commercial" civilization and "broke with the tradition which, regarding a preoccupation with economic interests 'beyond what is necessary for subsistence' as reprehensible, had stigmatized the middleman as a parasite and the usurer as a thief."[20] The self-indulgent and ostentatious use of riches was abhorrent to Calvin; but he was

[20] Tawney, *op. cit.*, p. 93. However, André-E. Sayous, who has studied the question of capitalism in Calvin's Geneva, in "Calvinisme et capitalisme: L'Expérience génevoise," *Annales d'histoire économiques et sociale,* Vol. VII (1935), pp. 225–244, finds that the Calvinist discipline so restricted capitalism as to keep it in a primitive form.

A fascinating debate has raged in recent times over the exact relationship of the reformers and the Reformation to the development of our modern economic system. The starting point is Max Weber's *The Protestant Ethic and the Spirit of Capitalism,* which, although first published in 1904 and now much outdated, opened up a whole new avenue of inquiry for many historians. Weber's thesis is that the material conditions for capitalism were present at many times in the past and in many places. However, it only came into existence in the West, in modern times, because "of the development of the spirit of capitalism." What Weber calls the "rationalistic economic ethic" was prepared by Luther's notion of the "calling" (which might very well be commercial) and by the Calvinist-Puritan idea of "worldly asceticism."

The Weber thesis was developed and partially modified by R. H. Tawney, who stressed the Calvinist rather than Lutheran teachings as being more sympathetic to capitalism and claimed, indeed, that the real acceptance and glorification of the New Economics was the work of seventeenth-century Puritanism. In Tawney's view, "To think of the abdication of religion from its theoretical primacy over economic activity and social institutions as synchronizing with the revolt from Rome is to antedate a movement which was not finally accomplished for another century and a half" (Tawney, *op. cit.*, p. 77). According to Tawney's analysis, other religions or sects might gradually and grudgingly make their agreement

not opposed to the accumulation of riches. The excessive exaction of usury from the poor was roundly condemned by Calvin; but he accepted the fact that the merchant ought to pay interest on the borrowed capital which made his profits possible. It should be little surprise, therefore, to discover that Calvinism appealed greatly to the rising bourgeoisie and, as Arnold Toynbee asserts, braced it for the coming struggle for power in the same way that Marxism later served the proletariat.

---

with the mundane necessities of the emerging economic world (the Jesuits, for example, tried to adapt Catholicism to the New Economics). Puritanism alone clasped that world to its breast.

The Weber-Tawney thesis, as it has come to be called, has been attacked and supported from many sides. A good account of the debate is given in the preface to the 1937 edition of Tawney's own book, *Religion and the Rise of Capitalism,* and n. 1 on p. 237 gives a good bibliography. We cite one attack on the Weber-Tawney thesis as a mere sample: Weber had emphasized the "spirit of capitalism" molding the material circumstances. H. M. Robertson, in his *Aspects of the Rise of Economic Individualism* (Cambridge, 1935), denounced Weber's "sociological method" and claimed that the opposite was true. Robertson insisted that the rising capitalism changed the religious ethic. He claimed that the "spirit of capitalism" rose not from religious impulses but from the "material conditions of civilization." Similarly, in the preface to Robertson's book, the economic historian J. H. Clapham asserts that "We see . . . the Puritan spirit not begetting capitalism but coming to terms with it."

What is our view in this lively controversy? Although it has become fashionable today in some academic circles to decry the Weber-Tawney thesis as either naïve or malicious, our opinion is that this is to miss the point of Weber's and Tawney's work and to overlook the limitations they themselves imposed on the application of their theories. As Weber pointed out, his own work was exploratory; he did not wish "to substitute for a one-sided 'materialistic' an equally one-sided 'spiritual' interpretation of civilization and history."

Our position is balanced in the same way. We believe that what men think and feel is of the utmost importance; but thoughts and feelings must always be related to the circumstances surrounding them. Thus, changes in a religious ethos are complementary to and go hand in hand with developments in other areas of men's lives. Cause and effect in such matters are interwoven and hard to distinguish, and the specific nature of the religious-economic link in the seventeenth century is a subject for empirical investigation.

But that some such connection existed seems beyond rational doubt. We find support for this in the fact that seventeenth-century thinkers—contemporaries of the twin developments of the reformed religious groups and of capitalism—linked the two movements closely. As Sir William Petty remarked, "Trade is most vigorously carried on, in every State and Government, by the Heterodox part of the same." By historical circumstances, the "Heterodox" in seventeenth-century England, for example, were generally Puritans; it was their heterodoxy, in our view,

The movement which Calvin had set going also contributed to the development of political individualism; but this was no part of Calvin's conscious intention. It came about accidentally, as a result of particular, historical pressures, and not in Geneva which the Calvinists dominated but in countries like France and England where they were a minority.

In Geneva, Calvin rejected both religious and political individualism, i.e., the freedom of the individual to make his own choice in these matters. Geneva was a Calvinist theocracy, and no deviation was tolerated. It was in this spirit that Calvin, at his university, trained men, such as John Knox and the English reformers, who returned to their own countries and tried to introduce the Calvinist reform.[21]

---

rather than specific Calvinist doctrine which linked them to the new economic activity. (In the same way, Parsees in India or Jews in various countries have played an economic role out of proportion to their numbers.) Weber and Tawney perhaps overemphasized the Calvinist doctrinal affiliation with capitalism; but at least they were on the right track. We prefer to emphasize the sense of isolation and of mission which inspires a group, as being a key value of heterodoxy. (The issue is complex, however; heterodoxy, almost by definition, closes off certain positions and employments but opens others; in some cases, the openings might be military rather than economic. Also, the past traditions and culture of the heterodox must be taken into account: Irish Catholics in nineteenth-century New England were heterodox; but they went into the police force rather than into business. To go further into this intriguing subject, however, would require a book in itself; we must rest content here with a mere glance at the subject.) It is also interesting to note that a significant by-product of the link between heterodoxy and economic development was the growth of tolerance. Sir William Petty went so far as to suggest that, from the economic point of view, schismatics had a positive value. In the seventeenth century, it seemed obvious that persecution was incompatible with prosperity, since it was the nonconformists who were in the forefront of economic progress. One result of this belief, in western and central Europe, was greater toleration for the Jews; especially was this a result of the Dutch example. See further Eli Heckscher, *Mercantilism,* tr. by Mendel Schapiro, 2 vols. (London, 1935), Vol. II, pp. 304–305. See, too, G. P. Gooch, *Political Thought in England from Bacon to Halifax* (Oxford, 1950), pp. 174–177, for the poet Andrew Marvell's defense of toleration as an economic measure. Emphasis on the commercial arguments for toleration may also be found in the thought of William Penn. This economic motivation for religious tolerance is overlooked by some historians, who assume that tolerance emerged largely from the inconclusive wars of religion, or was a result of the triumph of altruism and humanitarianism in the eighteenth century.

[21] Lutheranism had been too tied to German institutions (for example, the territorial princedoms) and nationalism to appeal fully, for example, to Frenchmen or Englishmen. Part of Calvin's accomplishment was to fashion a reformed theology which suited other than German conditions.

However, in their own countries—at least in France and England—these reformers were for long in a minority. They could not impose their will on others; hence, they were, initially, in the position of having to ask religious toleration for themselves. As it became clear that religious toleration was dependent on the disposition of political power, the Calvinists sought to achieve political power. Once again, their minority position frequently forced them onto the side of tolerance. The Calvinists became "anti-absolutists," and, occasionally, vaguely akin to democrats.

Thus, by historical accident more than by doctrinal tendency (although this did exist), Calvinism became related to "free" government and associated not only with economic but with political individualism. We can see this clearly if we study the French Huguenots.

The French Huguenots

I

France, unlike Germany, was a country in which the Reformation originated in a sect rather than in a national movement. It began, long before the entrance of the Calvinist reformers whom we have just been discussing, as a native development drawing strength from the French Renaissance rather than from Luther. In fact, Jacques Lefèvre d'Etaples anticipated Luther by publishing a Latin translation of St. Paul's Epistles (1512) with a commentary asserting that there was no merit in human works without the grace of God.

Around Lefèvre and his pupil, Briçonnet, a group of preachers (including Farel) gathered at Meaux under the protection of a remarkable woman, the king's sister, Margaret of Navarre. However, after 1521, when the Lutheran doctrines were condemned by the Sorbonne, suspicion began to fall on the little group at Meaux, and there were the first hints of persecution. The policy of Francis I toward the reformers was vacillating; he did not support them, but, on the other hand, because of exigencies of foreign policy, such as an alliance with the German Protestant princes, he at times tolerated them.

This uneasy toleration gave Calvin his chance. Lefèvre had died in 1536, the same year in which Calvin had written the *Institutes,* and direction of the Reformation in France quickly passed into

Calvin's hands, for want of any other dynamic leadership. Calvin preached a doctrine of passive obedience to the king and asked only for religious toleration. His appeal, at first, was mainly to the bourgeoisie and to the artisan class.

Unexpectedly, however, Calvin's doctrine was taken up by another class—the nobles. It is difficult to say just why Calvinism appealed to them at this time; perhaps they saw in it the same possibilities of particularism that the German princes had seen in Lutheranism (as against the universalist policy of the Holy Roman Emperor). In any case, by 1560 a major, if not the main, support of the sect came from the nobles. Even such lofty names as the Prince of Condé and Antoine of Bourbon (King of Navarre) were numbered among the Calvinists, and this association had momentous consequences.

It was at this point that French Protestants took the name of Huguenots (probably derived from the German word for "confederates"). Calvinism became a dogma held by many of the nobility, and was thereby involved in their political opposition to the Catholic Guise family and its pretensions. Further, the Calvinists who succeeded Calvin, who died in 1564, began to abandon his doctrine of passive obedience and to appeal to force. The result was the desultory civil wars that raged in France for about the next thirty years.[22]

II

Throughout the civil wars, the Huguenots remained a minority: even the most favorable estimates do not place their numbers higher than one-fourth of the French population. And, as a minority, their political theory tended to take the form of opposition to absolute rule, which gave no guarantees of protection to minorities or to particular groups. What the Huguenots wanted was government exercised by mutual agreement rather than by an absolute sovereign. In a sense, they were appealing to the medieval concept of government limited by traditional law and by the privileges of particular groups, as against the new version of absolute sovereignty introduced by Machiavelli and his followers.

[22] Jean H. Mariéjol, "La Réforme et la Ligue–L'Edict de Nantes (1559–1598)," Vol. VI (Paris, 1904), in the *Histoire de France,* ed. by Ernest Lavisse, is a standard French account. In English, there are the detailed volumes of Henry M. Baird, *History of the Rise of the Huguenots of France,* 2 vols. (New York, 1879), and *The Huguenots and Henry of Navarre,* 2 vols. (New York, 1886). A later work dealing with a slightly lesser time span is James Westfall Thompson, *The Wars of Religion in France, 1559–1576* (Chicago, 1909).

The Huguenot protest against absolute rule took two characteristic forms. Appeal was made to the protection either of natural law—and this was generally expressed in a theory that the state is based on a contract—or of historical rights.

The first kind of Huguenot political theory is illustrated by the work of an anonymous author who wrote under the name of Stephen Junius Brutus. We may recall that the Junius Brutus family in Rome was the family which was always on the popular side, and appears on that side in Shakespeare's plays *Julius Caesar* and *Coriolanus.* The writer, therefore, took the name of Stephen Junius Brutus in order to make it clear that he was against tyranny. In 1579 he published a book in Latin, called *Vindiciae contra tyrannos* (*The Vindication Against Tyrants*), which appeared in a French version two years later.

The *Vindication,* a fairly systematic treatise on the origins of sovereignty and the limitations to be placed upon it, was largely concerned to justify rebellion against kings. Junius Brutus held that there were two covenants or contracts: the first, between the king and his people, being conditional on a second—a sacred covenant between the king, the people, and God. Only so long as the king kept the sacred covenant did the people owe allegiance to him. The right of resistance, however, was not placed in the people but only in their magistrates; thus the *Vindication* gave an essentially aristocratic rather than democratic theory of political disobedience or rebellion.

The other justification of resistance to tryanny, that on historical grounds, is expressed in François Hotman's *Franco-Gallia.* This book was written a few years earlier than the *Vindication,* in 1573, and tried to undermine absolute royal power by claiming that the ancestors of the French kings had always been elected and their powers checked by the aristocrats, whose present-day heirs were the Estates General. Hotman's book, which we now know to have doubtful historical grounds, also placed the right of resistance in the hands of the magistrates or estates.

III

What we have said about Stephen Junius Brutus and Hotman illustrates the fact that, then and now, political views are in part held for the practical reason that they support one's real, i.e., power, position. Political ideologies ought to be understood, therefore, not merely as abstract, logical systems, but in relation to the people and the party who hold them as rationalizations for their sectarian desires.

Thus, the history of ideas must also be the history of the situations in which ideas have developed. The fascination in a study such as we are attempting here is to be found in this interplay between the ideas and the concrete circumstances. Huguenot political theory does not consist solely of a collection of books but of this collection seen within the ambiance of the sixteenth-century politico-religious struggle.

As further evidence for this view, we cite the opponents of the Huguenot political writers, the Catholic theorists known as the Catholic Monarchomachi.[23] They, as well as the Protestants, alternately justified and attacked regicide and supported and denounced hereditary monarchy. For example, when Henry of Navarre, the Protestant prince, became next in line for the succession to the French throne, the Catholics and the Sorbonne talked of election by the Estates General while the Huguenots, at this point abandoning Hotman and his arguments, accepted the unrestricted right of royal birth.

Nevertheless, the fact that political ideologies are advanced for sectarian reasons does not prevent them from having nonsectarian results. Thus, the occasional Protestant and Catholic attacks on unrestricted monarchy prepared the way for democratic theories of government. So, too, the sixteenth-century political situation in France led to the formulation of programs, especially by the Protestants, which, in an attempt to obtain support from the uncommitted groups, had to transcend the limited aims of the sect and take on a universal quality. In the same way, Calvinist political views, which led to a theocracy in Geneva, tended, for reasons that we have adumbrated, to democracy in France and, later, in America.[24]

[23] The name "monarchomach" was apparently invented by William Barclay, in his *De regno et regali potestate* (1600), as a description of those theorists who justified the right of resistance. As Sabine comments, "It did not imply an objection to monarchy as such." See George H. Sabine, *A History of Political Theory* (New York, 1937), n. 1, p. 374, and, in general, Chaps. 18 and 19.

[24] A similar point has been made about Calvinist doctrine and the economic organization of society. R. H. Tawney, *op. cit.*, p. 100, comments that "Both an intense individualism and a rigorous Christian Socialism could be deduced from Calvin's doctrine. Which of them predominated depended . . . above all, on the question whether Calvinists were, as at Geneva and in Scotland, a majority who could stamp their ideals on the social order or, as in England, a minority living on the defensive beneath the suspicious eyes of a hostile government."

IV

Let us now pass on from the political theory to the political practice of the French religious wars. A few years before Stephen Junius Brutus wrote his *Vindication Against Tyrants,* Catherine de' Medici, the last great descendant of the Medici, as mother of the reigning king, had attempted to heal the breach between the Catholics and Protestants by marrying the two dissenting sides. She made a marriage between Henry of Navarre, the leader of the Protestant side, and her own royal family.

The arrangement of the marriage, however, failed in its object of winning over the Protestants. Thereupon, Catherine elaborated a different scheme: the massacre of St. Bartholomew. This took place on Augut 24, 1572, during Henry of Navarre's marriage celebrations. The date is one remembered by all Huguenots to this day. Admiral Coligny, one of the outstanding leaders, and 2000 to 3000 Protestants were killed in Paris alone, and it is estimated that at least 6000 and perhaps 20,000 Huguenots were assassinated in the country at large.[25] Many others left France soon after, in an exodus of fear. Henry of Navarre, himself, was captured and had to adopt Catholicism in order to save his life.

But four years after the massacre, in 1576, Henry of Navarre, who had managed to escape, joined the Huguenots again, and reaffirmed his Calvinism. As he was a military leader of talent and energy, he was successful in conquering virtually the whole of France in the succeeding years. Thus, in 1593, he was at the gates of Paris. But his siege of the city failed. In the position of having to withdraw from Paris or to make a peace, he took an extraordinary step. He, the figurehead of all the Huguenots, returned to Roman Catholicism. According to the story, his famous words were "Paris is worth a Mass." At any rate, he succeeded in establishing himself on the throne of France as a Roman Catholic.

In becoming a Roman Catholic in order to reunite the French kingdom at a crucial moment, Henry was accepting what Machiavelli would have called a reason of state. We shall see the same reason accepted again when Charles II in 1660 came back to England; it is almost certain that he was a Roman Catholic, and it is equally certain that the Restoration would have been impossible if he had insisted on publicly remaining so.

[25] Cf. Smith, *op. cit.,* Chap. 4.

Henry of Navarre's conversion helped to end the long religious wars. Proclaiming himself restorer of France, he set to work, with the aid of a great administrator, Sully, on reconstruction. He placed the power of the state behind the building of roads and canals, and he encouraged the development of new industries. In a remarkably short time, the economic ravages of the French religious wars were repaired.

Then, in 1598, Henry took a religious step of great importance in French history. He issued an edict of toleration, usually called the Edict of Nantes. This Edict, on paper at least, expressed French religious policy for the next hundred years. It not only gave the Huguenots religious tolerance, but it virtually created a "state within a state" by allowing them to maintain their own armies and walled cities.

But just as, in the Edict of Nantes, Henry was anxious to establish freedom for the Huguenots, so he was also concerned to demonstrate loyalty to the Catholicism which he had readopted. In pursuance of this desire, he took a great interest in one of the large Jesuit schools, La Flèche. Descartes happened to be a student there. In fact, Henry willed his heart to the school; and Descartes was among those present when Henry's funeral rites were performed (he was stabbed to death in 1610 by a religious fanatic). This relationship is significant because it reminds us that Henry of Navarre achieved the stable, prosperous atmosphere of the seventeenth-century France which provided the background for the scientific work of men like Descartes.

V

After the death of Henry of Navarre, France by degrees became an increasingly autocratic country. True, a meeting of the Estates General (the French Parliament) took place in 1614, but it was the last until the French Revolution, 175 years later. Throughout the second half of the seventeenth century, the great Sun King, Louis XIV, dominated the French scene and wasted in his wars the wealth accumulated by Henry IV's minister, Sully, and his own minister, Colbert. The flavor of his government is given in his alleged statement: "*L'état, c'est moi.*"

In this atmosphere, the Huguenots suffered an uneasy tolerance. Increasingly, they, who were the industrious part of the French population, believing in the Calvinist virtues of thrift, frugality, and abstinence, found their commerce and industry hampered by the autocratic government of Louis XIV. Finally, in 1685, Louis XIV revoked the Edict of Nantes.

The result was another great exodus of Huguenots from France. Most of them emigrated to Holland, Prussia, or England. Many of them, when they came to England, lived in a place called Spitalfields. Here they worked as weavers, above all as silk weavers, and formed their own philosophical, literary, and scientific societies.[26] And here they did such fine weaving that the popes had them weave the white silk garment which the pontiff wears at the time of being made pope.

We can see in this story an image for the increasing subordination of religion to the interests of politics and economics in the sixteenth and seventeenth centuries. We have already seen this occurring in Luther's own practice. It showed itself more forcibly in the Thirty Years' War. This was a war which started in 1618 for religious reasons and ended in 1648 for political ones.; it was a war which saw a Catholic cardinal, Richelieu, maneuver France onto the side of the Protestant powers in order to defeat the Catholic Hapsburgs.[27]

VI

The Spitalfields story can also remind us that the Reformation, although a religious movement, also had lasting secular consequences. Let us try to summarize these briefly. One of the important results of the Reformation was that it strengthened the territorial state. Luther's protest broke the universalist ideal of the Catholic Church and then made religion subordinate to the state, reserving only the "inner life" to the individual. The religious wars which followed, in Germany, in France, and then throughout Europe in the Thirty Years' War, helped define the future political shape of Europe on the basis of independent, sovereign nation-states.

[26] The French Huguenots also must be credited with the popularization of English literature in Europe and especially in France. They compiled Anglo-French dictionaries and grammars, translated English works, and "created the first international periodicals in Europe. To them can be traced the Anglomania which was to seize the eighteenth century." H. M. P., "English Literature Seen Through French Eyes," *Yale French Studies*, No. 6, p. 111.

[27] C. V. Wedgwood, *The Thirty Years' War* (New Haven, 1939), is the best account.

Another important consequence of the Reformation was its encouragement of the rising middle class. The Reformation, at least in its Calvinist version, made religion a thing of this world and achieved the miracle of identifying good works with the accumulation of riches. The shame of profiteering was wiped away and what was formerly lust for wealth became the fulfillment of God's purposes on earth.

A strengthened nation-state and a rising middle class—these are characteristics of what we have learned to call modern history. If we add to these what may at first sight appear as contradictory elements, religious and political individualism, we are well on our way to the contemporary world. Thus, a movement which originated in a desire to purge a unified church and restore it to its pristine condition ended by having that church torn asunder and divided against itself, existing in a new world. A few years earlier, Columbus, intending to find a shorter route to India, discovered a New World accidentally. So, too, Martin Luther and his followers, intending a return to the old, helped create a new world, a new world not so much in space as in time—the world of modern times.

CHAPTER 7

The Scientific Revolution

IN RECENT years, historians have come to see that the most far-reaching change which grew out of the Renaissance was the evolution of the scientific method of inquiry. They have, therefore, given to the period of growth in science between 1500 and 1700 a new name, the Scientific Revolution. Professor Herbert Butterfield expresses the view of many contemporary historians when he says of the Scientific Revolution that it "outshines everything since the rise of Christianity and reduces the Renaissance and Reformation to the rank of mere episodes."[1]

1 Herbert Butterfield, *The Origins of Modern Science* (New York, 1952), p. viii. This is a broad, general, "humanistic" survey; perhaps the best available. A. R. Hall, *The Scientific Revolution, 1500–1800* (Boston, 1954), covers roughly the same period as Butterfield's book but in a more extensive and academic fashion; it is appropriate for detailed study. H. T. Pledge, *Science Since 1500: A Short History of Mathematics, Physics, Chemistry, Biology* (London, 1939), is rather a manual than a synthetic book on the subject, and presupposes a theoretical knowledge of the sciences treated. It contains, however, some interesting crotchets of the author as well as charts showing connections of masters and pupils and maps showing the birthplaces of scientists. Most of the book—at least two-thirds—deals with the period after 1800 and is, thus, later than the period with which we are dealing. S. F. Mason, *A History of the Sciences* (London, 1953), is a survey of science from ancient times to the twentieth century with only about one-sixth of the book devoted to the scientific revolution of the sixteenth and seventeenth centuries. Although by no means thorough, it is suggestive. It contains a bibliography. Herbert Dingle, *The Scientific Adventure: Essays in the History and Philosophy of Science* (New York, 1953), contains articles on Copernicus and Galileo as well as on other topics—including an interesting one on "Science and Ethics," a subject also treated by J. Bronowski in *Science and Human Values* (New York, 1957). A. Wolf, *A History of Science, Technology, and Philosophy in the Sixteenth and Seventeenth Centuries*, 2nd ed. (London, 1950), is especially valuable for its profuse illustrations, its concern with technology and its broad coverage. It is not a

Thus, historians recognize that the unfolding of scientific thought between 1500 and 1700 was critical in the creation of modern civilization. This recognition, which has been reached only recently, is itself the result of the powerful impact of science on the life of our generation. Science has made the world over in the twentieth century, root and branch—intellectually and physically. In doing so, it has transformed our understanding of the past as radically as our expectation of the future. Physically, we live in a new and a changing world. And intellectually we see the world differently, so that the processes of nature and even of history have a different logic for us.

In short, the Scientific Revolution between 1500 and 1700 was in the first place an intellectual revolution: it taught men to think differently. Only later was this thought put to a new practical use, in the Industrial Revolution about 1800 which gave our civilization its outward character. But the Scientific Revolution in itself mainly implied a basic change in the way in which people pictured the world. This was the profound change "from a world of things ordered according to their ideal nature, to a world of events running in a steady mechanism of before and after."[2]

I

The view of how the world works which was held before 1500 was quite different from ours. The difference is so great that it is difficult to think oneself back into the earlier view. Indeed, the most difficult of all historical tasks is to think as another civilization thought: to find its explanations reasonable and its view of the world natural.

We must grasp from the outset that a medieval painter was not merely ignorant when he made the figure of Christ larger in his picture than perspective would allow.[3] A medieval traveler was not un-

---

sustained treatment, as Butterfield's is, for example, but rather a series of topics handled more or less as separate units. In orientation, the book is somewhat outdated; for example, it still accepts without qualification the Burckhardt view of the Renaissance, and it treats religion as the blind enemy of scientific progress. Nevertheless, because of its comprehensive nature, Wolf's book is indispensable. The reader is also recommended to consult Thomas S. Kuhn, *The Copernican Revolution* (Cambridge, 1957), which contains up-to-date bibliographies.

[2] J. Bronowski, *The Common Sense of Science* (London, 1951), *passim*. The philosophy underlying the development of science is examined in this book.

[3] Cf. Lewis Mumford, *Technics and Civilization* (New York, 1934).

observant when he described what is obviously an elephant as an animal with five feet, one of which the animal uses like a hand. To us, the traveler's tale is ridiculous, because to us the elephant fits into the order of mammals and the unfolding of evolution; we see its connections with the horse and the tiger, with the whale and with man. But to medieval men, the world had a different set of inner connections; it was organized differently, as it is still organized differently today for other civilizations than ours. Today also, the primitive man who says that he *is* his totem animal, is not merely ridiculous;[4] and the islander who has a name for every kind of tree, but no general word for tree, is not merely ignorant. They are living in a different conceptual system of nature.

The conceptual system of nature which dominated the Middle Ages had been formed, or better had been welded together, by St. Thomas Aquinas about 1250. At that time, the works of Aristotle were already known to Arab thinkers, and had been translated by them from Greek into Arabic. By this roundabout path, Aristotle's system in physics and in medicine spread also to the West, and his works were translated into Latin. The great work of St. Thomas Aquinas was to join together the system of nature of Aristotle with Christian theology and ethics, and so to shape a single outlook for the next 300 years.

The Western rediscovery of Aristotle about 1250 was thus, as it were, a first revival of Greek learning. And there were isolated enthusiasts for classical knowledge at that time; Emperor Frederick II, whom his admiring age called *stupor mundi et immutator mirabilis,* was one. But in general, the discovery of Aristotle did not bring Greek thought and literature to the West whole. It brought only a first thin and single strand of humanism: an interest in the working of nature. Nature was still conceived, as it had been before 1250, to be kept going from moment to moment by a miracle which was always new and always renewed. But after 1250, men began to be interested in the form of the miracle; they wanted to know to what order, to what hierarchy it conformed.

All things in nature were thought then, as they are now, to be compounded from a number of fundamental elements. The number

[4] Cf. Ernst Cassirer, *An Essay on Man,* Anchor Books ed., (New York, 1953), p. 110.

which Aristotle had accepted, and which became general, was four. The elements were air and fire, water and earth.

The elements were held to follow what we should want to call laws; but it would be wrong to think of their pattern of movement as if it had, or was meant to have, the exact sequence which we now seek in a natural law. The elements did not so much obey a law as express and follow their own ideal natures. The elements of water and of earth moved downward when they could, because they were striving to reach their own natural centers. The elements of air and of fire moved upward, again in order to reach their natural centers. The dying Cleopatra in Shakespeare's play thinks of Mark Antony, who is already dead, and cries:

> Husband, I come.
> ....
> I am Fire, and Ayre; my other Elements
> I give to baser life.[5]

Here as elsewhere, Shakespeare is using the physical ideas of the Middle Ages. A man or a woman, like any other thing, was thought to be physically put together of the four elements. (And, as a result, these expressed themselves in his physique and in his temperament as four humors or inclinations—air as blood, fire as choler, water as phlegm, and earth as black bile or melancholy.) When a man or a woman died, the elements of air and fire left the body and sped upward to their centers in the sky; the elements of water and earth were left behind, and the body sank to the ground.

The striving of the elements to reach their true centers kept the world going. A piece of soil, for example, although it was not wholly earth, was mainly so; and therefore it strove to return to the earth. Each element had as it were a will of its own, and this will drove it to seek to fulfill its own essence. These strivings, these movements, kept nature going, not as a machine but as a hierarchy: a hierarchy in which each element was looking for its own fulfillment of itself. If ever that universal fulfillment were reached, nature would achieve completion; the universe would stand still and, in standing still, become God's perfect handiwork, no longer restless but fixed for ever.

[5] William Shakespeare, *Antony and Cleopatra,* Act V, scene 2.

II

In the system of the four elements, the heavenly bodies were not made of matter, in the earthly sense. In them, the elements of air and fire were dominant in a purer form, and indeed made a fifth element, the ether; so that the heavenly bodies were of a higher order than the earth, and it was reasonable that they should stay overhead. The thought which Newton had in his mother's garden in 1666, that the planets might be held in their orbits by a force of the same kind as that which draws the apple to the ground, was not possible to the Middle Ages; and this for two reasons. One reason is that they did not conceive the forces which act on the elements in this way: they did not see the starry sky as a machine. But the other reason, which is more often overlooked, lay deeper. It was that the Middle Ages could not have conceived the planets to be made of the same stuff as an apple.

The medieval picture of the movement of the heavenly bodies was also taken from the Greeks: from Claudius Ptolemy, an astronomer who belonged to the exiled school of Greek scientists in Alexandria in the second century of our era. This picture put the stars, which do not seem to move, on a fixed sphere round the earth. Between the earth and this fixed outer sphere there were thought to be, layer by layer, seven other spheres, each with its center at the earth. Each of these seven spheres carried what was then called a planet. That is, one sphere carried the moon and another carried the sun; beyond these, one sphere carried Mercury, one carried Venus, another carried Mars, another Jupiter, and another Saturn.

The motions which the planets carried out in their revolving spheres were complicated. For the planetary paths, when they are seen from the earth, go forward and then loop back on themselves from time to time. At long intervals, even these loops turn backward, so that at times two planets for a time seem to interweave their paths. (One rare conjunction of this kind sometimes intertwines the paths of Jupiter and Saturn, who are then said to play. History has it that Jupiter and Saturn played at the birth of Christ—and, incidentally, again at the English Revolution of 1688.)

Ptolemy described the paths of the planets, with their strange loops, by the rolling of one circle on one or more other circles. To the Greeks, and again to the medieval followers of Aristotle, the circle was in some mystic way a perfect path, and it was natural to build up other paths from circles. The circle was felt to be the path which a natural object would tend to describe of itself. By contrast, to the Greeks and to the medieval followers of Aristotle, the straight line was an unnatural path: Aristotle had supposed that an arrow, in order to go on flying on a straight path, had constantly to be pushed by the air from behind. This mistaken view was one of the main obstacles to the development of a realistic mechanics so long as Aristotle was accepted as an authority.

The curves which Ptolemy pictured, traced by circles riding on circles, give reasonable agreement with the timetable of the planets. It is, in fact, possible to build up most curves in this way, as epicycles made by circles riding on circles, if one uses enough different circles. The awkwardness of Ptolemy's system in the Middle Ages was that, in order to fit the improving observations, it called for more and more circles. Domenico Maria Novara, who taught Copernicus in Italy, was one of several astronomers who criticized the system of Ptolemy for its complication.[6]

Some of these critics saw that the complications of Ptolemy were unavoidable so long as the earth was thought to be fixed and all observations were traced back to the earth. The German astronomer Johann Müller, who used the Latin name Regiomontanus, had challenged the assumption that the earth must be taken as the center of the heavens, and he quoted the doubts of some Greek astronomers against it, early in the fifteenth century.[7] Regiomontanus died young, but others shared his doubts. When Leonardo da Vinci wrote in his notebooks, "The sun does not move," he was voicing the advanced opinion of others as well as his own.

III

The astronomical system of Ptolemy could not be overthrown merely by criticism. If it was to be displaced, then something else had to be put in its place; another system was needed. This system was formulated with admirable precision by the Polish astronomer Nicolas Copernicus.

To us, the system of Copernicus is coherent and satisfying; and we

6 Cf. Edwin Arthur Burtt, *The Metaphysical Foundations of Modern Physical Science* (London, 1949), p. 42.

7 Cf. Leonardo Olschki, *The Genius of Italy* (New York, 1949), pp. 374–375.

are tempted now to think it easier to conceive intellectually than the system of Ptolemy. But this was not so at the time, for several reasons.[8]

First, the motion by which Copernicus replaced the single motion of the sun is not easy: for it is made up of two separate motions of the earth. According to Copernicus, the earth swings round the sun once in a year, and at the same time the earth turns on its own axis once every day. This spin of the earth round itself outraged many ideas of mechanics which were then held—and which some people still hold. For example, if the earth is spinning under it, why does a stone fall straight? Why do things not fly off the spinning earth?

Second, the system of Copernicus did not chime with some of the detailed observations of astronomy. For example, it implied that the stars lie along different directions when seen at different times of the year. And this had not been observed then, and was not observed for another 200 years.

Third, the system of Copernicus offended the medieval sense that the universe was an affair between God and man, in the way that the Bible pictured it from the sixth day of Genesis. This fear of outraging religious tradition made Copernicus delay the printing of his work, as 300 years later it made Charles Darwin delay writing *The Origin of Species* for many years. Copernicus published his book, *Revolutions of the Heavenly Bodies,* only in the year of his death, 1543, and tradition has it that he held the first copy in his hands on his deathbed.

These objections made it certain that Copernicus could not expect to persuade the run of traditional minds of his time. Instead, he addressed himself to what he called the mathematicians. The mathematicians would appreciate the order and the essential simplicity of his vision. Copernicus had already announced to them in 1530 that he would show "how simply the uniformity of motion can be saved." Mathematicians could appreciate that, at the least, the new system reduced the number of circles which rolled in the epicycles of Ptolemy. Copernicus, of course, was not able to get rid of the epicycles altogether, because he still insisted on looking for uniform motion in a circle, and did not realize that the paths of the planets are ellipses.

In a sense, then, Copernicus was appealing to the aesthetic judgment of his fellow mathematicians. This aesthetic appeal makes a

[8] Cf. Burtt, *op. cit.,* pp. 23–25.

complex and important idea, which underlies all the intellectual advances since the Scientific Revolution. And it is a humanist idea. Copernicus, who was born in 1473, had had a characteristic humanist education, in law, in medicine, and in the classics, first in his native Poland and then in the Italian universities. He had been encouraged in part by the pope, who wanted to reform the irregularities of the calendar which had been accumulating, uncorrected, since Julius Caesar introduced it in 46 B.C. At bottom, Copernicus rejected the system of Ptolemy on the same ground on which other humanists rejected the work of the Schoolmen: because it lacked beauty and unity.

Certainly, Copernicus was inspired by the sense that there is a unity among the planets when they are seen in a single vision from the sun. As Leonardo Olschki comments, Copernicus' *The Revolutions of the Heavenly Bodies* "is written in the typical humanistic style, with a profusion of Latin and Greek quotations, and much solicitude for polished eloquence and rhythmic elegance. The conclusion is summed up in a solemn peroration praising the sun in poetical terms as master of the universe and the perfect work of a divine architect."[9] "The earth conceives from the sun," wrote Copernicus; and "the sun rules the family of stars."

Thus, one thought which moved Copernicus, and which has moved scientists ever since, is that nature has a unity, and that this unity expresses itself in the simplicity which we find in her laws when we have them right. The assumption that the laws of nature ought to be simple is usually justified by scientists as a useful working rule or procedure, and is quoted in the form in which William of Ockham had stated it long before, that we should not multiply hypotheses. But the demand for simplicity truly goes much deeper than this, and expresses a longing to find a unity in nature. There is, in fact, no criterion of what is simple, other than the demand for the greatest measure of unity. To take a modern example, Einstein and other searchers for a unified field theory have said that such a theory is simpler than the separate theories of gravitation and of electromagnetism which it combines. But when we look at a unified field theory, we find that the combination is simple only in the sense that it is presented in linked and coherent formulas; the formulas themselves are more complex than ever.

[9] Olschki, *op. cit.*, pp. 376–377.

Simplicity, then, is the expression of unity; and the search for simplicity as the expression of the unity in nature was a humanist idea. It owed much to the revival of Plato, for example, in the Academy of Florence, and with it to the interest in the mystical ideas of Pythagoras, that mathematics expresses the harmony and the beauty of the universe. This is why Copernicus made his appeal to mathematicians. This is why he wrote, "mathematics is written for mathematicians, to whom these my labours, if I am not mistaken, will appear to contribute something."

We see how complex were the ideas which underlay the belief of Copernicus, that his system of describing the movements of the planets embodied a profound truth in nature. To Copernicus and to those who later were convinced by him, his system had a unity which the system of Ptolemy lacked; and this unity expressed itself in the simplicity which the system gave to the heavenly paths. There is, however, as we have tried to show, no ready definition of simplicity; it is an aesthetic feeling; it inspired Copernicus with the sense of beauty, as it had inspired Pythagoras long before and has inspired scientists ever since. At the same time, the sense of unity in the world inspired him, as it has inspired other scientists, with the feeling that the truth of the processes within nature is being reached. In this way, science, like art, gives to those who practice it the humanist conviction that beauty and truth are two faces of what they seek, two faces which belong together: the conviction that beauty is truth which the poet John Keats felt, characteristically, when he looked at the figures on a Grecian urn.

IV

Copernicus had not expected to interest nonspecialists in his vision of the heavens, but two cosmic accidents ensured that he did so. In 1572, a new star blazed in the familiar constellation of Cassiopeia. This star was observed by the pioneer of accurate observation, the Danish astronomer Tycho Brahe. For two years the new star was brighter than any other star in the sky; and for a time it was so bright that it could be seen in daylight.

Then, on the turn of the century another new star appeared, once again dazzling and disturbing the accepted order of the heavens. This new star was observed with care by the astronomer Johann Kepler.

The traditional picture of the fixed stars had now been thrown out twice, each time by

> A strange new visitant
> To heavens unchangeable, as the world believed,
> Since the Creation.

The sky all at once seemed to be in flux, and the accepted system of astronomy was shaken by doubt.

These celestial happenings are remarkable, for they were both supernovae, that is, explosions of old stars; and supernovae are fairly rare in any one galaxy. In the whole of recorded history, there have been only three supernovae in our galaxy. The first of these, the supernova which is now the Crab Nebula, exploded in the sky on July 4, 1054, but no Western watcher has left a record of it; it was recorded only in China and Korea. The other two supernovae were those which have just been described, and flared up in the heavens as if providentially at the time when Copernicus had challenged the accepted vision of astronomy among professionals.

These events, of course, made an especially deep impression in an age when all men, and particularly educated men, believed that their personal fate was written in the sky. Almost everyone then believed in astrology; it was treated, as nothing else was, as an exact science; and it was, in fact, necessary, in order to cast a true horoscope, to know the positions of the planets precisely at any time. There was, therefore, a real interest among astrologers in an exact knowledge of the planetary movements, and in a system of astronomy which would describe them accurately. Kepler, who did more than any other man to establish the system of Copernicus in sound detail, made his living in part by casting horoscopes. And Pope Paul III, to whom Copernicus dedicated *The Revolutions of the Heavenly Bodies*, "never appointed a consistory without having one of his astrologers calculate a favorable 'conjunction.' "[10]

Kepler, even more than Copernicus, was carried away by the mystery of mathematics, by the strange relations between numbers and by the complex properties of natural space. His books have such titles as *Mysterium Cosmographicum* and *Harmonice Mundi*. He writes

[10] *Ibid.*, p. 380.

again and again that *Natura simplicitatem amat* ("Nature loves simplicity") and *Amat illa unitatem.*[11] He listened seriously for the harmony of the spheres: he tried to relate the speeds of the planets to the musical intervals, not as a piece of magic fancy but in strict mathematics—nothing less satisfied him. He tried to fit the five regular solids, which had been known since antiquity, into the orbits of the planets. And his mathematics was excellent; he was the first mathematician to go beyond the five regular solids which the Greeks had known, by discovering the two star-shaped regular solids which go by his name.

At the same time, Kepler had learned from his work under Tycho Brahe (who was never converted to the system of Copernicus) to take more accurate observations of the whereabouts and the movements of the planets than had been taken before. Working restlessly through Brahe's and his own figures, and seeking always for the arrangement that would fit them, as he was convinced, into the system of Copernicus, Kepler found three laws to describe the motions of the planets. He published these in 1609 and in 1619.

The first of Kepler's laws is, in a sense, the most revolutionary, for it breaks with the tradition of movement in a circle which had been taken for granted since the time of the Greeks. Instead, Kepler discovered that each planet moves on an ellipse, and that the sun lies at one focus of this ellipse. The second law describes the varying speed at which a planet travels along its ellipse: the line from a planet to the sun sweeps out equal areas in equal intervals of time. The third law relates the movement of one planet to another: the year in which a planet rounds its orbit is such that the square of the year is proportional to the cube of the average distance of that planet from the sun.

With the discovery of these laws, exact, compact, and remarkable, the paths of the planets were mapped once and for all. In his life's work (he lived from 1571 to 1630), Kepler had established the system of Copernicus in these formulas beyond challenge. There was no further step to take until the three laws could be shown themselves to be parts of a single unity, a single law holding each planet to the sun. Almost everything was ready for that step, taken roughly fifty

[11] Cf. Burtt, *op. cit.*, p. 46. The reader will find a more popular but graphic account of these matters in Arthur Koestler, *The Sleepwalkers* (London, 1959).

years later by Isaac Newton. But only *almost* everything was ready. What Copernicus and Kepler had made ready were the facts and the mathematical relations. What was still lacking before Newton could go to work was a practical understanding of the mechanics of motion.

V

If the planets run as a machine runs, then they must be moved and controlled by mechanical forces. The laws which these forces obey must be laws of mechanics. These celestial laws might, of course, be different from the laws of mechanics which are obeyed by forces on earth. The Middle Ages would have taken it for granted that they are different. By the year 1600, however, minds which had been shaken by the new stars and the new speculations found it natural to ask whether the laws of mechanics in the sky might not be the same as the laws of mechanics on earth. At the least, they found it natural to ask, with more persistence than had been used before, what precisely are the laws of mechanics on earth.

A central question in mechanics, because it was a question of fact and not of theory, asked at what rate an object falls to the ground. Aristotle had said that a heavy object falls faster than a light one, and St. Thomas Aquinas had followed him in this as in other opinions. The opinion had not gone quite unchallenged after Aquinas. As Pierre Duhem has shown in his work on the school of Ockham, first Jean Buridan and then Nicholas of Oresme in the fourteenth century put forward the view that unequal objects fall equally fast; and their view was held in the University of Paris for the next century and longer. More recently, other scholars have found this view expressed even earlier, about 1335, by the Mertonians in Oxford.[12]

In spite of these historical debates, however, the established view until the seventeenth century remained that of Aristotle, that heavier objects fall faster than light ones. Then, about 1600, the accepted view was challenged and defeated rather quickly. The story goes that it was defeated by Galileo, who is said to have dropped two cannon balls of unequal masses from the Leaning Tower of Pisa, and to have shown that they reached the ground together. Alas, there is little evidence for this story: if cannon balls were dropped from the Tower

[12] For a brief account of these views, see John Herman Randall, Jr., "Leonardo da Vinci and Modern Science," *Journal of the History of Ideas* (April, 1953), as well as the relevant pages in Butterfield, *op. cit.*

of Pisa, they were not dropped by Galileo, and it is unlikely that they reached the ground triumphantly together.[13]

It is, in fact, foolish to suppose that the doctrine of unequal rates of fall proposed by Aristotle was as brazenly out of step with the facts as this, and might have been overthrown at any time in 2000 years simply by dropping two unequal masses. Objects which are dropped from any height, if they are unequal, also meet unequal resistance from the air, and they do not, therefore, fall equally fast. On the contrary, if the new mechanics claimed that they should, then the old mechanics could point to the fact that they did not. Deeper thinking was needed to challenge Aristotle, and more subtle experiments. One reason why such experiments had been difficult hitherto, and why all experiments in mechanics were difficult, was that there did not exist until this time any reliable clockwork to measure the passing of time in small intervals.

The work of Galileo, therefore, begins rightly with a more searching discovery: how to measure small intervals of time. This is the robust approach which marks off Galileo from his contemporaries, and makes him a first leader of practical science and pioneer of the empirical method. Copernicus and Kepler were theoretical minds: Copernicus appealed to mathematicians, Kepler to the courtly interest in astrology which then drew educated men. By contrast, Galileo was grounded in the practical outlook of the trading republics of northern Italy; he appealed particularly to the interest of Venice in navigation and in gunnery. He spent his most active years, from 1592 to 1610, at the University of Padua, which was under the protection of the Republic of Venice.

This was the age of Mediterranean commerce and, fanning out from the Mediterranean and the Atlantic coast, of the voyages of discovery by Italians, by Spaniards, and by Englishmen. Until the sixteenth century, the records of earlier travelers had been treated as curiosities. For example, Marco Polo had written a book about his travels in the East in 1298; yet, it was not till almost the sixteenth century that an Italian cosmographer Paolo Toscanelli, a friend of

[13] Lane Cooper, *Aristotle, Galileo, and the Tower of Pisa* (Ithaca, 1935), deals with the evidence concerning Galileo's supposed dropping of weights from the Tower of Pisa, and disproves the whole story. Another familiar story about Galileo is corrected in Edward Rosen, *The Naming of the Telescope* (New York, 1947); see p. 122 below.

Christopher Columbus, used Marco Polo's book seriously as a source for geographical and navigational information.

Astronomy was now an important aid to navigation; but astronomy is not complete without accurate means of keeping time. For example, the latitude of a ship at sea can be found simply by observing the astronomical heavens; but the longitude can be found only if we know what time it is on some fixed longitude—say, on the longitude of Greenwich. For lack of a clock which could keep time at sea, navigators had to guess their longitude by what they called "dead reckoning" until well into the eighteenth century.

When Galileo Galilei was born, in 1564, there was not even a clock which could keep time accurately on land. The medieval clocks were not designed to be precise instruments which would run uniformly day in and day out. Medieval clocks were, in the first place, monastery clocks, whose purpose was to divide the day of prayer into equal parts or canonical hours, from the first matins to the evening vespers and the last complin. The day that was so divided ran essentially from dawn to dusk, and it was, therefore, a different day in summer and in winter. The clocks of the Middle Ages were a convenient means for dividing a day, but not for measuring time in any absolute sense. And these clocks were not capable of measuring time absolutely; they were clockworks, but they had no means of control to make them run uniformly—they had no escapement.

Galileo was not yet 20 when, as he reports, he noticed the regular swing of a lamp during a service in the cathedral at Pisa in 1583. How could he test its regularity? Galileo put his finger on his pulse, and he observed that the swinging pendulum and the beating pulse kept equal time. He had established an underlying uniformity in nature, on which many clocks run to this day: the uniform motion of a pendulum, which keeps virtually the same time whether it swings in a large or a small arc.

Galileo was not the first to guess that a pendulum keeps almost perfect time: it had been guessed by Leonardo, a hundred years earlier. And Galileo did not strictly prove his guess: that was done mathematically by Christian Huygens a hundred years later. What Galileo did was to use the pendulum. He probably made it a basic means for measuring time in experiments in mechanics, and thereby he brought time and the clock into the practical business of carrying out experiments on earth. This was many times more useful than any demonstration on the Leaning Tower of Pisa.

Galileo was, therefore, able to measure the fall of objects through quite small distances: more precisely, he measured the time which they take to roll down a slope. He was able to show that, whatever the slope, a ball rolls in such a way that the total distance it has gone from rest is proportional to the square of the time it has been traveling. For counting time in this way, of course, the beats of a pendulum are ideal. And Galileo proved his law to be true whatever the mass of the ball: Aristotle was defeated, as it were by the way. As a result of this work, Galileo was able to show, between 1604 and 1609, that a thrown or falling object travels to earth along a parabola.

The Paris school of philosophers had produced arguments long before to cast doubt on Aristotle's mechanics. For example, they had asked the following question. Suppose that three equal balls are dropped together; then it is clear that they will fall steadily side by side. Suppose now that two of these balls are joined: surely all three will continue to fall side by side. Is it then, they asked, not clear that a single ball, and an object made of two balls joined together, will fall side by side? And does it, therefore, not follow that Aristotle is wrong in believing that heavier objects fall faster than light ones?

Now these elegant but abstract speculations, and the arguments on the Leaning Tower, were at an end. Galileo had done more than compare the fall of one object with another: he had formulated a precise law which governs the fall, in a form which was mathematically beautiful and convincing. His practical ingenuity with the pendulum and his mathematical skill in writing the law combined to give his work a finality which nothing before had reached. Galileo had perfectly joined logic to experiment.

At the same time, Galileo's work also implied a breach with Aristotle's belief that an arrow is kept in flight only so long as the air pushes it. This belief also had been doubted long ago by Jean Buridan and Nicholas of Oresme in Paris. Now Galileo proved it false by experiment. He proved that a mass which is moving will go on moving until some force acts to stop it. This demonstration was equally important in the development of the new mechanics.

VI

About the year 1609, Galileo had news from Holland that people there had put two lenses together and had been able to see distant objects as if they were close at hand. Galileo had worked in optics, and this and the information he obtained enabled him to make a telescope for himself. He went on improving the telescope, and he also made what we should now call a compound microscope.

Galileo showed his telescope in Venice and at once created a sensation. Its usefulness to the merchant seamen there, for example, to identify distant ships, was patent.

But Galileo used the telescope more spectacularly to look into the sky with fresh, sharp eyes. He saw the craters of the moon, the spots on the sun, the phases of Venus, and what later turned out to be the rings of Saturn. The heavens were suddenly opened and transformed.

Characteristically, Galileo later proposed a still more subtle use of the telescope. Early in 1610, he had observed that Jupiter has four moons, and that they change their positions from one side of the planet to the other in a rapid and complex sequence. Therefore, Galileo proposed that a table of the positions of the moons of Jupiter should be made, to show where they will be at any time on any future night. In this way, it would be possible to read the time simply by looking at the moons of Jupiter through the telescope. Jupiter was to be turned into a divine clockwork, and the mariner at sea would be able to fix his longitude simply by looking at its face.

This suggestion did not turn out to be very practical, but it is revealing. It shows again how important a part the clock played in Galileo's thought, and in the ambitions of the merchant seamen of his time. Galileo's thought, here as elsewhere, was practical: no one else could have turned the casual observations of a lens grinder in Holland into a reasoned scheme for telling the time at sea.

The telescope was hailed with equal enthusiasm when Galileo went to Rome in 1611. Nevertheless, there began to be some critical voices. Some of Galileo's academic colleagues from the University of Padua refused to look through the telescope. Their ground was an absolute and, as it were, pious faith in Aristotle. The telescope might indeed reveal things which the eye did not see. But it revealed them, said the critics, by the agency of the devil: it was a form of conjuring and therefore at bottom an illusion.

This is in itself a puzzling objection. But under it there surely runs a deeper uneasiness. Scholastic and clerical minds were growing uneasy at the erosion by the new science of the evidence of the senses in the visible world. The Bible and Aristotle had expressed what seemed to that generation to be the evidence of the senses. Now Copernicus, Kepler, and Galileo were turning the visible world into a shadow play: the sun did not move, the earth did, the sky held hidden visions. The Scientific Revolution was creating an invisible world behind the world, and those of an older generation were afraid of it, as in our lifetime many people have become afraid of the invisible world of atoms and genes.

From this time, Galileo became embroiled with the Holy Office because he publicly avowed the system of Copernicus. The new sky that he had uncovered made him impatient with the astronomical conservatism of the church. Meanwhile, the work of Copernicus had become an issue in theology, and in 1616 it was condemned "until corrected."

The events which followed have been disputed, but are now well established. Galileo was instructed by Cardinal Bellarmine that the theory of Copernicus was contrary to Scripture and was not to be taught or defended. Cardinal Bellarmine seems to have been as friendly as his office allowed, and he had been one of those who had acclaimed Galileo's work with the telescope. There is no certainty from the Vatican records that Galileo was forbidden altogether to write about the system of Copernicus. Indeed, in 1620 the authorities decided to make only minor corrections in Copernicus' book, after which it became allowed reading.

In 1625, Galileo began to write a *Dialogue on the Two Principal World Systems,* in which their merits were argued at a sort of Renaissance court. The *Dialogue* did not openly advocate the system of Copernicus, but it put the arguments in favor of Ptolemy into the mouth of a man, Simplicius, who clearly did not have the respect of the author. In a way, then, the *Dialogue* was a subtly offensive form of siding with Copernicus; and it was certainly a popular and effective form. After some awkwardness, it was printed in Florence early in 1632. Within six months, the printer of the *Dialogue* was ordered to stop its sale; and two months later Galileo was called to Rome for trial, in spite of the protests of his protector and of his doctors.

A principal accusation against Galileo was that he had flouted the specific prohibition laid on him in 1616 not to discuss the Copernican system at all. No original of this prohibition exists in the Vatican records. What does exist is a copy which has not been witnessed. This copy may have been a forgery, which may have been smuggled into the record after 1616 by an enemy who was not content with the mild rebuke which Cardinal Bellarmine had given Galileo then. In the face of this document, Galileo had little defense. The pope had been one of his admirers in 1611, but he now thought that he recognized himself as the foolish figure in the *Dialogue*. Galileo was threatened with torture; ill and nearly 70, he was forced to recant in abject terms. Legend has it that he ended his recantation with the words *eppur si muove* ("and yet it moves"). But he was a broken man, was treated with great harshness and, when John Milton visited him under house arrest near Florence in 1638, he was blind. He died in 1642, and the pope insisted that any monument to him must contain no words which would "offend the reputation of the Holy Office."[14]

VII

The humiliation of Galileo is a climax in the Scientific Revolution. It makes it clear also why the Revolution petered out in Italy. In the same way, the contempt of Luther for the system of Copernicus[15] had early strangled the Revolution in Germany: in 1596, Kepler, who was a Protestant, had had to take refuge with the Jesuits. The climate in which the Scientific Revolution flourished when Galileo died, in 1642, was England in revolt against a dictatorial king; and in that year Isaac Newton was born in England.

The book in which Copernicus had put forward his system was called *Revolutions of the Heavenly Bodies*. Since then, the word "revolution" has come to be used in a social and a political sense,

[14] *Galileo's Dialogue on the Great World Systems*, in the Salusbury translation, revised, annotated and with an introduction by Giorgio de Santillana (Chicago, 1953), is the best edition, with a brilliant historical introduction by Professor de Santillana. See further the same author's *The Crime of Galileo* (Chicago, 1955). Also to be consulted for Galileo's writings is *Discourses and Opinions of Galileo*, tr. with an introduction and notes by Stillman Drake (New York, 1957). The reader should also consult Galileo's *Dialogues Concerning Two New Sciences*, trans. by H. Crew and A. De Salvio (New York, 1952), which presents his fundamental discoveries in mechanics.

[15] Cf. Smith, *op. cit.*, p. 618.

which has grown out of this astronomical use.[16] The new science saw revolution as the natural movement of events, in which the old is constantly turned over and replaced by the new. The long allegiance of the Middle Ages to established authority was over.

Galileo himself had said this at the time of the great struggle before the work of Copernicus was condemned. In a letter which he wrote in 1615 to the Grand Duchess Christina of Tuscany, *Concerning the Use of Biblical Quotations in Matters of Science,* he wrote:

> Methinks that in the discussion of natural problems, we ought not to begin at the authority of places of scripture, but at sensible experiments and necessary demonstrations. For, from the Divine Word, the sacred scripture and nature did both alike proceed. . . . Nature, being inexorable and immutable, and never passing the bounds of the laws assigned her, . . . I conceive that, concerning natural effects, that which either sensible experience sets before our eyes, or necessary demonstrations do prove unto us, ought not, upon any account, to be called into question, much less condemned upon the testimony of texts of scripture, which may, under their words, couch senses seemingly contrary thereto. . . . Nor does God less admirably discover himself to us in Nature's actions, than in the Scripture's sacred dictions.[17]

These sentences mark the breach between science and religion, in the claim that the book of nature is as authoritative a work of God as are the books of the Bible. And they do more. They set up nature as an activity outside men, in which man can trace beauty and truth together, not by projecting himself—not, as it were, by looking from the earth as center—but by entering humbly into her inexorable design. Science has become an impersonal way of looking at the world, a reading of the language of the universe in a new and absolute symbolism. This is the mathematical mystery of nature which had inspired Copernicus and Kepler, and Galileo summarized it in his work and in his words:

16 See further *Oxford English Dictionary,* the word "revolution." Sigmund Neumann, in his article "The International Civil War," *World Politics* (April, 1949), declares that, as used by Galileo, revolution "connoted a *natural* phenomenon outside of human control. . . . In transferring this *scientific* observation into the field of human history, the Renaissance meant to recognize in revolution 'the power of the stars,' i.e. the interference of super-human forces within world affairs" (p. 336).

17 Burtt, *op. cit.,* pp. 72–73.

Philosophy is written in that great book which ever lies before our eyes—I mean the Universe—but we cannot understand it if we do not first learn the language and grasp the symbols, in which it is written. This book is written in the mathematical language, and the symbols are triangles, circles, and other geometrical figures, without whose help it is impossible to comprehend a single word of it; without which one wanders in vain through a dark labyrinth.

CHAPTER 8

The Elizabethan Age

I

IN TALKING of Leonardo, Machiavelli, and Galileo, we looked upon the North Italian states as the center of the intellectual universe. Their supremacy, however, began to fade toward the end of the sixteenth century and their place to be taken by Holland, France, and England.

The reasons for this displacement are numerous; but they are almost all connected with the breakout from the Mediterranean into the Atlantic and toward the American continent. At that moment—back in 1492—the world's center of gravity began to shift. The reality behind the word "Mediterranean," i.e., center of the world, disappeared. No longer were the major seafaring powers those whose ships shuttled to and fro across the Mediterranean, such as the Venetians and the Genoese. Instead, the powerful seafaring nations were those whose ships carried the men of Brittany, or Holland, or England across the great oceans.

The decline of North Italy, started by the transfer in sea power and trade, was hastened by the "Barbarian" invasion of Italy, so poignantly deplored by Machiavelli. If we add to the several reasons given above the suppression of free thought by the Counter-Reformation church, as illustrated in the case of Galileo, the loss of leadership by the North Italian states is readily understandable.

That leadership passed northwest. It passed, at the end of the sixteenth century, especially to England. We shall see that in English history of the sixteenth and seventeenth centuries a movement remarkably like both the Renaissance and the Reformation played itself over again; that the Renaissance had its counterpart in the Elizabethan Age, and the Reformation in the subsequent Puritan Revolution.

II

The Elizabethan Age is a difficult period to describe fairly because it is so rich, so wide in its cultural reach, and so full of great men that, instead of giving one chapter to it, we ought plainly to give it a book. We can only attempt the task by a sideways movement, by trying to catch sudden glimpses of the age in some of its more neglected manifestations. For this, however, we need to give a little of the historical setting.

We have already mentioned Henry VIII's determination to establish the Tudor succession after the civil wars of the fifteenth century. His heir, Edward VI, a boy-king, lived only a few years and did not enjoy an equally successful reign (in fact, he did not really rule; a regency exercised power in his name). Edward's successor, Mary, who bears in English history the unhappy name of "Bloody Mary," was a Roman Catholic, who married the Spanish king, Philip II, and tried to undo Henry's reformation. At her death in 1558, her half-sister Elizabeth, the daughter of Henry and Anne Boleyn, came to the throne. She ruled for forty-five years, until 1603. It is this period (at its richest from about 1580, and running on perhaps till Shakespeare's death in 1616) which is known as the Elizabethan Age.

England's greatness at this time derived from a number of factors. One of the less showy ones was the presence of a Puritan element (like the Huguenots in France), which now helped to transform England into a business and manufacturing community. But the spectacular, and obvious, cause of England's greatness lay in the navigational achievements of men like Drake, Hawkins, Cavendish, and Raleigh.

These adventurers set out westward in order to explore and seize territory, to bring home new products (Raleigh is credited with having brought potatoes and tobacco to England) and, especially, to gain treasure. The latter they achieved simply by holding up the treasure ships of the Spanish fleet, and bringing the gold and silver back to England. Indeed, expeditions were fitted out expressly for this purpose, and even Queen Elizabeth occasionally took an open financial part in them, with a handsome rake-off as recompense. One example may give an idea of what was involved. Drake's expedition of 1577–1580, to America and the Pacific Coast (involving, incidentally, the circumnavigation of the world) resulted in a "take" of around 1½ million pounds sterling at a cost of 5000 pounds; the equipment involved consisted of 4 boats, with a total tonnage of 375 tons, manned by 160 men.[1]

Episodes like these inevitably led to a Spanish reaction. That reaction took the form of the great Armada which, although Philip had sound commercial reasons in mind too, he launched in the main as a religious campaign on behalf of Roman Catholicism against the heretics. In any case, the destruction of the great fleet which Philip II sent against England in 1588 was a blow against both Catholicism and Spanish economic and political power.

The English victory was due, in part, to Drake and his lighter, superior-sailing ships. Another reason, however, for the defeat of the Armada was a great storm which arose, according to the English view, as by the providential hand of God, and swept the Spaniards from the sea. Whatever the reason—Drake's ships or God's hand—the scattering of the Armada and the loss to the Spaniards of a great many of their warships left Elizabeth fundamentally unchallenged in her reign. In large part, therefore, the Elizabethan Age's great literary adventure sprang from the optimism and sense of destiny which the defeat of the Armada, almost by divine intervention, inspired in the minds and hearts of Englishmen.

III

Elizabeth, the woman after whom the age is named, was the last of the Tudors. The issue, in fact, went by default, for she never married. Thus, although she defended her throne by executing Mary, Queen of Scots, her successor was Mary's son, a Stuart, who became James I of England.

[1] Cf. Henri Sée, *Modern Capitalism: Its Origin and Evolution* (New York, 1926), p. 54. A. L. Rowse, *The England of Elizabeth* (London, 1950), puts the "take" at £600,000 (p. 152), while G. R. Elton, in *England Under the Tudors* (London, 1955), claims three boats, not four, were used (p. 346). These, however, are all minor details and do not affect the importance of Drake's expedition. For further information on English expansion, see James Williamson, *A Short History of British Expansion*, 2nd ed. (New York, 1931).

The circumstances surrounding Elizabeth's refusal to marry are diffcult to fathom. It seems inconceivable, in light of the strong feelings that the Tudors—especially Henry VII and Henry VIII—had as to their unique role in achieving English unity, and, furthermore, after the execution of Mary, that Elizabeth should have voluntarily gone to the grave without children, leaving the succession uncertain or to a Stuart. One feels that, surely, the bar to her marriage and childbearing must have been a deep physical or psychological one.[2]

In spite of the fact that she did not fulfill the natural functions of marriage and may even have had deep masculine traits, Elizabeth certainly affected an intense femininity. She wore feminine dress in an excessive way; the Spanish ambassador at times complained to his government that he was being called to interviews with the Queen at which she was so indecently dressed that he . . . really did not know where to look. She liked to bedeck herself with jewels and decorations of all sorts. In fact, some historians claim that her taste for jewels had state repercussions; Elizabeth spent so much money on them that little was left over to pay for ships.

Elizabeth insisted on getting adulation from everyone, particularly from her court favorites, and had a tendency to take strong likes and dislikes to various members of the court. All this behavior was undoubtedly rooted in a wish for a kind of adoration which, without being physical in character, had a deep sexual undertone. Of this the famous and tragic story of Essex serves as an illustration.

Elizabeth's need for veneration and her showy affectation of

[2] See the discussion on this matter in Christopher Morris, *The Tudors* (London, 1955), p. 161. The best single study of Queen Elizabeth is J. E. Neale, *Queen Elizabeth* (New York, 1934). For a detailed treatment of the Elizabethan period, consult J. B. Black, *The Reign of Elizabeth* (Oxford, 1936). A. L. Rowse, *The England of Elizabeth,* is an elaborate and detailed examination of the structure of Elizabethan society; an excellent work. See also his *The Expansion of Elizabethan England* (New York, 1955). For treatments of the Tudor period as a whole the following may be found useful: G. R. Elton, *England Under the Tudors,* is a splendid survey of the period 1485–1603, with the emphasis on political and constitutional matters; it contains a first-rate critical bibliography. S. T. Bindoff, *Tudor England* (Penguin Books, 1950), is a shorter attempt along the same lines. Conyers Read, *The Tudors: Personalities and Practical Politics in Sixteenth-Century England* (New York, 1936), and Christopher Morris' book mentioned above attempt to deal with the period by organizing the material around personalities; the latter book also contains some interesting illustrations.

femininity were not merely personal traits; they had important national consquences. They determined the tone of the court and the type of court favorite; and they influenced the sort of poetry and play that was written. As is acknowledged by the age being named Elizabethan, her character was almost all-important in influencing the "spirit" of the times. One of her contemporaries, aware that she held together the "Elizabethan" world order, compared her to the *primum mobile,* the master-sphere of the physical universe.[3]

IV

Among the many men who paid homage and rendered adoration to Queen Elizabeth, Sir Walter Raleigh was outstanding. He seems to have collected around himself (like a satellite having its own satellites) a strange set of characters, to whom the name "School of Night" was given, it is said, by Shakespeare in *Love's Labour's Lost.* Shakespeare, it is alleged, made fun of Raleigh under the name of Don Armado, and used the term "School of Night" to mock the group which surrounded him.

The School of Night was certainly not a formal group; in a formal sense, it has rightly been questioned whether such a "School" can be said to have existed. Yet, in an informal sense, the friends and followers who shared Raleigh's unorthodox views were a coherent group; and this group linked a courtier and adventurer like Raleigh with a respectable mathematician like Thomas Hariot and with writers, on the way up in social status, like Christopher Marlowe and George Chapman (of Chapman's Homer).[4] It reflected, therefore, the life at

3 E. M. W. Tillyard, *The Elizabethan World Picture* (London, 1950), p. 6. For the Elizabethan *Weltanschauung,* cf. Theodore Spencer, *Shakespeare and the Nature of Man* (New York, 1942).

4 M. C. Bradbrook, *The School of Night: A Study in the Literary Relationships of Sir Walter Ralegh* (Cambridge, 1936), is the fundamental, and almost the only, study of this group. Miss Bradbrook believes that the School probably included Raleigh, Hariot, the earls of Northumberland and Derby, and William Warner (p. 8). To this list she later adds Lawrence Keymis, a scholar and sailor (p. 40). The list, however, is quite conjectural; and, as we have said, there is even some question as to the actual existence of the School of Night itself. For our purposes, however, it does not matter whether or not the School existed in any formal way; it is as a focus, real or symbolic, of the Elizabethan spirit that we wish to consider it. For Shakespeare's supposed parody of the School, consult Frances M. Yates, *A Study of Love's Labour's Lost,* Shakespeare Problems Series (Cambridge, England, 1936). The case against the existence of the School of the Night is argued by

court, the thrust of discovery and colonization, the "dangerous," new ideas of the Scientific Revolution, and the vaulting periods and stops of Elizabethan prose and poetry. It is in this group, perhaps, that we can find the "spirit" of Elizabethan times.

Let us start with the cynosure of the School of Night, Sir Walter Raleigh. Raleigh, it is said, made his fortune with the queen by throwing his cloak before her at the right moment, when she was about to step over a puddle. There is no evidence that this is a true story. Psychologically, however, it is entirely true. It was exactly the action that would have captured Elizabeth's imagination; and it was the sort of extravagant gesture that Raleigh would make. It could even be the thing for which he would stand by the side of a puddle for a week. Raleigh was an extremely shrewd man in his own advancement and knew perfectly how to handle such a situation.

Walter Raleigh, like Drake and other rising men of the time, came from the western, i.e., the Cornish and Devon, part of the country. His family had been enriched by some of the loot after the destruction of the monasteries, but, as a younger son, Raleigh was expected to make his own way. Thus, after a short time at Oxford, he left around 1574 to fight in France (on the Huguenot side). On his return to England, he studied law at the Middle Temple. This completed his "gentleman's" education.

His progress at court was made simpler by the fact that he was a handsome man. He was also helped by his excellent intellectual accomplishments. For example, he read not only the learned tongues

---

Ernest A. Strathmann, *Sir Walter Ralegh, A Study in Elizabethan Skepticism* (Columbia, 1951).

On Raleigh the following may be consulted. Two recent short lives are Hugh Ross Williamson, *Sir Walter Raleigh* (London, 1951), and Sir Philip Magnus, *Sir Walter Raleigh* (London, 1952). Both are written in a popular style. It is interesting to note that a manuscript by Henry David Thoreau on Raleigh, written around 1840–1844, was discovered among his unpublished journals. It has been published, in a modern edition, by Henry Aiken Metcalf, with an introduction by Franklin Benjamin Sanborn, called *Sir Walter Raleigh* (Boston, Bibliophile Society, 1905). More standard works are Martin A. S. Hume, *Sir Walter Raleigh* (London, 1897), which concentrates on the empire-building and political aspects of its subject; Milton Waldman, *Sir Walter Raleigh* (London, 1928), one of the best and with a good bibliography; Edward Thompson, *Sir Walter Raleigh* (London, 1935), good and with a lengthy bibliography; and David B. Quinn, *Raleigh and the British Empire* (London, 1947), a volume in the Teach Yourself History series edited by A. L. Rowse.

but French and Spanish fluently; and none of this was lost on Elizabeth, who was also proficient in languages (it is said that she knew five or six fluently, and read Machiavelli in the original).

More important, Raleigh was helped by his first-rate literary gift. He wrote very good poetry at a time when advancement at court depended, in part, on such an accomplishment. As one writer states it, "The delicate games of tokens and poetry were essential to a world where personal relationships were never merely private, and where public relationships were never merely formal."[5] To mock or spoof one's rival successfully in a poem might be to laugh him out of court.

Many of Raleigh's poems were written in worship of the queen. He frequently addressed her in the figure of Diana, the moon. At other times he appealed to her as Cynthia (another name for Diana) or the shepherdess (her private name for him was "Water," a pun on his Christian name and on his seafaring exploits). Even when Raleigh wrote a poem whose overt meaning was apparently religious, about a pilgrimage to Walsingham, he was really adoring Elizabeth in the figure of the Virgin. In line with this, Virginia, the area in America which was first explored and colonized under Raleigh's instructions, was named in honor of the Virgin Queen.

There were, of course, other reasons why Raleigh was preferred at court. He was, like Drake, a fighter and a seafaring man. He participated, with notable heroism, in a number of military actions both on land and at sea, and he had to be restrained by Elizabeth, who feared to lose him, from others. He also took part himself in a good deal of what was simply high-sea piracy. Indeed, his piracy helped pay for many of his colonizing attempts, and ships which bore colonists out to America raided Spanish vessels on the way home.

Raleigh served England well in his colonizing attempts, although many of them initially failed. It was not, however, solely patriotism which motivated him in these activities. He was an entrepreneur, whose attempts at colonization were business ventures, made in an effort to profit from the lands granted him by the crown. Thus, with his half-brother, Sir Humphrey Gilbert, he undertook the planting of a settlement in America and, by himself, a plantation in Ireland.

Some of the money to finance his enterprises came from the grants and monopolies given him by Elizabeth. For example, in 1583, she

[5] Bradbrook, *op. cit.*, p. 33.

gave him the monopoly of issuing licenses for the sale of wine; this Raleigh promptly farmed out for 800 pounds a year. Somewhat later, he was given a pension of 2000 pounds from the customs.[6] Thus, we see that, while Elizabeth demanded admiration and masculine attention, she rewarded well those who catered to her needs.

Occasionally, Raleigh fell out of grace. He lost Elizabeth's favor in 1592 by marrying one of her maids of honor without authorization, a blow to the queen's vanity which did not go unavenged. To regain his place at court, Raleigh attempted a voyage to Guiana in 1595 and took part, heroically, in the 1596 attack on Cadiz, where he was wounded in the leg. His courageous behavior once again brought him the favor of the queen.

Raleigh was a true Elizabethan. With the death of the queen, his star began to wane. When the new king, James I, came to the throne, Raleigh was accused of treason and sent to the Tower. The real issue was that Raleigh's advocacy of an active anti-Spanish policy ran counter to James's desire to pursue a peaceful relationship with England's former enemy. In the Tower for thirteen years, Raleigh wrote, among other things, his *History of the World*, which had a strong influence on future historiography. Finally released by the king on the condition that he bring back gold from an expedition to Guiana, he failed in his quest for Eldorado. He was beheaded shortly after his return to England, in 1618.

Raleigh's death might be one way of marking the end of the Elizabethan Age. He embodied in his personality the elements of vigor and adventure that marked the period so strongly. He was a courtier and a poet, an entrepreneur and a pirate. Perhaps it was because he was so much of a microcosm of his age that around him gathered the group we are calling the School of Night. Raleigh's multiple personality could serve as a focus around which his friends, who practiced more specialized activities, could gather and find a unity of purpose and spirit.

[6] Quinn, *op. cit.*, pp. 37–38. Raleigh's "monopolies" did not endear him to the London populace, who made ballads about him such as the following:

> He seeks taxes in the tin,
> He polls the poor to the skin,
> Yet he vows 'tis no sin,
> Lord for thy pity!

Cf. Bradbrook, *op. cit.*, p. 31.

V

One of Raleigh's closest companions of the School of Night was Thomas Hariot. Hariot was an excellent mathematician, an astronomer, and something of an alchemist—the usual mixture of a scientist at that time. He forms part of the scientific side of the Elizabethan Age, which, although too often overlooked, reflects, as much as the literary side, the challenging attitude of sixteenth-century Englishmen.

Hariot, who occupied a position in Raleigh's household as a tutor, was a famous mathematician by the time he was 25. He made major contributions to algebra, especially in notation and symbolism; for example, he introduced the signs > and < for "greater than" and "less than." There is evidence, too, that Hariot either procured or constructed a telescope and discovered spots on the sun before or contemporaneously with Galileo; however, he kept his discovery quiet.[7]

Hariot served Raleigh in other capacities than that of a mathematician. Thus, in the expedition sent by Raleigh to Virginia in 1585, Hariot went along as a naturalist. The account which he wrote, called *A Briefe and True Report of the New Found Land of Virginia* (1588), was a model of description, and remained the best account, until Jefferson's writings, on the natural history of America.

In his *Report,* Hariot also helped Raleigh as a propagandist. Thus, the pamphlet, besides being a description of Virginia, was a "promotion" tract, designed to lure more settlers to Raleigh's holdings in America. Hariot did not hesitate to pay fulsome tribute to his patron:

> And the dealing of *Sir Walter Raleigh* so liberall in large giuing and graunting lande there, as is alreadie knowen, with many helpes and furtherances els: (The least that hee hath graunted hath beene fiue hundred acres to a man onely for the aduenture of his person): I hope there remaine no cause whereby the action should be misliked.

Hariot was led by his science in the direction of materialism and, thus, to what were held to be atheistic doctrines. For instance, he did not believe in the Creation as written down in the Bible because, he said, quoting Lucretius, *Ex nihilo, nihil fit* ("From nothing, you can

[7] The papers of Hariot, in which his work with the telescope is recorded, were not discovered until 1784.

make nothing"). This is, fundamentally, the basis of modern materialism in science. Hariot's biographer, Aubrey, gave a harsh, mischievous twist to this; after quoting the phrase, *Ex nihilo . . .* , he said, "But a nihilum at last killed him." In fact, what happened to Hariot was that a little spot on his nose developed into a cancer of the face which caused his death. There is almost a Shakespearian quality to Hariot's defiance of the norms and his punishment for it.

VI

Contemporary with Hariot was another mathematician and alchemist, John Dee, who received various commissions from the queen and her consultants. Dee, himself, although not a great scientist, does illustrate once again that science and scientists played a formative role in Elizabethan times. His name appears in the most unexpected ways. Not only did he give scientific advice about New World expeditions, but he backed a plan to organize a fishing fleet of 400 large ships to compete with the Dutch in the North Sea herring trade (the scheme was too costly and was never acted upon). When Jean Bodin, the French political thinker, was received at Elizabeth's court, he was met by the learned Dr. John Dee, who recorded the meeting in his diary.

Dee is also said to have warned Queen Elizabeth of the supposed plot on her life, attributed to the Spanish-Jewish physician, Dr. Lopez. The facts are vague, and it may well be that Dee himself was involved in the plot: one of those curious plots and counterplots which go on around the person of a sovereign and involve his medical advisers.

There was an odor of charlatanism about Dee, and it was his type of scientist which served as the butt of Ben Jonson's play *The Alchemist*. This play is a satire on the unseemly aspects of the science of the time.

One of the episodes in *The Alchemist* refers to magnetism. In 1600, William Gilbert, then physician to the queen, published a book on the subject. This book was the first important scientific work published in England to enjoy a European reputation. It had a great effect on the discussion about the world's rotation and similar problems, for at that time scientists were not clear in distinguishing between magnetism and gravitation. Part of the success of Gilbert's book was that it seemed to give some explanation as to why things did not fall off the earth as it spun.

In *The Alchemist,* Jonson coupled this exploratory work on magnetism with another new discovery, tobacco. We have already remarked that Raleigh introduced tobacco into England; in fact, one of the things which confirmed people in the belief that Raleigh was an atheist was that he smoked a pipe of tobacco just before he went to the block. In *The Alchemist,* Jonson has a character, named Drugger, want to open a tobacco shop. To ascertain the most auspicious location and the best time at which to open his shop, Drugger consults the Alchemist (the sixteenth-century version of our "location" economists and product psychologists?). The interesting thing is that, after giving him some of the usual verbiage about the necessity that the conjunction of the planets be right, the Alchemist gives Drugger an unusual piece of advice. He says,

> Beneath your threshold, bury me a Loade-stone
> To draw in Gallants, that weare spurres.

To us, today, this hardly seems very knowledgeable. If anything, it seems rather a funny and ridiculous use of science; and so it is. Jonson himself was making fun of it. Yet, the amount of scientific knowledge it represented, at a time when magnetism was a new phenomenon and very ill understood, is quite striking. In fact, Jonson was no wilder in his estimate of magnetic attraction than Gilbert and the other scientists who supposed that it accounted for the gravitational effects of the earth.

Jonson combined the idea of tobacco, with its evil reputation, and the new theory of magnetism in order to make fun of Gilbert, Dee, and the other scientists of the time. There is, however, another remark in *The Alchemist* which may have been intended more seriously. When the Alchemist, characteristically named Subtle, wishes to transform some base metal into gold, he insists that a great deal of base metal be supplied; for, as he says, or in words to that effect, "You cannot transmute from nothing. Unless you originally have metal you cannot make anything." This was a very important scientific remark at the time, and again shows Jonson's awareness of what scientists were saying. It is exactly like Hariot's remark, quoting Lucretius, that from nothing you can make nothing.

VII

Jonson did not belong to Raleigh's circle, although he served, like Hariot, as tutor to his son; but another famous playwright, Christopher Marlowe, did. A brilliant young man, born of a shoemaker's family, Marlowe attended the Canterbury Cathedral School and then Cambridge University, helped by scholarships. Somehow, it appears, he was involved even at that time in spying on behalf of the government. There is a letter in existence from the Privy Council to the university asking them to give Marlowe his degree and not listen to derogatory stories about him; the letter explains that, whatever he had done, he had done on behalf of the government.

While at Cambridge, Marlowe played the gentleman. However, as a playwright in London (hardly the career, at that time, for a university graduate), he "went down" in the world and lived an almost picaresque existence. He belongs, therefore, to the tradition of Villon and Molière, to the line of wandering poets and players, picking up a living on the seamy edge of a great civilization.

Living in this fashion, Marlowe wrote his plays. Most of them, like *Tamburlaine* and *Doctor Faustus,* dealt with people like himself, who "overreached."[8] Tamburlaine wanted boundless power; Faustus desired ultimate knowledge; and the Jew of Malta thirsted after illimitable riches.

To men of the Elizabethan period, all life seemed to be heroic. Endless new worlds awaited discovery, and no boundaries existed to knowledge. Marlowe, with his tumultuous and bombastic poetry, caught the spirit of the age in its exaggerated form. Figures like Raleigh, Drake, and Hariot were fit models for Marlowe's "superhuman" protagonists. What Marlowe did was to heighten some aspect of these complex figures and portray it—lust for glory, knowledge, wealth—in stark singleness. Like an alchemist, separating pure gold from the base metals, in the crucible of his art he extracted the "pure" drives of human nature.

Life for Marlowe was crowded and short. His end came in a quarrel in a public inn near Greenwich. The details of his death are shrouded in mystery and intrigue. There is, for example, a serious possibility that Marlowe was the victim of a political assassination.

[8] See Harry Levin, *The Overreacher* (Cambridge, 1952). Marlowe's death is discussed by Leslie Hotson, *The Death of Christopher Marlowe* (London, 1925).

The man who killed him was, like Marlowe, also in the pay of Walsingham, the head of Elizabeth's "intelligence" service. The two men who testified that the murder was in self-defense were both shady characters, and one a well-known spy who may have been involved in the plots around Mary, Queen of Scots. Probably we shall never know what really happened at Marlowe's death.

In any case, the circumstances of Marlowe's death can remind us of a prominent characteristic of his age: it was an age of tempestuous passion. For example, Raleigh was sent to jail around 1580 for fighting and brawling; in general, he enjoyed the reputation of an unruly gallant. Ben Jonson killed a fellow actor in a duel. As they lived, so they died: Raleigh was executed for treason and Marlowe was murdered in a tavern argument. Like Renaissance Italy, Elizabethan England was characterized by a strong note of violence.

There is another circumstance attached to Marlowe's death which tells us something significant about his age. He was, at the time, about to be called to the Privy Council to answer charges of atheism; indeed, Raleigh too was up on charges, having rather foolishly baited with impious remarks a local clergyman. In fact, the entire School of Night was under suspicion of holding atheistic views.

We are told by one modern scholar that atheism, not agnosticism, was the prevalent heresy of Elizabethan times, because the orthodox scheme of salvation was so pervasive that one could revolt from it but not ignore it.[9] We are not so sure of this explanation. When the new science, with its Copernicanism and materialism, cast doubt on the orthodox explanations, it gave rise to a form of disbelief or skepticism which, to the dogmatist, might seem like atheism. For example, in the case of Raleigh, his philosophic skepticism led him to religious toleration; he defended the Brownists (a Puritan sect) from persecution, in a speech of 1593 in the House of Commons. To the orthodox, to whom a Brownist was (in modern terms) a combination of Communist and Jehovah's Witness, this was further evidence of Raleigh's impiety and atheism.

Some authentic atheism did exist in Elizabethan times. Indeed, the Elizabethan Age may very well have been more atheistic than the ages which immediately preceded it, although this same charge (never fully

[9] Tillyard, *op. cit.*, p. 16. On the general subject of atheism, see George T. Buckley, *Atheism in the English Renaissance* (Chicago, 1932).

substantiated) was, and is, also made against the Italian Renaissance. During an age of "overreaching" even God might be by-passed by a few daring spirits. When men themselves became godlike in their powers and attributes, there seemed no need for other gods. Yet, it is doubtful whether atheistic beliefs were ever held by more than a small circle of "advanced" thinkers during the Elizabethan Era. Atheism was a part of the protest and extravagance of the time, but only a small part.

However, for this and for other reasons—the violence, the vaunting ambition, the stress on personality, the sense of discovery in the sciences, and the zest for exploration in the world—we may say that in Raleigh and the School of Night, in the new intellectual setting of England of 1600, we catch an echo of that Renaissance which played itself out in Northern Italy just a hundred years earlier.

Marlowe's closeness to Hariot marks the union of science and literary life, and Hariot's admiration for Raleigh is of a piece with the admiration Leonardo felt for the condottiere's son, Lodovico Sforza. Thus, as in Italy, Elizabethan England witnessed the union of the man of ideas and the man of action, and permitted its aristocracy of intellect to remain in close touch with its aristocracy of wealth and power.

VIII

The ideal of the "all-round" gentleman, espoused by Castiglione in *The Courtier,* which set the tone for Italian court life from its appearance in 1528, also dominated the court of Elizabeth.[10] It meant that the medieval separation between the education of a "clerk" and that of a knight was abandoned. An Elizabethan gentleman was educated in both writing and fighting.[11] He was expected to read Greek and to know modern languages, to dance and to make music, to read and to write poetry. We have already seen an excellent example of this man in Sir Walter Raleigh.

[10] Originally published in 1528, *The Courtier* did not appear in English until Sir Thomas Hoby's translation in 1561.

[11] Raleigh's half-brother, Sir Humphrey Gilbert, projected a scheme for a "Queen Elizabeth's Academy," where her wards and other noble youths would study "matters of action meet for present practice, both of peace and war." See further Sir Ernest Barker, "The Education of the English Gentleman in the Sixteenth Century," in *Traditions of Civility* (Cambridge, 1948), Chap. 5.

However, the English added to this "Italian" ideal of the courtier the notion of the "governor," i.e., the local magistrate. It was thought that the Tudor gentleman, along with literature and philosophy, should know something of law so that he might serve the commonwealth in the capacity of a magistrate, an officer, or an administrator.

This idea, which reflected the strong tradition of local government in England, was expressed as early as 1531 by Sir Thomas Elyot, in his *Book of the Governor* (the difference in concept between Elyot and Castiglione is significantly indicated by the titles of their books, which appeared within three years of one another). Elyot's book was enormously popular in England, and went through seven editions by 1580.

One feature in the ideal picture of the gentleman was hotly debated by Elizabethans: was gentle birth necessary to a gentleman? Or was it sufficient to have the virtues and talents of a Raleigh without his good ancestry? Elyot took the conservative position that gentle birth was a requirement of the gentleman, and he bolstered this by the view that society was composed of hereditary "degrees" in which men were fixed; to change those hereditary degrees, to permit social mobility, was to shake the order of society. This philosophy, fully set out in Elyot's work, was used by Shakespeare in the famous speech on "degree" uttered by Ulysses in Act I, scene 3, of *Troilus and Cressida.*

Puritans and nonconformists tended to oppose the distinction between those who were born and those who were made gentlemen—and, often, the entire idea of the gentleman. For example, an early reformer, Thomas Cranmer, in 1541 declared that God "giveth his gifts both of learning and other perfections in all sciences unto all kinds and states of people indifferently . . . wherefore if the gentleman's son be apt to learning let him be admitted; if not apt, let the poor man's child, that is apt, enter his room."

Whatever the merits in this controversy, and whether born or made, the Tudor gentleman had to be a man of versatile accomplishment. He was to be interested in art, literature, and music, as well as in warfare and civil rule. And, we must not forget, he was in some degree to be interested in science. This is nicely illustrated in the well-known connection of Sir Francis Bacon with the developing science of his time; and it is made dramatic by the fact that he died of a cold brought on during an experiment with the antiseptic properties of snow. So, too, during his imprisonment in the Tower, Raleigh not only wrote the *History of the World,* but engaged in chemical experiments and compounded drugs.

Carried to its high point, the Elizabethan ideal of a gentleman as an all-round, versatile person later created the "virtuosi." The gentleman, however, was to be a virtuoso not only in the arts and sciences but also in the art of life. Thus, exuberant figures like Raleigh and Marlowe not only wrote poetry and plays but lived dramatic lives of passion and extravagance. In this, the Elizabethan concept of personality was similar to that of Renaissance Italy (as we remarked when we discussed Leonardo da Vinci, his contemporaries regarded his life as in itself a work of art). Both centered on the idea that life consisted in the development and expression of individual personality. A Leonardo and a Hariot, a Machiavelli and a Sir Walter Raleigh might have inhabited the same court.

IX

It was not only the desire for science and poetry which animated the Elizabethans. There was another, more pervasive passion which drove them forward: the lust for gold. Behind the privateering and colonization and the attempts at alchemy stood the same desire. Indeed, the alchemist's dream, as Jonson put it, to "turn the age to gold," had come partially true with the discovery of the New World and the mines of Potosí.

In the gullible character of Sir Epicure Mammon, Jonson was satirizing the moneymaking passion of his age. For example, Sir Epicure raves and rants about what he will do with the gold the Alchemist is making for him. In his imagination he conjures up harems of beautiful women and extravagant feasts. When warned that all this will turn to dust, he has a ready answer. He will take an elixir of youth, "and so enjoy a perpetuity of life and lust."

Much of this in Jonson, as we have said, was satire. But in Marlowe it was presented in great seriousness, in terms of high-sounding tragedy. Dr. Faustus, too, conjures up harems of beautiful women, but they are real and not in his imagination. Tamburlaine and the Jew of Malta, with their boundless ambitions for power and wealth, were drawn by Marlowe with a deadly earnestness (which, it is true, almost passes into unintended parody). Marlowe, not Jonson, comes closest to portraying the Elizabethan lusts—for knowledge, for power and for gold—in their true colors.

This particolored lust it was which served as the driving force for entrepreneurs like Raleigh and encouraged them to undertake expeditions, to plant settlements, and to hunt for treasure in Guinea. The motive behind the projects and promoters of Elizabethan times was not, therefore, a pretty one.

What started, however, as unbridled lust changed in time into rational capitalism. It was this which linked the Elizabethan Age with the Puritan times that followed. Men like Raleigh and Drake broke open the routes to wealth, which were then pursued by more sober men, employing more economical means.

Indeed, although it has been overlooked until recently, there was an outburst of industrial activity in the Elizabethan Age of sufficient intensity to deserve the name of a "first industrial revolution."[12] This movement of growing industrial activity shows a combination of both the Elizabethan and Puritan attitudes; as such, it was a transition movement in both thought and action. Thus the traditional picture of the Elizabethan Age must be still further broadened by the addition of industrial expansion.

Actually, the two expansions—industrial and overseas—were connected. A good deal of capital was needed to start the new industries; and some of that capital came from the gold and silver acquired through piracy. Also, English industry was helped by the enormous price rise which resulted from the influx into Europe of New World treasure. This was so because the inflation in England, unlike that in Spain, for example, was not disastrously rapid. England experienced a threefold rise in prices from 1500 to 1600, whereas in Spain it was

[12] J. U. Nef, in "War and Economic Progress, 1540–1640," *Economic History Review* (1942), calls it an "industrial revolution." See, too, his *Cultural Foundations of Industrial Civilization* (Cambridge, 1958), pp. 50–62; this book is a broad treatment of economic history and the rise of industrialism, seen in terms of the general cultural conditions surrounding their development; it is a stimulating book. On the whole subject, cf. Rowse, *The England of Elizabeth*, p. 112, and Elton, *op. cit.*, Chap. 9. Elton thinks the term "industrial revolution" is too grandiose.

fivefold. As A. L. Rowse comments, there occurred "a long wave of profit inflation and capital accumulation which bore up the expansive achievements of the age in every sphere."[13]

This helpful inflation was especially important because capital accumulation was a difficult matter in the sixteenth century. For example, it was not until 1567 that the building of an English bourse, modeled on that of Antwerp, was accomplished by Sir Thomas Gresham. The joint-stock company did not come into existence until around 1555, with the formation of the Muscovy Company for trade with Russia, and reached its impressive development only when merchants of the Levant Company got together with other merchants to form the East India Company, chartered in 1600. Generally, to exploit a copper mine in Cornwall or a lead mine in Derbyshire, two or three gentlemen got together and formed a company.

Once accumulated, in regulated companies or joint-stock ventures, the capital was invested in such things as the mining and manufacturing of iron, copper, and brass and in the making of soda. Or it was used to increase the production of coal; for example, the contention is made that between 1565 and 1625 coal shipments from Newcastle increased at a more rapid rate than at any other period in history. Or it was used to back Huguenots setting up glass manufacturing in Sussex or a German immigrant setting up the first paper mill in England.

In these expanding industries, the first "industrial revolution" came into existence. The economic growth of England at this time was so rapid and extensive that a present-day economic historian, J. U. Nef, suggests that not until the eighteenth century was the rate of expansion again as fast as in the period around 1600. It is for this reason that he calls the expansion of Elizabethan industry the "first industrial revolution."

X

The Elizabethan Age, before its end, witnessed a great English surge in a broad sweep of life—political, economic, literary, and scientific. Yet, like all periods of history, it had to end. With the death of Elizabeth and her principal courtiers, the idea of the gentleman and the whole way of life which accompanied it began to wane. The free expression of personality and the violent, tempestuous mode of

[13] Rowse, *The England of Elizabeth*, p. 109.

life gave way to a more restrained and conscientious manner of existence. Alongside or in the place of the Elizabethan spirit arose a new ethos, the Puritan ethos. It was the Puritan ethos which served as the English counterpart to the displacement of the Italian Renaissance by the Reformation.[14]

In one sense, we may mark the culmination of the Elizabethan Age

[14] The subject of Puritanism, Puritan values, and the Puritan ethos is, of course, large and complicated. The term "Puritan" itself seems to have been first used sometime around 1563–1567, to describe Thomas Cartwright and his followers, who were demanding further reforms in the Elizabethan Church. Cartwright, among other things, wished that ministers be chosen by the people and not by bishops appointed by the queen. Unsuccessful in their efforts to amend the scheme of church government, the Puritan preachers succeeded in indoctrinating the people with a new view: it is this "revolution" in attitudes, spirit, and modes of behavior which makes up the Puritan values. As William Haller, *The Rise of Puritanism* (New York, 1938), comments, out of this conflict "came modern civilization" (p. 46).

Puritanism, of course, was not a fixed thing but rather a continuous movement of men and ideas. It changed with the changing times. For example, the early Puritans believed in a doctrine of passive obedience to government; later ones were willing to resort to the sword. So, too, at any one moment there were left and right wings and a center party. As M. M. Knappen, *Tudor Puritanism: A Chapter in the History of Idealism* (Chicago, 1939), comments, "As circumstances changed, so did the party attitude" (p. 339). Nevertheless, there were certain dominant attitudes and values which people called "Puritans" held with some consistency and which separated their holders from what we have called "Elizabethan man."

The best treatments of the Puritan movement of ideas are Haller's and Knappen's. Haller's book starts in 1570 and carries the story to 1643. (There is also a sequel, *Liberty and Reformation in the Puritan Revolution*, New York, 1955, which brings the story down to 1660.) It is a brilliant work, based solidly on the primary materials, and showing great insight into the psychology of Puritanism. It also contains a fascinating discussion of the literary style used in Puritan preaching and writing as well as an analysis of how the preachers, addressing large audiences, brought about "democratic" consequences unintended by themselves. Throughout, the emphasis is on the preachers and their pamphlets, but as seen against the background of the whole movement of Puritanism.

Knappen's book starts its story earlier than Haller's: in 1524, with Tyndale's decision to go to Germany and to translate the Bible into English. Knappen views Puritanism as "a transitional movement linking the medieval with the modern," and, by restricting his book to the Tudor period, Knappen consciously emphasizes the medieval characteristics of Puritanism. Thereby, he disowns the Weber-Tawney thesis as well as our own connection of the rising seventeenth-century science with Puritanism; and he ought therefore to be read as a corrective to our view. In Chap. 17, "The Spirit of Puritanism," an attempt at a synthesis of the main Puritan values is made, and the whole of Book II deals with special areas of Puritan thought and activity. There is also a useful appendix (II) on terminology ("Puritan," "Independent," etc.) and one on "The Historiography of Puritanism" (III), as well as a select bibliography.

by the 1611 version of the Bible, translated under the patronage of King James. Elizabethan literature, before this, had been unmistakable; its forms were fluid and experimental, and its language was constantly enriched by metaphor. After 1611, this style lost its dominance over English prose and was replaced by that of the 1611 Bible itself, which was written in a well-balanced way, correctly punctuated, with clearly delineated sentences (whereas the Elizabethan sentence tended to be a paragraph). The new style, at once the product and the end of Elizabethan literature, dominated subsequent literary and scientific prose.

In another, broader sense, however, we may symbolize the end of the Elizabethan Age with the Puritan closing of the theaters around 1642. Indeed, the conflict between Puritan and Elizabethan values is nowhere more dramatically (a forgivable pun) illustrated than in the field of aesthetics. This conflict had already been foreshadowed at the very beginning of the Elizabethan Age. The figure of Sir Philip Sidney, and his book *The Defence of Poesie,* may fittingly serve as an example.

Sidney was a nobleman, born in 1554, and named after his godfather, Philip II of Spain. Sent to Oxford at 14, he remained there for three years before going to Paris in the train of the English ambassador to France; he was in Paris at the time of the Massacre of St. Bartholomew. Proceeding to Frankfurt, Germany, Sidney met the famous Protestant divine Hubert Languet, and was greatly strengthened in his Protestantism. After Germany, Sidney, in an early version of the Grand Tour, traveled to Austria, Hungary, Italy (where he studied astronomy and absorbed the Italian aesthetic theories), and back to England by way of the Low Countries.

Back at court, Sidney became a favorite. He was a perfect courtier, a consummate horseman, and a gifted poet: in short, the ideal of his age. But his life was as short as it was brilliant. In 1586, he was mortally wounded at the battle of Zutphen, in the Low Countries. The story is told that he refused a last drink of water, and asked that it be given to a dying soldier, whose need, he said, was greater than his own. Thus, his death, as his life, accorded with the tone of grave gallantry which filled the court of Queen Elizabeth.

We have said that Sidney was a gifted poet. His own work, for he also acted as patron to the poet Edmund Spenser, was, however, published posthumously: *Arcadia,* a romance, written in 1580, appeared in 1590, and the sonnets, *Astrophel and Stella,* in 1591. Sidney's poetry is moving and important, but we shall not say anything about it here. Rather, we are interested in the dispute about the nature of poetry in which Sidney became embroiled almost by accident and which led him to write *The Defence of Poesie.*

In 1579, a hack writer, Stephen Gosson, who is of no importance except for this tiny place he occupies in the history of literature, wrote a book called *The School of Abuse.* He dedicated it to Sir Philip Sidney because Sidney, although given to writing poetry, was regarded, through his friendship with Languet, as a man with Puritan inclinations.

The School of Abuse was directed primarily against the rising theater of the time. It contended that people ought not to go to the plays, and listed a number of reasons why not. Most of these reasons resolved themselves into two complaints. One was that, on the whole, a play was likely to prompt you to that kind of thought and that kind of action which Puritans considered undesirable. The other was that if you went to a play you were likely to find yourself sitting next to people (particularly women) whose reputations left something to be desired. Thus, Gosson's simple objections were that plays led men into bad company and dealt with lascivious subjects.

Sir Philip Sidney replied to this in a book which, written in 1583 and published in 1598, has become famous; it was called, in its two editions, *The Defence of Poesie* and *An Apology for Poetry.* Sidney was goaded into writing this book because *The School of Abuse* had been dedicated to him, of all people; and Sidney wished to show that a man could have proud motives and noble ideas, such as he certainly had, and still be on the side of the new arts, the new poetry, and the new plays.

Most of his arguments he borrowed from the Greeks, the Romans, and the Italians. He admitted that, roughly speaking, playwrights did deal with subjects that nice people might not talk about. But Sidney claimed that the playwright did so in order to show virtue triumphant in the end—now the hackneyed excuse for gangster films in which, after ninety minutes of splendor, the gangster-hero realizes that "crime doesn't pay." Going further, however, Sidney claimed that poetry was superior even to philosophy, science, and history in that these simply teach us what is right while poetry alone makes us will it.

Of course, there are many further arguments in Sidney's book; but our purpose is not to enter into the detail of either Gosson's *School of Abuse* or Sidney's *Defence of Poesie*. The odd, prophetic thing for us about these two books is that they were engaged in a debate about plays and poetry at a time when little good poetry and virtually no plays had yet been written in England. The great period in English drama was, roughly speaking, from 1590 to 1615; by 1590, Sidney had been dead four years. Yet, his *Defence* validates in advance the coming plays of Marlowe, Shakespeare, and Jonson and the poetry of Spenser, Donne, and Milton.[15]

When the great age of Elizabethan and Jacobean literature was hardly on the horizon, a portentous and prophetic battle about poetry and drama was waged by Sidney and Gosson. On the very eve of the outburst of what is certainly the most prolific and exciting period in the history of English literature, the conflict between Elizabethan and Puritan values is clearly put before us in terms of aesthetics. It serves as a forerunner of the blaze of dissent and discussion which will go on in the political, religious, and economic fields, as well as in the aesthetic, and which will culminate in the war of Puritans versus Cavaliers.

XI

What link is there between the Puritan opposition to the Elizabethan Age of literature, as exemplified in Gosson's *School of Abuse,* and the growing industrialism of Elizabethan times, increasingly controlled by the Puritans? Perhaps, if we glance at the attitudes of the Elizabethan writers themselves, we may perceive an answer.

For instance, in Shakespeare's *Coriolanus,* the subject is the quarrel between the upper classes and the laboring classes in Rome. Coriolanus is a man who takes every possible opportunity to alienate the representatives of the plebeians, including, incidentally, Junius Brutus, to whom we have already had occasion to refer.

The most interesting thing about *Coriolanus* is that, as one realizes

[15] Sidney vindicated the coming plays, in general terms, even though he insisted on the observance of the classical unities of time and place and objected to the mingling of tragic and comic elements in a play.

after reading a few pages, Shakespeare is openly in sympathy with Coriolanus; and this because his hero is defending the idea of hierarchy and order, so dear to the medieval world and so dear to Shakespeare. The play makes it clear that Shakespeare believed in a world where everything had its right "degree" and everyone his right place in society.[16]

By 1600 and thereabout, the medieval ideas on degree were as outmoded as the medieval ideas on economics. The "industrial revolution" of that period, with its consequent social mobility, coupled with the growth of Puritan sentiments, had shattered both sets of ideas. Yet, Shakespeare, and many of the other playwrights, held to the old vision. Why?

One reason is that many of the playwrights were employed to write plays by noblemen; an obvious sympathy might be expected. A second reason lies in the fact that many of the Puritans were opposed to ideas of "degree" as well as to the theater and the writers of plays. The playwrights, who had every reason to be thoroughly out of humor with the rising group of Puritans so anxious to shut the theaters, retorted with a defense of hierarchy and an attack on the Puritans.

We can see this antagonism to the Puritans clearly reflected, for example, in Jonson's *Alchemist,* where great fun is made of them. We can see the Puritan opposition to the playwrights in the very location of the theaters: they were all built outside the confines of the city of London. They were built there because, by 1600, the city was intimately associated with Puritanism and was asserting its independence.[17]

Puritanism, although intimately involved with expanding Eliza-

[16] *Coriolanus* is only one of several plays in which Shakespeare is openly on the side of patricians and the upper class and in favor of the established order. There is a famous speech on degree in *Henry V,* delivered on the night of Agincourt; and we have already mentioned the speech of Ulysses in *Troilus and Cressida.* Some scholars hold that the speech on degree in *The Play of Sir Thomas More* was also written by Shakespeare.

[17] The independence of the city of London was jealously guarded; for example, it was symbolized whenever the king, whose court was outside the city, wanted to enter the city. When the king reached the entrance to the city at Temple Bar, this gate (rebuilt a number of times, and finally by Wren in 1672) was closed, and the royal procession was stopped until the king received the Lord Mayor's permission to enter. This ceremony still takes place every year.

bethan industrialism and trade, as we have seen, rejected and criticized the Elizabethan and Stuart fostering of luxury production and monopoly; one such "luxury" was the theater. By rejecting the theater, the Puritans also rejected the courtly culture for which it stood. This antagonism had profound consequences; it ended one sensibility and led to another. It meant that Puritanism turned its back on the Elizabethan Age.

Here, we see the unity between Gosson's attack on the theater and the industrial development of Elizabethan times. The next step was to suppress not only plays extolling degree and hierarchy but degree and hierarchy itself.[18] With the taking of this step, we enter fully into the Puritan Revolution.

[18] The Puritans were neither all nor consciously opposed to degree and hierarchy; in fact, it was only a very small number—mainly the Levelers—who actually turned to democracy. However, this development was latent within much of the logic of the entire Puritan position. As William Haller points out, "If plutocracy may be said to have first reared its head among the orthodox Calvinists, democracy may equally as well be said to have stolen up behind the orthodox in the guise of the heretics, the mystics and enthusiasts of all sorts whom the preachers summoned to bear part under the banners of Christ in the war of the spirit on wickedness in high places" (*The Rise of Puritanism*, p. 215).

PART II

THE AGE OF REASONED DISSENT:

From Cromwell to Rousseau, 1630-1760

CHAPTER 9

The Puritan Revolution

I

WHILE ELIZABETH lay dying, in March of 1603, one of her relatives, Robert Carey, wrote to James of Scotland that he had horses posted along the way and would give the king, as soon as possible, the eagerly awaited message to come down to England and assume the crown. Speed was important. The change from the Tudor Elizabeth to James and the Stuarts opened the way for all sorts of plots. In one such plot Raleigh was supposedly implicated, and for this he was sent to the Tower.

Any king would have found it difficult to follow upon Elizabeth. She had been an astute manager of men and never neglected the obvious tricks which make it easier for a woman to control men than for a man to do so. James, however, was especially handicapped. He had been brought up in a Scottish court which, modeled on the French court, was greedy and prodigal. His tutors, however, had been Scottish Calvinist divines. They, in opposition to the court, drummed into him the lesson that tyrannical kings may be deposed by their people and that the king's power was greatly circumscribed.

James reacted strongly. He became a violent believer in the divine right of kings and wrote the *True Law of Free Monarchy* in his own defense. He held that monarchy was a divinely ordained institution, that the king was accountable only to God, and that, therefore, the king's power was above the laws.[1]

[1] Interestingly enough, the divine-right theory of kingship had originally been adopted by the Protestant reformers, in an attempt to strengthen the claim to independence of the national state against the papacy. As G. P. Gooch suggests, "Divine right thus began its career as a defensive weapon against militant Cathol-

This brought James into conflict with Parliament. In the very first session of his reign, in 1604, king and House of Commons fell out on the question of the Commons' right to determine the disputed elections of its own members. Discord was not restricted merely to matters of constitutional import; on national affairs, such as James' treatment of the Puritans and the Catholics, as well as on foreign policy, the king and his Parliament were also at odds. To James' assertion that these were questions of "king's craft" and "far above their reach and capacity," the Commons retorted with a grim defense of constitutional limits on the power of the king.

By 1611, James had had enough of Parliaments. With the exception of a so-called Addled Parliament, which sat for a few months in 1614, he ruled England alone until 1621. The rights of Englishmen, unable to find voice in Parliament, sought refuge in the courts of common law. Here, opposition to James' assertion of divine right was led by the great constitutional lawyer Sir Edward Coke, who viewed the laws not as the instrument but as the boundary of the royal prerogative. He held that the king was under the law and that the law was to be found in the decisions of the courts and in common law. (This was to make judges the ultimate sovereigns; in many ways, it foreshadowed the claim later made by the French Parlements.) Even this barrier against the king, however, was broken down by the dismissal of Coke as lord chief justice in 1616. Thus, the defense of the rights of the subject once again was thrown back to Parliament, when and if the king saw fit to summon it again.

The fact that the Commons tended to be in sympathy with Puritanism exacerbated its controversies with James. The latter's hatred of the reformers had been intense from childhood on. Their opposition to him in Parliament added fuel to the fire. From the Puritan view, James' disapproval of their religious and political ideas was made worse by his abuse of the system of granting royal patents and monopolies to worthless favorites. From his side, James flaunted his partiality for certain courtiers and threatened the Puritans that he would "make them conform themselves or I will harry them out of the

icism" (*Political Thought in England from Bacon to Halifax*, London, 1950, p. 11). Gradually, however, the reformers realized that they were simply substituting one tyranny in place of another, and they abandoned the divine-right theory. Thus, by the seventeenth century only Anglican divines, who saw the power of the church and of the crown linked, upheld James in his theory of divine right.

land." It was the latter policy which prevailed. One result was the Pilgrim removal to America in 1620.[2]

In conflict with Parliament over constitutional issues and theory of state, out of sympathy with its Puritan leanings, James was nevertheless wise enough to keep affairs in England from boiling over. He defended his divine-right views vigorously with his pen, but he did not push beyond a certain point in practice. It was not for nothing that he was known as the "wisest fool in Christendom."

II

James died in 1625 and was succeeded by his son, Charles I. Although he inherited various of his father's quarrels with Parliament, for Parliament was now sensitive to extensions of the king's prerogative, Charles held initial advantages over his father's position. He was personable and dignified in comportment, whereas James had been uncouth; he was temperate, whereas James' court had been marked by vice and loutishness; and he threw over his reign a cloak of beauty by patronizing the artistic work of Van Dyck and Rubens. On a less subjective level, he won popularity by his anti-Spanish policy and he sponsored a government which was benevolent and not without efficiency. Indeed, it was a paternalistic government, for Charles stood on the side of the common people and tried to protect them, through his special courts, from the worst rigors of enclosures and industrialism.

With all this in his favor, Charles was really in a worse situation than his father, for he lacked the latter's willingness to give way at the crucial point. He was prepared to die for his beliefs. And like his father, Charles believed in the divine right of kings. His courtiers and officials confirmed him in the righteousness of his actions and

[2] G. M. Trevelyan, in *England Under the Stuarts* (London, 1904), points out that James I, and Laud, must be given the credit for establishing "Anglo-Saxon supremacy in the new world." Between 1628 and 1640, about 20,000 nonconformists fled to America (pp. 173–174). Trevelyan's work covers the entire period from 1603 to 1714; although written more than 50 years ago, it is still one of the best books in the field. Over half the book deals with the origins, events, and effects of the Puritan Revolution. Its bibliography, however, is seriously outdated. The standard authority is S. R. Gardiner, with his *History of England, 1603–1642*, *History of the Great Civil War*, and *History of the Commonwealth and Protectorate*. Clarendon's *History of the Great Rebellion* is a classic, written by a partisan eyewitness.

belief. His courts, from which the judges were removable at his pleasure, obediently told him that he could command the persons and money of his subjects without let or hindrance if he thought a sufficient necessity existed. One of his chief advisers, Strafford, counseled him that "The King is loose and absolved from all rules of government."

Charles joined to his extensive view of the royal prerogative an even more pro-Catholic and anti-Puritan bias than his father's; his French wife, Henrietta Maria, had a strong influence on him in this respect. The weapon he used, as we shall see later, was William Laud; but the policy of anti-Puritanism was Charles' own and it coincided with his opposition to Parliamentary assertions of right.

The spark which set off his first major explosion with Parliament came when he entrusted the conduct of affairs to his favorite, Buckingham. Incensed at the latter's mismanagement of policy, the Commons sought to impeach him. Blocked in this by the king, they refused to vote supplies. Charles dissolved Parliament and resorted to a "forced loan." It was, of course, illegal, and, although he was able to arrest some of those who refused payment, he had effectively to abandon the attempt to collect the loan. He was not only unsuccessful, however; he was also impolitic, for he had openly violated the "fundamental laws" of the nation.

Eventually, shortages of money, largely occasioned by his military needs (for he was variously engaged in war against Spain and France), forced him to call Parliament again into session. This Parliament, of 1628, voted the king supplies but, in turn, it made him accept the famous Petition of Right. This asked that the king henceforth observe the rights of his subjects. It demanded that billeting of troops and trial of civilians by martial law cease and it declared that arbitrary imprisonment, and taxation without the consent of Parliament, were both illegal. By his acceptance of the Petition, Charles admitted his forced loan to be unconstitutional.

Immediately after the Petition, however, events seemed to show that Charles had not been sincere in spirit. A second session of Parliament, in 1629, sought to bind him more closely to the wishes of Parliament and to the rights of his subjects. Firmly led by Sir John Eliot, with a member for Huntingdon, named Cromwell, supporting him in a very minor role, it passed three resolutions. Whoever, it was declared, should bring in innovations in religion should be considered an enemy of the kingdom (this was aimed at the king's alleged attempt to reintroduce popery in England). Whoever counseled the levying of tonnage and poundage tax (i.e., custom duties), without a Parliamentary grant, should also be held an enemy to his country; and whoever willingly paid these taxes was proclaimed to be a betrayer of the liberties of England. As Sir Charles Firth comments: "The significance of the resolutions lay not merely in their challenge to the King, but in the union of political and religious discontents which they indicated. Elizabeth's policy had called into being a religious opposition. James had created a constitutional opposition. Under Charles the two had combined, and from their alliance sprang the Civil War."[3]

In short, Charles had lacked statesmanship; and this was one of the fundamental causes of the Puritan Revolt. He had given occasion to the leaders of the Puritan opposition to his policies to identify themselves with the defense of the constitution and the rights of the subject. Another king, with a different personality or conception of his position, would have meant, in all probability, a different outcome. In the words of one recent writer, "In this respect at least the accidents of personal history were more important than either social tinder or political spark."[4]

III

What, however, of the "social tinder" and "political spark"? The character of Charles I is pivotal in any causal analysis of the Puritan Revolution—his rejection of any real settlement in 1642 forced Parliament to a war it really did not desire—but without the circumstances of political and religious strife, his character would not have affected the course of history in such a momentous fashion.[5] It is, therefore, the constitutional and religious issue and, in the eyes of some his-

[3] Sir Charles Firth, *Oliver Cromwell and the Rule of the Puritans in England* (Oxford, 1953), p. 18. This is one of the most interesting books on Cromwell.

[4] H. R. Trevor-Roper, "The Outbreak of the Great Rebellion," in *Men and Events* (New York, 1957), p. 194.

[5] On the pivotal role of Charles I, see further J. H. Hexter, *The Reign of King Pym* (Cambridge, 1941), pp. 16–17. Hexter's book is a model of how history should be written; it not only deals with the subject in hand but it is, as well, a brilliant essay in historiography.

torians, the social and economic question which must be examined for any real understanding of the events of 1640–1660.

In many ways, such an understanding emerges best from a narrative treatment of the period.[6] The Puritan, first among modern revolutions, was the least consciously prepared for by any body of systematic ideas: political, religious, or social. Indeed, so much was this the case that the Puritans, once in power, were at a loss to know what government to install in the absence of the king. They had not intended a revolution; and they had always assumed the coexistence of king and Parliament. Because of this "event tumbling into unintended event" quality of the Puritan Revolution, we shall devote some space to the narration of its events.

Before that, we need a synoptic view. So overlapping are the civil and ecclesiastical spheres in the seventeenth-century polity that a real separation of them is impossible; nevertheless, the issue first in importance in our judgment is the constitutional issue. The generally accepted view of government was in terms of a "balanced polity." In sum, this meant that the king's powers were limited by the so-called "law of nature," by the common law, as administered in the courts, and by the rights of Parliament. Head of the commonwealth, the king was not absolute in his powers but constrained by the rights of his subjects, the latter protected by law and in Parliament.

Between them, James and Charles brought this view of government into jeopardy. In theory, they arrogantly asserted the divine right of kings. In practice, they destroyed an independent judiciary and thus the appeal to law; they eroded property rights by "forced loans" and extra-parliamentary taxes; and they threatened personal rights by arbitrary and illegal arrests. Therefore most Englishmen wanted quite simply to restrict the king's despotic rule and to restore the "balanced polity." Even at the outbreak of civil war, the Houses claimed to fight in the cause of preserving the king's just prerogative, the privileges of Parliament, and the liberties of the subject.

On the other hand, it must be admitted that Parliament itself was

[6] C. V. Wedgwood, *The King's Peace, 1637–1641* (New York, 1955), maintains exactly this thesis: that a narrative history restores the "immediacy of experience" and is more interesting because it treats the "behaviour of men as individuals" and not as groups or classes. In a curious way, however, Miss Wedgwood's book is disappointing; perhaps the forest is lost for the trees. Both analysis and narrative seem necessary parts of the historical craft.

asserting greater rights than it could claim by tradition. Charles was correct in his estimate of affairs in 1629 when he remarked that the Houses "hath of late years endeavoured to expand their privileges by setting up general committees for religion, for courts of justice, trade and the like." However, the demands of Parliament, though large, were specific; they wanted an increased say in economic, foreign, and religious affairs. In the Grand Remonstrance of 1641 (which was carried only by a majority of 11), they indicated with pride what they had wanted and what they thought they had secured. As one modern writer summarizes it:

> The Houses have put down the prerogative courts used by the malefactors around the King for curtailing the liberties of the subject; they have re-established the credit of a bankrupt monarchy and at the same time, by outlawing ship money and subjecting the collection of tonnage and poundage to the consent of Parliament, have made arbitrary exactions by the King once and for all illegal; they have partially reformed the Church; they have provided machinery for the regular summoning of Parliament; so henceforth the people may always voice their grievances in the assembly of their chosen representatives; finally they have brought to justice some of the wicked councillors to whom all England's woe is due.[7]

In two areas—financial and administrative—the king was at a disadvantage in asserting his prerogative. He could raise large sums of money, consistently, only through Parliament; thus, he was constrained by the gentry who controlled Parliament. And he could assert his administrative power locally only through the local gentry—the justices of the peace and sheriffs and vicars—who resented attempts of a central government to filch away their authority.

Only in one area—the religious—did the king have a centralized,

[7] Hexter, *op. cit.*, p. 163. For the Grand Remonstrance itself, see *The Constitutional Documents of the Puritan Revolution, 1628–1660*, selected and edited by Samuel Rawson Gardiner (Oxford, 1889). In the petition to the king that preceded the long list of grievances, the House of Commons demanded that "for the future your Majesty will vouchsafe to employ such persons in your great and public affairs, and to take such to be near you in places of trust, as your Parliament may have cause to confide in." They also wished to deprive "the Bishops of their votes in Parliament," a more radical and novel step than the others we have mentioned. On seventeenth-century constitutional history, consult Francis D. Wormuth, *The Royal Prerogative, 1603–1649* (Ithaca, and London, 1939); Margaret A. Judson, *The Crisis of the Constitution* (New Brunswick, 1949); George L. Mosse, *The Struggle for Sovereignty in England* (East Lansing, 1950); and J. W. Gough, *Fundamental Law in English Constitutional History* (Oxford, 1955).

relatively efficient, and powerful organization to do his bidding: the Anglican Church. And it was here, in this area, as we shall see, that his assertion of right touched off the Puritan Revolution. To understand the background of the events we shall describe, we must therefore recall the religious situation in general.

The Anglican Church was a product of the Reformation. The constant issue, therefore, after the break with Rome, was: how far should the church be reformed? As we have seen, under Elizabeth's reign, a group known as the "Puritans" arose who wished a further degree of change in the English Church than the queen was prepared to institute. Led by Thomas Cartwright, they at first concentrated on changing the method of governing the church; unsuccessful in this arena, they turned to preaching and to the indoctrination of the people with new views and attitudes.

There were many sort and shades of "Puritans," but they all shared the basic Calvinist ideas, especially of predestination. In common, too, they had the desire either to "purify" the usage of the established church from stains of popery or to worship by forms so "purified" in separate congregations. Within their ranks, however, there was what might almost be called a "trend to the left." Thus, there were Episcopalian, Presbyterian, and Independent or Congregationalist Puritans, with the latter trailing off into Separatists. The Presbyterians wanted a system of church government on the model of Geneva, with an ascending hierarchy of "elders" (the Greek for "elder" is *presbyteros*) in control; a similar system of church government existed in Scotland. The Independents wished each congregation to be legally independent of every other, with no ecclesiastical authority higher than the individual church; they were prepared, however, to remain within the established church. Those who wished to separate from the established church, its state control and financial support, were known as Separatists.[8] United in their desire to change the established church, the Puritans thus split into two major groupings: those who stressed reform of church structure and ceremonies and those who stressed liberty of conscience.

Against all the Puritans, Charles and his Archbishop of Canterbury, Laud, pressed a campaign in favor of order and uniformity in

[8] For all questions of Puritan terminology, consult the already cited work by M. M. Knappen, *Tudor Puritanism* (Chicago, 1939), and especially Appendix II, "Terminology."

the Church of England. In the realm of theory, the attack was made on the doctrine of predestination. In the realm of practice, the attack was made in such terms as the insistence on altars in the east end of all churches, and the repression of Puritan "lecturers." The aim was to harry the opposition into conformity. The assertion of royal ambitions in the matter of religion took on the same tone as assertion of prerogative in matters of state. As we shall see, the two "matters" became intertwined in the events of 1639 and led to the Puritan Revolution.

Was the Puritan Revolt not only a constitutional-religious controversy but also a class conflict, waged in religious terms? Suggestive answers to this question have recently been brought out in a controversy over the gentry. The gentry, it is agreed, can be defined as that class which drew the largest part of its income from the exploitation of their land, and whose members did not belong to the peerage. Were they rising or falling in power? Rise or fall—what was the significance of their movement? A controversy has raged over these questions in what one critic has called a "Storm over the Gentry."[9]

One thesis is put forth by R. H. Tawney. He claims that the gentry, which he equates with the new, emerging middle class, was on the rise economically, displacing the aristocracy, and that the Revolution of 1640 was merely the political expression of what was already a social fact. In its baldest terms, the Cavaliers were aristocrats and the Roundheads were bourgeois; the Revolution enshrined the middle-class gentry as the new rulers of England.

Another thesis, similar in centering attention on the gentry and on economic-social criteria, has been put forward by H. R. Trevor-Roper. He claims that the price revolution of 1540–1640 spelled the doom of the mere gentry, because they lacked the credit or cash to meet the new conditions by turning to large-scale estate management (i.e. "capitalistic" farming) or by rescuing their fortunes through favors received at court. These impoverished squires turned to unsuccessful anti-court rebellions before 1640 and to successful revolt in 1640. In his words, the Great Rebellion "is not the clear-headed self-assertion of the rising bourgeoisie and gentry, but rather the

[9] J. H. Hexter, "Storm Over the Gentry," *Encounter* (May, 1958). This is a brilliant and incisive analysis of both the Tawney and the Trevor-Roper theories.

blind protest of the depressed gentry [among whom Trevor-Roper places Cromwell]," and, "it was a protest by the victims of a temporary general depression against a privileged bureaucracy, a capitalist City."[10]

A more old-fashioned interpretation is by one of Cromwell's earlier biographers, Sir Charles Firth. He stoutly maintains that the Revolution was not primarily a social or economic movement, and cites some interesting information in support of his position. According to Firth, in the House of Commons of 1642, "About 175 members followed the King's flag while nearly three hundred remained at Westminster. In the Upper House the preponderance was overwhelmingly on the King's side. Rather more than thirty peers threw in their lot with the popular party, while about eighty supported the King, and about twenty took no part in the struggle." Indeed, many families were split over the issue: in Cromwell's own family, an uncle and a cousin ardently espoused the Royalist cause. Firth's conclusion is that "though the bulk of the upper classes was on one side, the war never became a social war, but remained a struggle of opinions and ideas."[11]

In any synoptic view, the various motives—political, religious, economic, and social—must all find their place. However, the stress must be put, in our judgment, where Firth puts it: on the political and religious "opinions and ideas." Indeed, the Puritan Revolt was not a "social war" exactly because it lacked a systematic and theoretical ideology of classes, which would have made men conscious of their social and economic interests. These interests existed, of course, but they were dwarfed in men's minds by other interests and only found their expression indirectly, in religious and constitutional guise.

This is understandable. Men have been wont, since 1517, clearly

[10] H. R. Trevor-Roper, "Social Causes of a Great Rebellion," in *Men and Events, passim.* See, too, his "The Gentry, 1540–1640," *Economic History Review* (1953). For Tawney's views, see his "Harrington's Interpretation of His Age" (British Academy Lecture, 1941), "The Rise of the Gentry," *Economic History Review* (1941), and "The Rise of the Gentry: A Postscript," *ibid.* (1954). On the general subject, George Yule, *The Independents in the English Civil War* (Cambridge, 1958), may also be consulted.

[11] Firth, *op. cit.,* pp. 70–71. The latest work on the composition of Parliament is D. H. Pennington and Douglas Brunton, *The Members of the Long Parliament* (London, 1954). In general, it supports the notion that political divisions in the House of Commons did not follow clear-cut class lines.

and consciously to fight about religion. In the seventeenth century, the conflict around political and constitutional issues also became open and acknowledged; it was an age of "reasoned dissent" in politics and religion. It was only in the late eighteenth and nineteenth centuries, however, that economic and social conflict became consciously motivated and provided with an ideology.

In the seventeenth century, therefore, a man's economic and social position was normally expressed not as such, but as a religious and political position. Emphatically, this does not mean that the religious and constitutional issues were unimportant or mere "fronts." Nor does it mean that an Anglican cavalier or a Puritan merchant was a hypocrite. It simply means that Puritans and Cavaliers thought of themselves more as religious-political groupings than as economic-social ones; their "ideology" expressed this dominant orientation of mind. And it behooves the historian, while discerning other aspects of the conflict, to realize that this cast of mind determined that the Puritan Revolt would be primarily a religious-political and not an economic-social struggle.

IV

Let us turn, once again, from our abstract, synoptic analysis of the Puritan Revolt to recounting its events in a sequent narrative. We had suspended our story with the Petition of Right of 1628 and the Three Resolutions of 1629. After the latter setback, Charles ruled for 11 years without a Parliament. His taxes were levied without Parliamentary grant, and obsolete fines and dues were revived. One way to obtain funds was ship-money. In 1634, it was levied only on the coastal towns; in 1635, it was extended to all England.

When this occurred, a man named John Hampden, who lived in Buckinghamshire, refused to pay it. His case became a *cause célèbre* in England and raised him to the status of a hero in the Parliamentary party. We shall not go into details about the case, except to say that Hampden's opposition to taxation wihout consent was linked with the fact that he was also a leader of the Puritans. It is interesting to note, further, that Hampden was Cromwell's cousin. Later, we shall show more fully how interconnected were the "Revolutionary families" of 1642.

The fact that Hampden came from Buckinghamshire is significant.

It tells us something about the areas of the country which were to support the Revolution. Buckinghamshire, for example, owing to its extensive growth of beech trees, was a center of the furniture-making industry—an industry first carried on by Puritans and later by Quakers.

The newer sheep-grazing and cloth-making districts also supported Parliament's assertion of right against the king. These districts were mainly on the east coast. Their prosperity was made evident by the fine brick houses built by the Puritan clothiers, which still stand.

Thus, in very general terms (i.e., overlooking pockets of resistance), it was the east coast which supported Parliamentary pretensions. On the other hand, it was the west, that is, from Oxford on westward, which tended to rally to the king's standard. In detail, of course, no sharp geographical division corresponding to political alignments can safely be made; the very counties themselves were split.[12]

If Hampden can be taken to represent the new Parliament-Puritan attitudes, a fitting representative of what they were fighting against may be seen in Archbishop Laud. The leading spirit in Charles' government, Laud, like Sir Thomas More before him, opposed speculation, enclosures, and other economic malpractices (at least so considered by Laud) of the time. As a member of the High Commission and of Star Chamber, he attempted to halt what he considered the increasing exploitation of the English people and to use equity, rather than positive law, on their side. Essentially a medievalist, Laud refused to countenance the growing reluctance of the church to judge economic life and social organization by ethical criteria.

Further, following Hooker, who, in his *The Laws of Ecclesiastical Polity* (1593), had asserted that in a Christian state, "one and the same people are the Church and the Commonwealth," Laud insisted that religious unity was essential for state unity. Hooker, however, had had a spirit of moderation and tolerance that Laud lacked. James I saw this when he remarked that "Laud hath a restless spirit, and cannot see when things are well, but loves to bring matters to a pitch of reformation floating in his own brain."

Laud's "restless spirit" led him to try to introduce Anglican (i.e., "popish") elements into the externals of worship in Scotland, as well

12 Cf. Maurice Ashley, *The Greatness of Oliver Cromwell* (London, 1957), p. 93. Ashley discusses the geographical division of England in the Civil War in detail in his Chap. 6.

as to curb the Puritan preachers. The Scots retorted with a Covenant and a resort to arms. The result, in 1639, was the so-called Bishops' War, for Charles, with his sense of divine right and his high-flown pride, backed his archbishop. "So long as this Covenant is in force," Charles declared, "I have no more power in Scotland than a Duke of Venice, which I will rather die than suffer." His words were prophetic.

Rapidly now, the chain of events uncoiled. Charles had to call Parliament into session to finance his Scottish wars. Annoyed at its demands, Charles dissolved the first Parliament of 1640. He was soon forced to call another in the same year; this was the famous Long Parliament, under whose auspices the Puritan Revolt broke out.

The Long Parliament first impeached Charles' favorite, Strafford, and, when this failed, took the irregular step of sending him to the block by attainder. Next, it abolished Laud's favorite courts of Star Chamber and High Commission. To gain time, Charles accepted these and other curbs on his authority, while he tried to gather troops with which to overawe Parliament. His attempt to arrest arbitrarily five leaders of Parliament, at the beginning of 1642, however, misfired.

The failure of this attempted coup d'état was the signal for the outbreak of civil war. Charles, who had sent his queen abroad with the crown jewels (to buy arms and ammunition), moved north. He planned to gather his friends about him and to get possession of Hull, an east coast port where he could land French troops from the Continent and which was also an arsenal of arms (collected for the Scottish war).

Charles approached Hull from a little cathedral town, called Beverley, where everybody was in his favor. He expected, therefore, to be warmly received in Hull. The lord mayor and the commander of the garrison, however, held a conference and decided to shut the gates to Charles; significantly, Hull was a big clothing-export town. This action, like the shots at Lexington in the American Revolution, was the first open act of defiance which set off the Puritan Revolution.

V

Parliament, from the beginning, held certain long-term advantages: it controlled London and thus the wealth of the capital; it could raise taxes by assessment on the counties and levy customs duties on the ports; and it commanded the fleet. But it was divided in its leadership and ambivalent in its desires. Many of those who wished to limit the king's prerogative had no desire to extend that of Parliament and the people; and they certainly had no active desire to go to war if the king could be "treated" into coöperating with Parliament. Thus, at first, the prosecution of the war dragged on the Parliamentary side, and it was only the political genius of Pym which forged a real weapon of war against the king.[13]

The initial advantages on the side of the king were his unquestioned authority and leadership over his forces; his military leaders, such as Prince Rupert, who had had training and experience in Continental warfare; his gentlemen horsemen, who could make excellent cavalry; and the intangible though important "divinity that doth hedge a king." In one sense, however, the very professional training of his officers made them less flexible than the Parliamentarians in meeting the special conditions of English civil war.[14]

The total effect of these factors was that the Parliamentary forces got the worst of it at the outset of the war. Gradually, however, they improved their position and by 1644, at Marston Moor, severely checked the king's forces. "New-modeled" by Parliament in 1645, unified and with regular pay, they effectively defeated the king at Naseby and Langport in that year.

The inspiring personality and military leader in the Parliamentary army was, of course, Oliver Cromwell. He had raised his own troops —the Ironsides—on a new principle: conviction in the cause for which they were fighting. Amalgamated into the Eastern Association, or league of counties, they eventually also became the core of the New Model Army; their leader, the dominating figure in the military forces of Parliament.

What sort of man and what sort of leader was this champion of the Parliamentary cause? He was, first of all, a member of the gentry. Indeed, all his life, he retained the tastes and habits of a country gentleman. He drank ale and wine. He hunted and hawked, and he was an

[13] For Pym's critically important role in the early stages of the rebellion, see Hexter, *The Reign of King Pym.*

[14] C. V. Wedgwood, in her little book, *Oliver Cromwell* (New York, 1956), pp. 42–43, has some illuminating comments on the special conditions of warfare in the English setting.

expert jumper. Like many other gentlemen of the period, he had a pronounced taste for music. Educated for a year at Cambridge, he enjoyed mathematics and history; one of his favorite books was Raleigh's *History of the World.* Finally, to round off the preparation required of an English country gentleman, Cromwell is thought to have studied law in London.

Two things distinguished Cromwell's background: he was brought up in a Puritan atmosphere (his own spiritual conversion occurred when he was 28) and he belonged by birth to the English ruling class.[15] The latter fact did not necessarily mean great wealth; it did mean local power in the county. Cromwell, himself, experienced great vicissitudes of fortune, at one point having to sell his own lands and become a tenant farmer. In 1636, however, he became heir to his maternal uncle and took possession of a considerable estate and income. Thus, in spite of shifts, Cromwell's position as a country gentleman remained assured.

Cromwell's status was derived from that of his general family (even though his father was a relatively impecunious second son), and his family was one of those which had made its fortune in the Reformation. It was Henry VIII's minister, Thomas Cromwell, who had laid the foundations of the family's prestige, and the original grants of land and money made by the king were added to by Thomas Cromwell's nephew, Richard.

A "new-rich" Reformation family, the Cromwells had freely intermarried with other such families. We have already mentioned that one of Oliver Cromwell's cousins was John Hampden. When Cromwell was elected to Parliament for his home town of Huntingdon in 1628, and for Cambridge in 1640, he sat next to another of his cousins, St. John (a very able lawyer and Hampden's counsel in the ship-money case). Indeed, it is estimated that about twenty of Cromwell's kinsmen sat in the Long Parliament. During the war, one of the Regicides was Major-General Edward Whalley, another of Cromwell's cousins.

[15] For Cromwell's social position, cf. Ashley, *op. cit.,* and Wedgwood, *Oliver Cromwell.* The prime source for Cromwell is W. C. Abbott, *Writings and Speeches of Oliver Cromwell,* 4 vols. (Cambridge, 1937–1947), which has largely replaced Thomas Carlyle's *Letters and Speeches of Oliver Cromwell.* Both Ashley and Wedgwood give good bibliographies on Cromwell and should be consulted for further works.

Cromwell himself married the daughter of Sir James Bourchier, a wealthy leather merchant. Cromwell's eldest daughter was married to General Ireton, Cromwell's right-hand man, and after Ireton's death, she became the wife of Lieutenant-General Fleetwood. The latter was succeeded, as Lord Deputy of Ireland, by Cromwell's son Henry. The other son, Richard, succeeded Oliver himself as Protector.

Cromwell's family background may help to explain why he was opposed to leveling movements within the Puritan Revolution, and worked to support authority and property. He held that the distinction between class and class was the cornerstone of society. "A nobleman, a gentleman, a yeoman," he said, "that is a good interest of the land and a great one." It is evident that Oliver Cromwell's support of the Revolution was not in the interests of a social renovation.

There were two main motives to Cromwell's objection to the government of Charles I. The first was that if Charles could tax his subjects as he willed, their property was not safe. Cromwell, therefore, opposed Charles' exercise of authority on the ground that it was unjust and without restraint. The second motive for Cromwell's resistance to Charles' government was a positive one and much more powerful. He believed that there should be freedom of conscience in religion in England. In short, Cromwell was a confirmed Independent. It was this which distinguished him from the "Cavaliers" around Charles I.

Cromwell's Puritanism explains, also, his character as a leader. Most of Cromwell's contemporaries, and many later writers, such as Pope, Voltaire, and Hume, considered him a consummate hypocrite. We do not subscribe to such a view. Cromwell's attitude to God was that of a mystic. He believed firmly that God had chosen him, Cromwell, to lead His people out of the wilderness. Therefore, he attributed all victories in battle to God rather than to his own strategy and tactics. The consequences of this view of himself, as God's chosen instrument, were twofold: it girded his loins to win battles, but it also, as Ernest Barker points out, enabled him "to hide the exigencies and the calculations of politics behind a plea of the intention of God. It enabled a man who loved mercy to practice sometimes a sad cruelty."[16]

[16] Ernest Barker, *Traditions of Civility* (Cambridge, 1948), p. 186. It must be remarked, however, that all elect Calvinists felt directly guided by the Divine Will.

It permitted him to pursue a policy of *raison d'état,* to plead "Cruel Necessity" in justification of his actions. Thus we have Cromwell saying, "It is easy to object to the glorious actings of God, if we look too much upon instruments. Be not offended at the manner; perhaps there was no other way left."

His belief in "Cruel Necessity" led Cromwell to justify an offensive war. Convinced that Scotland meant to invade England in 1651 and that war was inevitable, he wrote to Fairfax, the commanding general at the time, "Your excellency will soon determine whether it is better to have this war in the bowels of another country than our own." When Fairfax refused to lead such a war, Cromwell arranged to have him replaced as general of the armies. In his own defense, Cromwell declared, "I have not sought these things, truly I have been called unto them by the Lord, and therefore am not without some assurance that He will enable His poor worm and weak servant to do His will."

Those who were, and are, unsympathetic to Cromwell saw in the disguise of "poor worm" a proud and haughty hypocrite. It is our belief, however, that Cromwell really felt himself a "poor worm" in the sight of God; any pretension to power and might on his part was only as the humble instrument of an all-powerful God. Difficult as it may be for us to enter sympathetically into such a frame of mind, it is even more difficult to avoid the conclusion that the Puritan Revolution (like Luther's, which preceded it by over a hundred years) was led by a religious spirit who drew sustenance from what he believed to be intimate converse with God.

VI

The execution of Charles I, in 1649, was the work of the Army. Cromwell at first opposed the desire to bring Charles to account for his evil doings but later consented and took a leading part in pressing on the trial of the king. The general abhorrence of the Puritan Revolt during the next 150 years resulted from this execution, which scandalized even those who carried it out.[17] In fact, when Cromwell

[17] One ought, however, to remember that the execution of Charles I followed such punitive actions of his as imprisoning Sir John Eliot in the Tower, where he died in 1632, for his part in the Parliament of 1628–1629; and sentencing, by his Star Chamber court in 1637, William Prynne, a barrister, Henry Burton, a divine, and John Bastwick, a physician, "to be fined £5,000 apiece, to lose their

tried to implicate every important family in England by making one member of it sit on the bench of judges, only half of the "judges" showed up.

Charles never accepted the jurisdiction of the court which tried him. How could he who believed in the divine right of kings do so? In 1629 he had declared that "Princes are not bound to give an account of their actions but to God alone," and he repeated this idea at his trial in 1649. In addition, he made a reasoned attack on the Puritans which explains why their Commonwealth was not accepted by the bulk of Englishmen. Charles said: "It is not my case alone, it is the freedom and liberty of the people of England; and do you pretend what you will, I stand more for their liberties. For if power without law may make laws, may alter the fundamental laws of the kingdom, I do not know what subject he is in England that can be sure of his life, or anything that he calls his own." How ironic to see Charles accusing the Puritans of exactly the tyranny that they had charged against him! Yet, he was right. Charles' tyranny had helped bring about the Puritan Revolt. The arbitrary and illegal government of Cromwell and the Army was to cause the English people to turn back to the Stuarts in the Restoration of 1660.

With the death of the king, monarchy was abolished and a republic set up in England. What followed can be told briefly. The attempt to govern by the Long Parliament broke down in 1653, and Cromwell, after the Barebones Parliament, resorted to rule as a Protector (very much like Napoleon as First Consul in a later revolution). He might have solidified his rule by making himself king, a step which probably would have been popular with the nation, but he knew that this would lose him the support of the republican-minded army. Thus, he was forced to live out his reign without "the divinity that doth hedge a king."

The personal nature of Cromwell's rule was shown at his death in 1658. Although his son Richard was proclaimed Protector, he lacked the force and driving will of his father; unrooted in kingly legitimacy, he was soon out of power. By 1660 almost everyone in England had

ears, and to be imprisoned for life for attacks on the bishops and on ecclesiastical innovations" (Firth, *op. cit.*, p. 22). Unless one thinks of the king as really being divine, his execution was a judicial (though extreme) punishment of one man by other men. The psychological importance of executing a king is, of course, another matter.

had enough of the Puritan experiment in government. The return of Charles II from the Spanish Netherlands was arranged by General Monck without difficulty or bloodshed. So ended the Puritan Revolt.

VII

The significance of the revolt, of course, stretches outside the years 1640 to 1660. For example, the voyage of the Pilgrim Fathers in 1620 was one Puritan response to the Stuarts, played out in another country, while the 1640 revolution was another, acted out at home. The 1620 exodus, therefore, is integral to the perspective which permits seeing the depth of the Puritan movement.

The profound impact of the Puritan Revolt on subsequent American and English history hardly needs underlining. The effect of the Revolution on the European continent, however, was relatively slight because, unlike the Jacobins of the French Revolution, the Puritans did not "universalize" their message. Nevertheless, the Puritan Revolt was one of the most influential and boldest movements in modern history, for despite everything that had been said about liberty of the subject and freedom of the people, no one had ever cut off a king's head until 1649.

The weapon which accomplished that act was also something new in history—the first mass, democratic army. The origins of the New Model Army—and the name suggests its novelty of conception—really dated from the Irish Rebellion of 1641.[18] At that point, Parliament realized that to give Charles I an army with which to crush the rebellion meant that their own resistance would also be overawed. Therefore, Parliament resorted to the revolutionary move of taking into its own hands the control of the Army. This step marked an assertion of new right by Parliament rather than a defense of old privilege against Charles. It also marked a break in the tradition linking crown and army.

At first, as we have suggested, the Parliamentary army fought a losing battle because it lacked a moving spirit, unified leadership, and a military genius. The Parliamentary support secured by Pym, the emergence of Cromwell and his small group of "Ironsides" (the name given to Cromwell by Prince Rupert after the battle of Marston Moor) corrected this defect. A "new" spirit animated Cromwell's

[18] C. H. Firth, *Cromwell's Army* (London, 1912), is the standard book.

troops. They were all godly men who fought, not for pay or through compulsion, but because of belief; to paraphrase Hooker, "one and the self-same people were the Church and the Army." As the soldiers themselves declared, "We were not a mere mercenary Army, hired to serve any Arbitrary power of a State, but called forth and conjured, by the severall Declarations of Parliament, to the defence of our owne and the people's just rights and liberties." Eventually, around this nucleus arose the New Model Army, uniformly clothed in russet coats, its officer ranks open to anyone with talent, numbering about 22,000 and enjoying an annual budget of about 1½ million pounds a year.

This mass army soon ran into trouble with the Parliament that supposedly controlled it. Parliament was dominated by Presbyterians who wished a national church on the lines of their faith. In accord with this, for example, a bill was passed in 1645, to suppress sectarianism. Unitarians and freethinkers could be put to death and Baptists imprisoned. The Army, however, was largely controlled by Independents (for example, Congregationalists and Baptists) who saw that a Presbyterian national church might be as oppressive as an Anglican state church; as Milton said,

> *New Presbyter* is but *Old Priest* writ large.

The conflict between Presbyterians and Independents, between Parliament and Army (exacerbated by a quarrel over disbanding pay), was settled in favor of the latter in what is known as Pride's Purge. In 1648, Colonel Pride and a body of soldiers forcefully prevented the Presbyterian majority from entering the House of Commons; the minority whom they admitted sat on as the Rump Parliament.

The Army itself, however, was split on the question of what form of government England ought to have. On one side was the bulk of the officers, led by Cromwell and his son-in-law, Ireton. On the other side were the common soldiers, who chose Agents or Agitators (often officers) to air their views.

In 1647, Ireton drew up "Heads of Proposals." These retained government by king, lords, and Commons, but with the Commons controlling the militia, thus preventing a revival of despotism. The "levelers" within the Army responded with "The Agreement of the People." This demanded manhood suffrage, equal electoral divisions, and biennial Parliaments and asserted that certain "native rights," such as freedom of religion and equality before the law, were beyond even the reach of Parliament.

In the Army Debates which took place between these two sides, we hear almost all the arguments and positions involved in the modern notion of political democracy. Colonel Rainborough, the leader of the democratic party, declared: "For really I think that the poorest he that is in England hath a life to live, as the greatest he; and therefore truly, sir, I think it's clear, that every man that is to live under a government ought first by his own consent to put himself under that government." Ireton retorted that to give men without any property stake a vote was to endanger both liberty and property. As for natural rights, he argued, no two men agreed about them; it would mean the end of civil law. "When I hear men speak of laying aside all engagements, to consider only that wild or vast notion of what in every man's conception is just or unjust, I am afraid and do tremble at the boundless and endless consequences of it." Cromwell's attitude was that government should be for the people but not necessarily by them. "That's the question," said Cromwell, "what's for their good, not what pleases them."

A supposed compromise between the people's "good" and their pleasure was reached in 1649 when an act was passed establishing the English Republic and declaring England to be henceforth governed "as a Commonwealth, or a Free State." The House of Commons passed a resolution declaring that "The Commons of England, in Parliament assembled, being chosen by and representing the people, have the supreme power in this nation." But, in reality, England was neither a free state nor were the Commons "chosen by" the people. From 1649 to 1653 the Rump Parliament (hardly "chosen by" the people) ostensibly ruled, and after that Cromwell exercised the office of the Protectorate.[19]

The rule of Cromwell and his major generals was based on force,

[19] Trevelyan makes the point that the Army's democratic aspirations were foolish because the majority of Englishmen were opposed to religious tolerance and would have used universal suffrage to abolish it *and* republicanism (*op. cit.*, p. 283). The text of the Army Debates can be found in *Puritanism and Liberty: Being the Army Debates (1647–9) from the Clarke Manuscripts*, selected and edited with an introduction by A. S. P. Woodhouse (Chicago, 1951). Woodhouse's introduction is a valuable one. The previously existing text was *The Clarke Papers*, 4 vols. (Camden Society, 1891–1901).

represented by the Army (although, it must be remarked, the Army was really no less representative than Parliament), and not on the people's consent. It was, in fact, a dictatorial regime, opposed by the majority of the nation. One legacy of this crude military rule was to strengthen the English hatred of a standing army. Another consequence was to erode the ground from under the Puritan rule and to bring it to an end in 1660.

Nevertheless, as an innovation in modern history, the New Model Army must be numbered among the achievements of the Puritan Revolt. It introduced for the first time an army composed of ordinary citizens, fighting for ideological reasons and taking a part in politics. As such, it may be considered the fulfillment of the trend started in the Peasants' Revolt of 1525. From another point of view, it is the forerunner of the mass, democratic armies which have so strongly characterized modern times.

VIII

We have already suggested that the Puritan Revolt is not primarily significant as a social revolution; that would be to anticipate the meaning of the French Revolution. Nevertheless, there were faint hints of ideologies other than the constitutional or religious in the events of 1640–1660, and these are generally associated with the group of "levelers," who were responsible for the drafting of the Agreement of the People.

Actually, there were two kinds of levelers—the Levelers properly speaking, led by John Lilburne and by such men as Richard Overton and William Walwyn, and the so-called economic levelers (Diggers). The starting point for Lilburne, as for most of the Levelers, was religious. He had experienced the usual inner spiritual warfare of the Puritan in struggling to become a saint and to assure himself of grace abounding. Then, as one recent writer explains it, "Spiritual warfare in his version of the story became directed against the evil in human institutions rather than in the heart of the sinner himself. The end sought was not personal conversion but the general good. The struggle of Christ's redeemed ones became a struggle for the redemption of the state. The holy community, the New Jerusalem, came to be conceived rather as a going community of free citizens than a withdrawn though visible congregation of the elect."[20]

On this basis, of a secular "Jerusalem" emerging from belief in a spiritual one, Lilburne and the Levelers pressed their agitation. Although they were so accused by their enemies, they did not believe in socialism (they did, however, wish enclosed land to be returned to cultivation). Their demands were almost purely political: they opposed both royal and parliamentary government and wanted a more direct rule by the people. They held that Cromwell had substituted one tyranny for another.[21] Ousted from the army as early as 1648, Lilburne's Levelers continued their campaign against Cromwell in the country at large.

The real economic radicals were the Diggers, led by Gerrard Winstanley. They actually began to cultivate the common lands and to advocate communism. At first, a strong religious element pervaded this communism, as we see in a title of a 1648 pamphlet, *The Light Shining in Buckinghamshire,* but gradually the economic element became stronger in the Digger ideology. Their numbers, however, were never large and their role in the Puritan Commonwealth totally unimportant. Therefore, their ideas and writings are significant mainly as a primitive expression of later communist doctrine.

The Levelers and the Diggers, and such sects as the Fifth Monarchy Men, who believed in the imminence of the Second Coming and in their own right to rule until that event, were at all times peripheral and at no time successful in the period 1640–1660. It is, therefore, evident that we cannot list as one of the aims or accomplishments of the Puritan Revolt a social program or a new distribution of property. At most, this was only favored by a radical wing which never achieved any power. Nevertheless, the Levelers represented the democratic aspirations and the equalitarian strain present in Puritan preaching.

20 William Haller, *Liberty and Reformation in the Puritan Revolution* (New York, 1955), p. 271. This book does an excellent job of tracing the evolution of religious feelings into political radicalism. For the Levelers, see T. C. Pease, *The Leveller Movement* (Washington, 1916), Joseph Frank, *The Levellers* (London, 1955), and M. A. Gibb, *John Lilburne* (London, 1947). Also to be consulted are Wilhelm Schenck, *The Concern for Social Justice in the Puritan Revolution* (London, 1948), and Perez Zagorin, *A History of Political Thought in the English Revolution* (London, 1952).

21 This was the same charge made against the Puritans by Charles I. It is not surprising, therefore, that the Levelers intrigued later with the Royalists for the return of the king.

They stood for the secular extrapolation of what was otherwise a purely spiritual striving, and as such they carried the political message implicit in one aspect of Puritanism to many who were not Puritans.

IX

The major accomplishment of the Puritan Revolt was in the area of religion, but it also and necessarily drew after it major political consequences. Actually, that accomplishment—the idea of a free church—was the work of only a small segment of the Puritans: the Independents. It was they who maintained the notion of a free church against the Presbyterians and against thinkers like Hooker who believed in a single union of state and church.

The belief in a free church entailed the belief in tolerance, which the Presbyterians also denied. The Independents, in fact, demonstrated tolerance (although limited) at work during their own tenure of power, the Protectorate. Under Cromwell, all creeds and doctrines were publicly tolerated except Anglicanism and Roman Catholicism. And the adherents of these creeds, although denied the right of public worship, were not forced into attendance at other forms of worship by fines or punishment, as had previously been the case in England.

The tolerance achieved by the Independents was only a partial victory.[22] It lapsed again in 1660, and it was not until the Toleration Act of 1689 that the nonconformists again won the legal right to exist and to worship publicly. As individuals, however, the members of the free churches did not win rights of full citizenship until the repeal of the Test and Corporation Acts (which also operated against Catholics) in 1828. Nor did they gain the right of admission to Oxford and Cambridge until 1871.

Clearly, the history of toleration is a long and slow one. We have seen it take some of its impetus from the controversy surrounding

[22] Toleration, of course, was neither complete nor all the work of the Puritans. For example, the Cambridge Platonists should also be mentioned in the history of toleration, as paving the way within the Anglican Church for what is known as "latitudinarianism." For the early history of toleration in England, from the beginning of the English Reformation to the death of Elizabeth, see W. K. Jordan, *The Development of Religious Toleration in England* (Cambridge, 1932). Two later volumes carry the story to 1660.

Calvin's burning of Servetus. It is to the credit of the Puritan Revolt that it marks another notable point in the evolution of religious toleration.

The free church was also unlike the intolerant state church in another way: it was a voluntary group. It represented the free association of individuals in a society, prepared to accept other freely gathered groups. As such, it became the model of social organization on which England developed. According to the economic historian George Unwin, "The expansion of England in the seventeenth century was an expansion of society and not of the State. Society expanded to escape from the pressure of the State."[23] Thus, English colonization was originally undertaken by freely formed companies of gentlemen adventurers, such as that sponsored by Sir Walter Raleigh, or by the Puritan connections who took the lead in the Providence Island and Saybrook ventures.[24] And further, as Ernest Barker points out, "Lloyds and the London Stock Exchange [were] both originally associations based on the social life of city coffee-houses in the eighteenth century."[25]

The religious idea of a freely associated group also served as the model for political organization. It could lead to the formation of political parties, or partial associations, and, to cite Barker again, "Parties, after all, are social formations: The Whig party began its life in a city inn."[26] The result was that the armed conflict of Roundheads and Cavaliers died down to the party bickerings of Whigs and Tories.

To view government as an affair of parties was not necessarily to tie it to any particular form of government. Thus there was no theoretical requirement for the Puritans to be antimonarchical. It is true that they beheaded a king, installed a republic, and introduced into England its first written constitution, the Instrument of Government. But all this was the result of historical circumstances rather than of doctrinal necessity. The events which followed the Restoration in 1660 showed that the system of party government was con-

23 George Unwin, *Studies in Economic History* (London, 1927), p. 341.

24 The Massachusetts and Virginia Companies are sufficient testimony of the power of privately formed companies; they were able to lay the foundations of future growth in America.

25 Ernest Barker, *Principles of Social and Political Theory* (Oxford, 1951), p. 28.

26 *Ibid.*, pp. 28–29.

veniently adaptable to the monarchical as well as to the republican form of government.

The idea that the state was composed of various parties rather than fixed in a single monolithic unity had as its complement the notion of the limited state. One of the Army documents, probably the work of Ireton, uses the image of committing "our stock or share of interest in this kingdom into this common bottom of Parliament." Thus, instead of the state being Hobbes' Leviathan, it was a "limited" company whose ends were the maintenance of peace and order. Neither the free play of religion nor of politics was to be overborne by the government.

Furthermore, it was only natural that, abhorring the whole mechanism of ecclesiastical discipline and conformity, many of the Puritans turned away from a strict state regulation of economic matters. In the case of Lilburne, for example, we know that he was debarred from entering the wool trade by the Merchant Adventurers' monopoly power. Thus a belief in freedom of economic enterprise took its place alongside a belief in religious and political freedom as a legacy of the Puritan Revolt.[27]

X

Yet, with all its accomplishments, the Puritan Revolt was rejected in 1660 by almost all the English people. After about 40 years of continuous strife—parliamentary and military—Charles I's son, Charles II, came back to England. He returned without the firing of a shot or the striking of a blow—and to a population which initially did not care for him. He was not particularly an attractive king. He was too much absorbed in the luxurious life of his court and in his numerous mistresses. He was suspected of being, and probably was, in secret, a Roman Catholic. Indeed, the only obvious benefit that he conferred upon his country was that, within a short time of his return to London, the influence of French literature and drama made itself felt within England.

[27] For Lilburne's debarment from the wool trade, see further Haller, *Liberty and Reformation in the Puritan Revolution,* pp. 267–274. It must be emphasized, however, that not all of Puritanism's associations were with liberty; it was primarily the Independent branch of Puritanism which was affiliated with the various freedoms we have mentioned. Cf. Woodhouse's introduction to *Puritanism and Liberty,* p. 53.

Why did the English people accept the restoration of Charles II and throw away the fruits of victory, achieved at so much cost of blood and labor? Why, in the political and religious settlement which followed Charles' return, did England revert, more or less, to the state of affairs existing before the Civil War?

The answer appears to lie in the combined constitutional-religious nature of the 1640 revolt. At that time, the majority of the English nation—Presbyterian as well as Independent—opposed the king's despotic assertion of right against Parliament. The English people were largely united in their desire to insist on their political liberties. On this basis, the Puritan Revolt began.

But events led to the seizure of power by an army which, composed of Independents, placed religious liberty above political freedom. Cromwell and his followers were open in their assertion that the liberty of the "people of God" was more important than the civil liberty of the nation. This permanent limitation on the potential sovereignty of the people, in the interest of a minority, was acceptable neither to the left nor the right of the forces which had opposed Charles—neither to the Levelers nor to the Presbyterians. Thus it was these groups which combined with the Royalists in 1660 to secure the peaceful restoration of Charles II to England.

In the reaction against the Puritan tyranny, the English nation went too far in restoring autocratic power to the king. The result was that political and religious liberties had once again to be vindicated in 1689.

In itself, therefore, the Puritan Revolt had failed. Yet, it bore within itself the ashes whence might arise a new phoenix. As we have attempted to show, there were involved in the ideas and events of 1640–1660 a desire for tolerance, the belief in a free church within a free state, and the conception of a limited state. These beliefs—in religious, political, and economic individualism—the Puritan Revolt developed and extended from their origins in the Reformation and handed on to modern times in dramatic and unforgettable terms.

CHAPTER 10

The Royal Society

THE COMMONWEALTH which the military fervor of Oliver Cromwell had created in the 1640's came to an end after the perfunctory rule of his son Richard in 1660. In that year the son of Charles I, who had been beheaded in 1649, was brought back to England and became king, by universal consent and without bloodshed.

The restoration of Charles II opened a new age in England. Many of the gains of Cromwell's revolution remained, and became accepted in the outlook of educated men. What was rejected was the dictatorial aspect of the revolution; and Charles II became a popular king (although, for example, as we have said, he was probably a secret Roman Catholic) because he understood and shared the tolerance of his subjects. England, starved of the arts under the Commonwealth, recreated a lively literature and an imaginative architecture. Opposing theories of the state were argued, with much originality, by Thomas Hobbes on one side and by John Locke on the other. And the greatest of the world's scientific societies, the Royal Society, was founded in the year when Charles II was restored, and had John Locke among its first Fellows.

I

Groups of scientists began to meet to discuss their work, here and there in Europe, early in the seventeenth century. For example, in 1611 Galileo was made the sixth member of the Accademia dei Lincei, (Academy of the Lynx-eyed), which had been founded in Rome by young nobles who were enthusiasts for science. In the middle years of the century, a similar group set up its own laboratory for a short time in Florence. A group of scientists which at different times included Descartes, Desargues, Fermat, and Pascal met privately in Paris from about 1630. A little later, during the Puritan Revolution, at least two private groups of scientists began to meet regularly in London and Oxford.

One group centered round John Wallis, a powerful mathematician to whose work, *Arithmetica Infinitorum,* Newton was indebted for some of his major mathematical conceptions. This group met at Gresham College, which had been founded in 1598 by the merchant of that name, who is still remembered in Gresham's Law—the law that bad money will displace good money if both are in circulation at the same time.[1] By an odd irony, however, this law seems in fact to have been stated for the first time by Copernicus when he reported on the Polish currency in 1526.

Those who met at Gresham College with Wallis were interested in astronomy and geometry, and they included Christopher Wren, who was a professor in both subjects. Indeed, Christopher Wren's bold architecture was based in part on his geometrical genius. The Royal Society was formally founded on November 28, 1660, at a meeting in Gresham College which began with a lecture on astronomy by Christopher Wren; and the Society went on meeting at Gresham College until 1703.

Wallis' group had begun their weekly meetings about 1645, and about the same time a second group began to meet which centered on Robert Boyle. Boyle was a philosopher, a physicist, and a chemist; he revived, though in a form still far short of its modern form, the Epicurean theory that all matter is made of atoms; and he established the law which governs what he called "the spring of the air"—the pressure of a gas—which still goes by his name. Boyle himself called the group which met with him the Invisible College.

Both these groups consisted largely of men of Puritan sympathies. Robert Boyle's family had been on Cromwell's side; for although they were lords, they were upstart lords, created by Queen Elizabeth and

[1] Sir Thomas Gresham, whose father, Sir Richard Gresham, was the largest single beneficiary from Henry VIII's confiscation of the monasteries, acted as envoy for Henry VIII in Holland, and there found that the pound was falling rapidly because Henry VIII was coining bad money. He stabilized the pound about 1550, helped to found the Royal Exchange in London, and so set going the movement of the European money market from Holland to England.

holding lands which had once belonged to the monasteries. Boyle's father, the first earl of Cork, in 1602 had bought the house in which Boyle was born in Ireland from Sir Walter Raleigh, who had it as a gift from Queen Elizabeth.

In the other group, John Wallis earned the favor of Oliver Cromwell because, like mathematicians ever since, he turned his ingenuity to breaking enemy cyphers. He decoded for Cromwell messages from the king to his military units, which meant a good deal to both sides. Cromwell turned out Royalist professors from Oxford to make room for Wallis and men like him. In doing so, he replaced the out-of-date learning of the old universities by a fresh and forward-looking science. *The Ballad of Gresham College,* written when the Fellows of the new Royal Society were still more widely known as Gresham Philosophers or Collegiates, was right in its trenchant statement of the contrast.

> Oxford and Cambridge are our laughter,
> Their learning is but pedantry:
> These Collegiates do assure us,
> Aristotle's an ass to Epicurus.

Thus the new men differed in their outlook from the old, both politically and intellectually.

II

Charles II was a man of some intellectual curiosity, and he took an amateur interest in science. But he cannot have liked the Puritan groups of scientists, who had begun to meet during the Revolution, and who numbered among their men many who had risen under Cromwell. It has been estimated that, of the sixty-eight early Fellows of the Royal Society about whom we have information, forty-two were clearly Puritans.[2] Charles II can hardly have been enthusiastic about

[2] Dorothy Stimson, "Puritanism and the New Philosophy in Seventeenth-Century England," *Bulletin of the Institute of the History of Medicine* (May, 1935), pp. 321–334. Also see Robert Merton, *Social Theory and Social Structure* (Glencoe, Ill., 1949), who gives this fact, and whose *Science, Technology and Society in Seventeenth Century England* (Bruges, 1938) contains the fundamental research on this subject. On the history of the Royal Society, see Dorothy Stimson, *Scientists and Amateurs* (New York, 1948); Sir Henry Lyons, *The Royal Society, 1660–1940: A History of Its Administration Under Its Charters* (Cambridge, 1944), which is solid and "official"; and *The Record of the Royal Society of London, 1897* (London, 1897), which includes an account of the foundation and early history of the Society, copies of the first three charters, and much other interesting material.

giving his name to a society which was dominated by men whose political and religious views were distasteful to him.

Yet Charles II did give his name to the new Society: it was called *The Royal Society of London for the Promotion of Natural Knowledge,* and it received its first charter—and nothing else—from him in 1662. The man who persuaded Charles II to this step was probably the diarist John Evelyn. Evelyn belonged to a smaller group of scientists who had remained Royalists, and the first presidents of the Society were drawn from this group. Evelyn himself gave the Royal Society its motto;[3] and he displayed the practical bent of the early Fellows in his protests, now 300 years old, against the smoke which even then fouled the air and withered the plants of London.

Plainly the temper of the times was stronger than the political bias of men in power. Science had a new charm, and scientists had a new prestige. And part of this prestige may already have come from their sense of mission and the aura which they were beginning to carry of being dedicated men. Most of these men were Puritans by birth, and came from the families of merchants and smallholders who were thrusting their way into the world. But, intellectually, their Puritanism did give them a special devotion to the truth as they saw it for themselves, and a grave indifference to the authority of the past, both of which are still summarized in the word "nonconformist."

It may be that in England Charles II acknowledged these men, whose temperament was the opposite to his own, because tolerance was his political creed. But in France, the young King Louis XIV had no taste for tolerance. Yet, Louis XIV was willing, as Charles was, to give his name to a new society of scientists which contained many men with whom he was out of sympathy. The French Society was the Académie Royale des Sciences, and it was founded formally in 1666. The distinguished mathematician Christian Huygens became its secretary.

Christian Huygens also was surely a most unlikely man for Louis XIV to choose. He was a Protestant in a Roman Catholic country. He was a Dutchman in France, at a time when Louis XIV was constantly making war on Holland. And he was a Cartesian in his views, although the Académie Royale tried in the main to exclude Cartesian scientists; indeed, Huygens' family had befriended Descartes when

3 The motto was *Nullius in verba,* from Horace.

he was in exile in Holland, and this is how Christian Huygens had come under his influence.

In the end, these difficulties became too much for Huygens, and he left the Académie Royale in 1681. Not long after, in 1685, Louis XIV ended the uneasy toleration of the Protestants in France, and set going a new migration of Huguenots through Europe.

III

The Royal Society and the Académie Royale had at the outset a strong practical bias; their founders and their Fellows were convinced that science could be useful, and should be planned to be useful. For example, Huygens at the Académie Royale was much concerned with the clockwork: he has a claim to the disputed invention of the balance spring, and he was the first man to make a pendulum work practically in a mechanical clock.

One other example from the Académie Royale is worth singling out, because it is also a development from the work of Galileo. Two other foreign members of the Académie were Jean Cassini and the Danish astronomer Ole Roemer. Cassini had taken Galileo's hint and had made a table of the eclipses of the moons of Jupiter as a means for finding one's longitude. In 1676, Roemer set to work to watch the moons carefully, in order to improve the table. He found that the eclipses did not keep perfect time: and he noticed that they were early when Jupiter was near the earth, and late when Jupiter was far away. He concluded, rightly, that light travels at a finite speed, and he calculated its speed to an accuracy which, for its time, deserves respect. The moons of Jupiter would not, after all, do as a clockwork, but they prompted instead this remarkable and unexpected fundamental discovery.

In the Royal Society, systematic observation was not confined to astronomy and the problems of navigation. One of the first Fellows of the Society, Sir William Petty, was a pioneer of statistical methods. Like others in the Society, Petty was a many-sided man: he was professor of anatomy at Oxford and a student of medicine. (He was also a constant inventor, and one of the gadgets for the household with which he is credited is the water closet; but the Elizabethans had probably anticipated him in this.) Petty's most important work is in the subject which he named political arithmetic, that is, the statistical study of economics. In 1662 he published a *Treatise on Taxes and*

Contributions which was one of the first works to understand that the basic values of an economy derive not from its store of treasure, but from its capacity for production.

In the year 1662 there was also published the first study of vital statistics, under the title *Observations on the Bills of Mortality of the City of London.* This work was for long also ascribed to Sir William Petty, but it has now been proved to have been written, probably with Petty's help, by John Graunt, another early Fellow of the Royal Society.[4] The statistics collected by Graunt became the basis of the first life-insurance tables; and similar tables were constructed about the same time in Holland in order to work out a system of state annuities which would finance the war against Louis XIV.

There was reason for this interest in the Royal Society in economic matters. The seventeenth century and the early eighteenth century continued to derive much of their wealth from trade; the major growth of industry did not begin for another hundred years. The technical study of trade—of navigation, for example, and even of insurance—was well established; but its economic study was new. There was no such economic study in France: there, under an absolute king always greedy for taxes, economics remained a theoretical subject, dominated by an essentially mercantilist approach—the belief that all values derive from treasure, or from some other absolute and fixed source. By contrast, in England trade began to be studied empirically, and for this purpose there were created such offices as that of Inspector General of Imports and Exports—"the first special statistical department successfully created by any Western European state."[5] One of the early holders of an office of this kind was John Locke.

It is worth stressing the unexpected interest among the new scien-

[4] It is a pleasing point that one decisive test by which the *Observations on the Bills of Mortality of the City of London* has been shown not to have been written by Sir William Petty is a statistical one. G. U. Yule, in a paper in *Biometrika* in 1939, showed that the length of sentences in the *Observations* does not match the length which Petty used in his writings.

[5] G. N. Clark, *Science and Social Welfare in the Age of Newton* (Oxford, 1937), p. 138. Werner Sombart dates the beginning of capitalism from the invention of bookkeeping by double entry, which was made by Leonardo of Pisa in 1202. In the same sense, one can state that the work of Sir William Petty and John Graunt in political arithmetic helped prepare the way for the Industrial Revolution. Certainly the economics of Adam Smith would have been impossible without the statistics whose collection they prompted. Adam Smith himself was at one time a commissioner of customs in Scotland.

tists in the details of trade and economics, for two reasons. It shows both how practical and how broad was the sweep of the Royal Society. Men like Christopher Wren, William Petty, Robert Boyle, and John Evelyn were willing to be interested in anything. Like the men of the Renaissance and like the Elizabethans, they fired one another, and each gave his own enthusiasm and his own genius. Of course, they liked the distinction of being in the Royal Society, and of having a finger in every intellectual pie. But more than this, they felt themselves to be part of the movement of the times: they felt themselves to be driving into an expanding future. Even a poet like John Dryden wanted to be a member of the Royal Society, because science was alive and was making over the world. The Royal Society promptly put him on its committee to teach scientists a simple and direct style for writing English.

IV

The practical bias of the Royal Society was displayed above all in its experimental work. In this, the Fellows who founded the Society were deeply influenced by Francis Bacon and what they called his Experimental Philosophy. Francis Bacon had written a number of works which proposed a society of scientists, exploring the world together by experiment and paying little heed to theory—and no heed at all to traditional theory. Among these works of Bacon was the *Advancement of Learning* in 1605, the *Novum Organum* in 1620, and the *New Atlantis,* unfinished and published in the year of his death, 1626.

The vision of Francis Bacon inspired the founders of the Royal Society in three ways. It laid stress on the communal working of scientists: for example, in the *New Atlantis* there is a College of Natural Philosophy called Solomon's House or the College of the Six Days' Works, in which the philosophers work together "dedicated to the study of the works and creatures of God." It constantly presented science as an activity designed to serve mankind, and this injunction was much in the mind of such religious men as Robert Boyle. And it advocated throughout, and for the first time, what is essentially the modern scientific method: the making of experiments, the drawing of general conclusions from them, and the testing of these generalizations in further experiments. Bacon was, in fact, the first clear advo-

cate in theory of what we now understand to be the inductive method in science.

So great was the preoccupation of the first Fellows of the Royal Society with experiments that Christian Huygens—himself a clear-headed exponent of the inductive method—criticized Boyle for failing to do anything else, and for drawing no conclusions from his experiments. The early experimenters were too much carried away by the new method, too full of the wonders of nature and of the thirst to explore them, to be at all systematic. The very brightness of their minds, the many-sidedness of their skill, made them impatient of the detailed step-by-step logic in experiment which a younger generation evolved later. The preoccupation of the Fellows whose minds had been formed by the writings of Francis Bacon was practical—it has been estimated that nearly 60 percent of the problems handled by the Royal Society in its first thirty years were prompted by practical needs of public use, and only 40 percent were problems in pure science.[6] And they wanted the practical results of experiment to be immediate, they were full of inventions and gadgets, and if the experiment did not come right overnight, they were tempted by morning to move on to another.

The most remarkable of these inventive minds was Robert Hooke. He was appointed in 1662, at the age of 27, to be Curator of Experiments to the Royal Society, which at that time expected weekly demonstrations of new work. He spent the rest of his life, until his death in 1703, in the service of the Society, for the most part as its secretary. Hooke's inventions were numberless.[7] Even before 1662, he had worked for Robert Boyle in Oxford, and had there built the air pump which Boyle needed for his work on the pressure of gases. He proved, long before the eighteenth century rediscovered it, that anything which burns draws some gas from the air. He invented the balance spring of the modern watch—and quarreled with Huygens

[6] Merton, *op. cit.*, Chap. 15. For the general influence of Francis Bacon on the early Royal Society, see also G. N. Clark, *Science and Social Welfare in the Age of Newton*, Chap. 1.

[7] See Margaret Espinasse, *Robert Hooke* (London, 1956); also A. R. Hall, *The Scientific Revolution 1500–1800* (Boston, 1954). R. T. Gunther, *Early Science in Oxford*, 14 vols. (Oxford, printed for the subscribers, 1923–1945), includes a facsimile edition of Hooke's *Micrographia* (Vol. XIII), among other delectable items.

about it—and he formulated the law which relates the tension in a spring to its elongation, which still goes by his name. And he produced one of the first books of drawings of microscopic animals, the *Micrographia,* which has remained a classic.

After the Great Fire which destroyed much of London in 1666, Robert Hooke was appointed with Christopher Wren to draw up a new plan for the city; and he did more of the practical work of rebuilding than Wren could find time for. The tall monument to the fire which is often ascribed to Wren was, in fact, Hooke's; it is one of the few things built by Hooke which survives, for his buildings, like so much of his work, have mostly disappeared, by neglect and by bad luck. Like Wren, Hooke was also an able mathematician, and was professor of geometry at Gresham College. During the 1670's, Hooke had the idea—in common with others—that forces of gravitation extend all through the solar system, and he began to speculate that gravitation might fall off as the square of the distance. A discussion on this subject between Hooke, Wren, and the astronomer Edmond Halley in 1684 prompted Halley to go to Cambridge to ask Isaac Newton what would be the path of a planet if it were, in fact, held to the sun by a force of gravitation which falls off as the square of its distance. An ellipse, said Newton at once; and he told Halley that he had solved this problem as a young man in 1666, while the plague kept him away from Cambridge. Because Newton could not find his old papers again in 1684, Halley insisted that he must write out his work in detail, and this is how we come to have the *Principia,* whose publication Halley helped to finance in 1687.

With all these achievements, why has Hooke been neglected? The answer can be put simply: he quarreled with Newton. When in 1672, not yet 30, Newton had sent an account of his wonderful experiments in optics to the Royal Society, Hooke had been critical of the conclusions. Like others, Hooke found it difficult to believe that white light is compounded of different colors; more deeply, Hooke and other scientists then did not really grasp the unspeculative and strictly logical nature of Newton's thought. Later, he corresponded with Newton about gravitation, and in a letter written on January 6, 1680, Hooke proposed the law of inverse squares. Newton, who in spite of his unmatched originality was morosely suspicious of the claims of others, saw in these and other points an attempt by Hooke

to belittle his work. He never forgave Hooke. Newton did not accept the presidency of the Royal Society until after Hooke died, in 1703; and he did not publish his *Opticks* until 1704, long after the work was done. When Hooke tried to get Newton to make some acknowledgment to his speculations on the law of inverse squares, Newton threatened to withdraw the most important part of the *Principia* and not to publish at all.

V

The bitterness between Isaac Newton and Robert Hooke was personal. But it reflected also an opposition of temperaments which goes deeper, and which marks the shift from the early experimental method of Francis Bacon to the new systematic method of modern science. Hooke was a fertile, sparkling mind in the tradition of Bacon; he threw out a hundred suggestions, he invented the universal joint and the iris diaphragm, he laid out the meridian at Greenwich, and his mind never rested on anything. By contrast, Newton had two immense gifts which are still the twin arms of science. Newton had a grasp of mathematics many times more exact and more powerful than his contemporaries, so that he was able to turn his sensible speculations about gravitation into precise formulas which observation could verify. And Newton saw in experiment a systematic procedure, each step of which should be critical and should thereby decide between alternative explanations. In both these intellectual conceptions, Newton was an age ahead of the colorful and disorderly experimenters whom Bacon inspired.

The work of Isaac Newton is technically so powerful that we easily miss the lucid originality of outlook which underlies it. Most important in this is the point to which Newton returns again and again, that he does not make hypotheses. He means by this that he has no preconceptions about the mechanism which underlies the phenomena of Nature. He has no theory of how gravitation works, or of what causes light of different colors to be refracted differently. These are philosophical speculations, what Newton calls hypotheses, which he leaves to the abstract thinkers of the past. Newton is concerned with the purely factual laws which gravitation displays or which light of different colors obeys. He conceives the universe as a machine, but he does not ask what particular mechanism drives the parts of the machine. He is, for the first time, a mind without metaphysics, who is content to put into mathematics the inductions to which the facts seem to lead him, and to verify that the inductions match in their consequences the phenomena which they are intended to summarize. This is a conception of the method of natural science which was quite new.

Newton conceived his method in the same terms both in his experiments in optics and in his grand synthesis of Kepler's three laws under the single law of gravitation. Of the two works, that on optics is easier to read, and is still fresh and inspiring today. Here every experiment is thought out in advance, Newton knows what he is looking for, and above all he knows in what way the experiment will be decisive between one alternative explanation and another.

Newton was by no means lacking in practical interests. He himself made the instruments for his experiments, he invented the reflecting telescope in which the stars are seen in a parabolic mirror, and he spent the last thirty years of his life in the very practical administration and reform of the Royal Mint. In this sense, Newton is heir to the tradition of Francis Bacon and stands on the shoulders of Hooke and his inventive colleagues.[8] What is new in Newton is something else. He is able to use the experimental method to set up a new system of fundamental concepts—the kind of system which Aristotle had set up, and which the school of Bacon had dismissed. It had been necessary to break the hold of Aristotle and the rooted belief that nature could be explained from self-evident principles; and this the first experimenters of the Royal Society had done. But when this had been done, a new basic system was needed, founded this time not on self-evident notions but on notions which would turn out, in their consequences, to match the facts of experience. This Newton knew how to do, and this is the inductive basis of modern science.[9]

VI

The Royal Society had no government grant, and it was, therefore, always in money difficulties. For example, it was always behind in

[8] For all these matters, see Merton, *op. cit.*, particularly Chaps. 1, 14, and 15. See also Edwin Arthur Burtt, *The Metaphysical Foundations of Modern Science* (London, 1949), p. 209.

[9] See J. Bronowski, *The Common Sense of Science* (London, 1951), Chap. 3 on "Isaac Newton's Model."

paying Robert Hooke the small salary which his appointments carried. One way to make its work pay for itself was to publish a series of reports, and to make these the property of the secretary; in this way, the secretary earned his keep by printing and selling the Society's *Philosophical Transactions*. No doubt this method of doing business reflects the strong sense of Puritan independence, the conviction that success must be earned, and earned as money, which marks the first Fellows of the Society.

The publication of scientific work, as the natural way to make it known to others, was an invention of the seventeenth century. It begins with correspondence, first between scientists, and then between scientists and a few men who became, as it were, clearing houses for scientific information. The first secretary of the Royal Society, a devious and difficult man named Henry Oldenburg, who was responsible for much of the friction between Hooke and Newton, between Hooke and Huygens, and between Newton and Leibnitz, was a postal clearing house of this kind; Robert Boyle called him a "philosophical merchant."

Thus, the secretary of a learned society in the seventeenth century begins as a center of information, and it is natural that he should become more formally the editor of a series of reports. The *Philosophical Transactions* of the Royal Society made an outstanding series, and in a certain measure were, as G. N. Clark puts it, a first encyclopedia of new techniques and knowledge. In this sense, the *Philosophical Transactions* were the forerunners of the first official encyclopedia of the new knowledge, Chambers' *Cyclopaedia,* which was published in 1728 and which continues to be published, in up-to-date revisions, to this day. But more important, the *Philosophical Transactions* set the pattern, that a scientist made his work known by publishing it in a learned paper. This step, by which the letter to a colleague becomes transformed into a scientific paper, is seen very clearly in the communication in the eighteenth century to the Royal Society of the private accounts of his work in electricity which Benjamin Franklin sent to England from America.

The publication of results carries with it a demand for a plain and understandable symbolism which all scientists can share. Mathematics provides such a symbolism, and mathematical notation, therefore, settled down to standard and communicable forms. It is notable that Isaac Newton, because he was given to secrecy, wrote the infinitesimal calculus in a private notation which was less useful than the notation which Leibnitz used. Unhappily, because of a quarrel between them, English mathematicians felt that they must show their patriotism by clinging to Newton's notation, and in this way they delayed the development of pure mathematics in England in the eighteenth century so that it fell far behind the rest of Europe.

More important than any formal symbolism, however, scientific work, to be understood, needs a clear expression in words. This the Royal Society stressed from the outset, and its Puritan sympathies were well suited to produce a simple and clear language for science. The Fellows of the Royal Society were exhorted to report their findings "without amplifications, digressions, and swellings of style: to return back to the primitive purity, and shortness, when men delivered so many *things* almost in an equal number of *words*." What the Society wanted was to exact "from all their members a close, naked, natural way of speaking; positive expressions; clear senses; a native easiness; bringing all things as near the mathematical plainness as they can."[10] The Society was bringing to science the same simplicity, the break with the wild and metaphoric style of the past, which the translators of the Bible had brought to their work in 1611, and in much the same spirit. To the founders of the Royal Society, God operated through nature, and operated always in the simplest way; and speech and writing ought to match that simplicity. Their style has remained the aim of science ever since, and has proved as elusive an aim to their successors as it was to them.

[10] *The Social and Political Ideas of Some English Thinkers of the Augustan Age,* ed. by F. J. C. Hearnshaw (London, 1928), p. 18.

CHAPTER 11

Hobbes and Locke

I

We have seen earlier that the French religious civil wars of the sixteenth century gave rise to the writings of Stephen Junius Brutus, François Hotman, and other Protestant as well as Catholic political theorists. It should, therefore, be no surprise that the great development in political thinking of the seventeenth century took place in England either during, or as a result of, the Puritan Revolt.

Thus, Thomas Hobbes published his *Leviathan* in 1651, expressly stating that his "Discourse of Civil and Ecclesiastical Government" was "occasioned by the disorders of the present time."[1] And John Locke, in the preface to his *Two Treatises of Government*, published in 1690, openly declared that his work was to "justify" the 1688 continuation of the 1640–1660 struggle, and "to establish the throne of our great restorer, our present King William," against the Stuart claims.[2]

In their works, however, Hobbes and Locke rose far above the exigencies of contemporary politics and gave classic statements of two positions. Hobbes stated the case for absolute sovereignty and the Leviathan state, while Locke put forth the defense of parliamentary government and the limited, liberal state. They wrote two of the

1 Thomas Hobbes, *Leviathan*, ed. by Michael Oakeshott (Oxford, 1955), p. 467. This is the best modern edition, with a very interesting introduction by the editor.

2 John Locke, *Two Treatises of Government*, ed. with an introduction by Thomas I. Cook (New York, 1947), p. 3. This is a very convenient edition, which also contains Filmer's *Patriarcha*, the book which Locke attacked in the first part of his work.

great books around which political argument, to a large extent, henceforth revolved.

One reason for their success was that the *Leviathan* and the *Two Treatises* were based on a complex and carefully articulated philosophy and on a comprehensive view of science. It is this fact which plays a major role in lifting the two works out of the category of transient, polemic productions.

Actually, both men were connected with the movement of science and metaphysics which centered around the Royal Society. Locke, in fact, as we mentioned in the previous chapter, was a member. He contributed papers and made practical experiments. He also made barometric observations for Boyle and helped prepare the latter's book *A General History of the Air* for publication.[3] Hobbes was not a member, but he involved himself in similar speculations and tried to make contributions in the field of physiology and mathematics, which, although unsuccessful, show his interests.

It was not, however, as practicing scientists that Hobbes and Locke made their mark. It was as philosophers, incorporating the ideas and methods of the new science, that they made their original contribution to the development of European thought. For example, Locke's *Essay Concerning Human Understanding* was the first comprehensive statement of the empirical approach to epistemology, while Hobbes' contribution to philosophy was concerned with the whole theory of mechanism and causality on which so many of the achievements of the Royal Society rested.

[3] See H. R. Fox Bourne, *The Life of John Locke*, 2 vols. (New York, 1876), Vol. I, p. 125. A standard work, though marred by its uncritical attitude to Locke, and by its excessive anti-Catholic and pro-Protestant and Whig bias. Nevertheless, it includes a good deal of the detailed information available about Locke's life, as well as reprinting many of his letters and written compositions. It should be supplemented by *The Life of John Locke with Extracts from His Correspondence, Journals and Commonplace Books*, ed. by Lord King, 2 vols. (London, 1830). Both Fox Bourne and King have been superseded, to some extent, by Maurice Cranston, *John Locke, a Biography* (London, 1957), which has used a great deal of new material—the Lovelace collection—and a rather less awe-struck view of Locke than its predecessors. A good, short view of Locke is offered by an old book, Alexander Campbell Fraser, *Locke* (Philadelphia, 1890). R. I. Aaron, *John Locke*, 2nd ed. (New York, 1955), gives a brief treatment of his life and then a more lengthy treatment of Locke's doctrines; it is weak on the political but excellent on the philosophical side (it ought certainly to be consulted on the *Essay*). It includes a bibliography.

The philosophy and science were not simply pushed aside when Hobbes and Locke turned to the field of politics. The epistemology of Locke, the metaphysics of Hobbes, served as the foundation stones of the *Two Treatises* and *Leviathan*. It is important, therefore, that we deal with the philosophy in order to understand the political superstructure erected upon it.

II

Let us begin with Hobbes. By the time of the Puritan Revolt, Hobbes was a man past middle age. His time of birth went back to the very roots of English greatness; he was born in 1588, the year of the Armada. In his younger years he served briefly as amanuensis to the great Francis Bacon, the man who first persuasively proposed that science was an inductive study, to be used for practical results. Later, Hobbes also became acquainted with Galileo and Descartes. Altogether, at the time of the Puritan Revolt, he must have appeared as a monument from another age.

Most of his life Hobbes spent in the service of the Cavendishes (whose title of nobility was Earl of Devonshire). However, before becoming a tutor in the Cavendish family, he had studied the classics at Oxford. The result of this was the first fruit of his literary labors: a translation of Thucydides. It is worth speculating on the effect made on Hobbes by this immersion in Greek thought, in a marvelous exposition of power politics.

As for science, Hobbes really took no interest in it until he was over 40. At that time there occurred the famous incident, picturesquely told by Hobbes's friend John Aubrey.

> He was 40 yeares old before he looked on geometry; which happened accidentally. Being in a gentleman's library Euclid's Elements lay open, and 'twas the 47 El. libri I. He read the proposition. 'By G——,' sayd he, (He would now and then sweare, by way of emphasis) 'By G——,' sayd he, 'this is impossible!' So he reads the demonstration of it, which referred him back to such a proposition; which proposition he read. That referred him back to another, which he also read. *Et sic deinceps,* that at last he was demonstratively convinced of that trueth. This made him in love with geometry.

The 47th Proposition of Euclid is Pythagoras' theorem about the squares on the sides of a right-angled triangle, and Hobbes followed the steps one by one, back to the elementary axioms from which it is proved. So impressed was he by the demonstrations that he generalized his experience into the notion that all thought—not merely geometrical thought—ought to be presented as an axiomatic system.

This was Hobbes' first great philosophical contribution. His conviction that the axiomatic method applies to all thought helped make the influence of mathematics and mathematicians paramount in European philosophy from that time on. True, Descartes had had the same vision earlier, in 1619, when, in a great mystical moment, it had been revealed to him that the world was fundamentally mathematical. But Hobbes really popularized the idea in England by expressing it in nonmathematical language and by employing the axiomatic method in his philosophical constructions.

Under the stimulus of his interest in geometry, Hobbes familiarized himself with all the latest developments in science. He made the acquaintance of Galileo and his work, during the years 1634–1637, and became his disciple. Indeed, Hobbes applied Galileo's assumptions—such as those concerning bodies in motion, and the distinction between primary and secondary qualities—and methods to all fields. In fact, much of Hobbes' basic achievement was to turn Galileo's physics into a metaphysics.

Hobbes also became acquainted, through the friendly offices of the French philosopher Father Mersenne, with Descartes, Gassendi, and many other French scientific thinkers. Although much influenced by Descartes, Hobbes was very critical of his system. He opposed, especially, the Frenchman's dualism of mind and matter. Hobbes turned, instead, to a complete materialist position and treated the mind as simply another body in motion.

III

This materialistic view of the universe is Hobbes' second major contribution to modern philosophy. He himself tells the story of how he arrived at it. According to Hobbes, he was once in company, around 1640 (by which time he was a man over 50), when somebody asked what was the nature of sensation. Hobbes, who was unable to think quickly on his feet, did not pretend to answer the question. But he went away and thought about it a great deal. He finally came to the conclusion that nothing in the outside world can be perceived unless it is in movement.[4]

Hobbes stated, more or less as an initial postulate:

The world . . . that is, the whole mass of all things that are, is corporeal, that is to say, body; and hath the dimensions of magnitude, namely, length, breadth, and depth: also every part of body, is likewise body, and hath the like dimensions; and consequently every part of the universe, is body; and that which is not body, is no part of the universe.

Having once established the fact that the universe is solely corporeal and materialistic, he set it in motion.

That when a thing lies still, unless somewhat else stir it, it will lie still for ever, is a truth that no man doubts of. But that when a thing is in motion, it will eternally be in motion, unless somewhat else stay it, though the reason be the same, namely, that nothing can change itself, is not so easily assented to.

Hobbes concluded, "When a body is once in motion, it moveth, unless something else hinder it, eternally." Thereupon, according to Hobbes, these bodies in motion strike our sense organs and set up further motions within us. It is these internal motions which we perceive as sensations. In short, "The cause of sense, is the external body, or object, which presseth the organ proper to each sense" and sets up further motions.

In a world composed of bodies in motion, what is man? He, too, Hobbes believed, was merely a body, or, better, a machine, in motion.

For seeing life is but a motion of limbs . . . why may we not say, that all *automata* (engines that move themselves by springs and wheels as doth a watch) have an artificial life? For what is the *heart,* but a *spring;* and the *nerves,* but so many *strings;* and the *joints,* but so many *wheels,* giving motion to the whole body.

[4] This idea had a profound effect on scientific philosophy for at least two reasons. In the first place, it enshrined the idea of velocity henceforth as a central part of scientific thought. Secondly, it served as the basis for a picture of the universe as a leaping, shivering affair of atoms, so that in our time writers like Eddington and Jeans picture a table actually trembling under one's hands. All this proceeds from Hobbes' decision that, whatever one may feel to the contrary, one cannot perceive anything but motion.

Hobbes, clearly, had taken the physics of Galileo as his model for the construction of a metaphysics of materialism and mechanism. It was this metaphysical counterpart to the seventeenth-century physical world which served as a basis for a large part of future thought.

IV

Hobbes' espousal of the axiomatic method and of materialism would be quite sufficient to earn him a permanent place in the history of philosophy. However, he is worthy of attention for a third achievement which is even more important than the others: the development of the doctrine of causality. The combination of axiomatics with the idea of bodies in motion, and, therefore, of time, led him to the formation of a concept of causality. Of all philosophers, if one must be singled out, it is Hobbes who saw most fully and consistently the world as an endless chain of cause and effect.

Hobbes started by accepting the medieval principle of final causality in terms of the Supreme Being. Then, for all practical purposes, he put it to one side. With God to one side, Hobbes asserted that causality was always to be sought in terms of particular motions of particular bodies. "Whatsoever effects," he believed, "are hereafter to be produced, shall have a necessary cause, so that all the effects that have been or shall be produced have their necessity in things antecedent." Thus, causal explanations were to be made in terms of elementary parts entering into temporal relations; before and after became equivalent to cause and effect.

Having equated causality with temporal relations, Hobbes went a step further. He equated causality with our ability to generate a thing; once again, we have the axiomatic method present in Hobbes' thought. He insisted that, if we want to "know" anything, we must construct it from its ultimate parts. For example, if we want to generate a circle, we must first "define" it; this enables us to deduce all its properties. Thus, the "cause" of the circle is our definition of it, i.e., our ability to define and construct it.

Hobbes, according to Cassirer, was "the first modern logician to grasp this significance of the 'causal definition.' "[5] Through his development of the theory of causality into the "causal definition," Hobbes influenced strongly the logic of the seventeenth century. Indeed, one of the interesting consequences of his concern only with

[5] Ernst Cassirer, *The Philosophy of the Enlightenment* (Princeton, 1951), p. 254

things that can be generated was, as Robertson points out, that he "could at once get rid of the subjects that mainly exercised his Scholastic predecessors. God, as having no generation, and spirits, as having no manifest properties (or phenomenal aspects)."[6]

A far-reaching consequence of Hobbes' view of causality was a rigid determinism. This found its epitome in the dictum of the French astronomer Laplace, that an omniscient intelligence, knowing the position and motion of all the atoms in existence at a given moment, could predict the entire course of the future.

Standing alone, Hobbes' theory of causality would earn him respect. When to it we add his "causal definition," his popularization of the axiomatic system, and his powerful development of the materialistic view of the universe, we see the real extent of his contribution to philosophy. Perhaps we can summarize his accomplishment by saying that, single-handed, he translated the method and ideas of the new science of the seventeenth century into a general explanation of man and the universe.

V

Locke's contribution to the development of philosophical ideas was to employ his talents as "an under-labourer in clearing the ground a little, and removing some of the rubbish that lies in the way to knowledge." Claiming that he took a humble position beside "such masters as the great Huygenius, and the incomparable Mr. Newton," Locke removed the rubbish lying in the way to knowledge by bringing into the forefront of modern philosophy the problem of epistemology: of how we know. He set himself the task of investigating the limits of knowledge so that man might not waste his energy in vain pursuits.

If by this enquiry into the nature of the understanding, I can discover the powers thereof, *how far* they reach, to what things they are in any degree proportionate, and where they fail us, I suppose it may be of use, to prevail with the busy mind of man to be more cautious in meddling with things exceeding its comprehension, to stop when it is at the utmost extent of its

[6] George Croom Robertson, *Hobbes* (Edinburgh, 1886), p. 79. Robertson's book is a judicious treatment of Hobbes' life and work and is slightly preferable to a similar treatment by Sir Leslie Stephen, *Hobbes* (New York, 1904), whose volume is included in the English Men of Letters series, ed. by John Morley. A further volume to be consulted is Ferdinand Tönnies, *Hobbes: Leben und Lehre* (Stuttgart, 1896), which, as the title suggests, treats both of Hobbes's life and doctrines and does so in an illuminating fashion.

tether, and to sit down in a quiet ignorance of those things which, upon examination, are found to be beyond the reach of our capacities.[7]

The problem of knowledge, i.e., epistemology, had become important in seventeenth-century philosophy as soon as Hobbes' materialistic philosophy was fully stated.[8] In a world of pure body and motion, how does man's mind perceive and know what is outside itself? This was a problem which the medieval scholastics, with their faith in the intelligibility of a world inhabited and manipulated by God, simply had by-passed. Locke, however, had to start from a Newtonian world which moved in mathematical regularity, like a "great Clockwork," and was made up of atoms of mass. He had to explain how man's mind, enclosed spatially within the brain, obtained its idea of this world. It was in his *Essay Concerning Human Understanding* (1690) that Locke presented his solution.

The *Essay* contains an attack on the Platonic view that there are innate ideas, that we come into the world furnished with a view of it which we simply rediscover. Locke pointed out, for example, that there were no principles "to which all mankind gave an universal assent." His main point of attack, however, was on the grounds that there was no need to assume innate impressions; men "barely by the use of their natural faculties, may attain to all the knowledge they have." Thus, Locke vindicated the empirical or natural road to all knowledge against the rationalistic or supernatural approach. Knowledge was not imprinted, once and for all, on man's brain by God; it was to be discovered, by experience, in the world. This was the "way" of seventeenth-century science.

[7] John Locke, *An Essay Concerning Human Understanding,* abridged and ed. by A. S. Pringle-Pattison (Oxford, 1950), pp. 11–12. This is the most usable edition of Locke's *Essay*. For some of Locke's earlier ideas on subjects covered in the *Essay,* see W. von Leyden, *John Locke: Essays on the Law of Nature* (Oxford, 1954), which reprints the essays in their Latin original as well as in English translations.

[8] Hobbes fumbled at an answer by asserting that the motions of objects strike the sense organs and produce counter motions which we call sensations or ideas. The mind then reasons about these ideas by computing and reckoning. According to Hobbes, "when a man *reasoneth,* he does nothing else but conceive a sum total, from *addition* of parcels; or conceive a remainder, from *subtraction* of one sum from another; which, if it be done by words, is conceiving of the consequences of the names of all parts, to the name of the whole" (*Leviathan,* p. 25). However, Hobbes did not make epistemology central to his work; this task was left for John Locke.

Locke's doctrine, therefore, was a doctrine of empiricism. He held that the mind, in the small child, arrives upon the scene entirely blank. He used the phrase "white paper," or in the Latin, *tabula rasa* (blank tablet). Locke stated the problem and the solution together:

> Let us then suppose the mind to be, as we say, white paper, void of all characters, without any ideas; how comes it to be furnished? Whence comes it by that vast store, which the busy and boundless fancy of man has painted on it with an almost endless variety? Whence has it all the materials of reason and knowledge? To this I answer, in one word, from EXPERIENCE; in that all our knowledge is founded, and from that it ultimately derives itself.[9]

The mind, through experience, has sense impressions made upon it from the outside. These impressions (which Locke later calls ideas) are then organized by the mind in a way which Locke found it difficult to explain, although he made a bold attempt.[10] In reality, Locke added, experience may furnish the understanding with ideas derived either from sensation or from reflection (i.e., "the internal operations of our minds, perceived and reflected on by ourselves").[11]

9 *Essay*, p. 42. In another image, Locke compared the mind to a sort of photographic box, with the developing process built in. "For methinks," Locke said, "the understanding is not much unlike a closet wholly shut from light, with only some little opening left to let in visible resemblances or ideas of things without" (p. 91).

This view remained an outstanding doctrine of science from the time Locke wrote, in 1690, for well over 200 years. It lasted, although attacked by Kant and Hegel on philosophical grounds, until the emergence of Heisenberg's uncertainty principle and Einstein's relativity physics led to a new scientific view: the view that the observer plays an essential part in the discovery of nature—that the laws of nature cannot, in fact, be formulated without making the observer enter into the formulation.

Of course, long before Heisenberg and Einstein, Locke's empiricism was criticized by philosophers. For example, it was pointed out that Locke's dictum that "all our knowledge is founded in experience," was an axiom not itself derived from experience. It was also noted that, while Locke attacked innate ideas, he accepted without question innate operations of the mind, i.e., the reasoning involved in demonstrative knowledge. The line of development and discussion of empirical philosophy may be said to have stretched from Locke to Berkeley to Hume and, thence, to Kant, who put the subject on a new footing. The interested reader is referred directly to the writings of the men indicated or to a summary by Ernst Cassirer in the third chapter, "Psychology and Epistemology," of his book *The Philosophy of the Enlightenment*.

10 See *Essay*, p. 22.

11 *Ibid.*, p. 43. Following Galileo, Locke accepted the division of qualities into primary and secondary; the primary qualities of any object are mathematical and are in the object, while the variable, changing, subjective, secondary qualities—such as taste and smell—are in us. This was also Hobbes' view; see p. 196 above.

Supplied thus with ideas, the mind perceives, by intuition or by demonstration (i.e., reasoning), which we shall explain shortly, whether they agree or disagree. This agreement or disagreement is what we call knowledge. In Locke's own words: "Knowledge then seems to me to be nothing but the perception of the connexion and agreement, or disagreement and repugnancy, of any of our ideas. In this alone it consists."

In the course of his argument, Locke drew an important distinction between innate and self-evident ideas. Although there are no innate ideas, he asserted, there are some self-evident ideas; that is, ideas about which we are intuitively certain. "For if we will reflect on our own ways of thinking," Locke suggested, "we shall find that sometimes the mind perceives the agreement or disagreement of two ideas immediately by themselves, without the intervention of any other: and this, I think, we may call *intuitive knowledge.*" It is intuitive knowledge which is "the clearest and most certain that human frailty is capable of." It is on intuitive knowledge that all other ways of knowing are founded.

These other ways are demonstrative, or reasoned, and sensitive knowledge. Demonstrative knowledge is where "the mind perceives the agreement or disagreement of any ideas, but not immediately." Sensitive knowledge, which is of less certainty than the other two, is of "the particular existence of finite beings"; Locke is not really quite sure of himself here, and we need not try to follow him closely. The important thing is that intuitive, demonstrative, and sensitive ways of knowing are all "natural" ways of knowing.

Further, they are ways of knowing shared by all thinking men and involve, therefore, merely "common sense." Thus, after Locke, the Declaration of Independence resounded with words like "We hold these truths to be self-evident. . . ." The self-evident quality of "these truths" is not innate; yet it is intuitive, and is available to all of us.

There does exist a form of knowledge, Locke admitted, which is neither intuitive nor clearly demonstrable: probable knowledge. Generally, he said, we call it "belief" or "opinion." The grounds of probability are two: "conformity of anything with our own knowledge, observation and experience," and the "testimony of others, vouching their observation and experience."

Revelation, according to Locke, is probable knowledge, and it is of the latter order of probability. On this difficult and dangerous matter, Locke took the following position. He declared that revelation which accorded with reason must be accepted. However, revelation which clashed with the "clear evidence of reason" was to be rejected. Locke subtly explained that "whatever God hath revealed is certainly true; no doubt can be made of it. This is the proper object of faith [i.e., even if it clashes with reason]: but whether it be a divine revelation or no, reason must judge." Handling revelation after this fashion, Locke, although a deeply religious man, hardly gave comfort to those who turned to it for certainty.

Indeed, it was this cutting edge of the Lockean epistemological razor which made enemies for him and his followers, among the orthodox and the conservative. His solution to the problem of how we know left the mind shorn of its received ideas—religious and political. Man was now left with a blank tablet, on which new ideas might be written and inscribed. As a result of Locke's "clearing the ground . . . of rubbish," the new man of the seventeenth century might start from scratch to construct his own, new world.

VI

Hobbes and Locke were animated by the same view of nature. They saw the world as made up of bodies in motion, which were arranged in an orderly pattern and followed well-defined, causal laws. Their view of the world was reflected in their view of human society. Through their thoughts run what we may call the atomic view of man and his social relations. They really thought of human societies as starting from a mass of atoms, and they were trying to discover what we might call the "gas laws" of human conduct. Or, to phrase it another way, they both wished to reconstruct human society from its component parts by means of a "causal definition." In accord with this, they began their political philosophies with man in an original state of nature.

Starting from the same point, Hobbes and Locke, however, built up quite different societies. Let us deal with each one separately, treating Hobbes' first.

Perhaps because he lived through the Puritan Revolution as a more mature man, and was uprooted by it, the occurrences of the civil war caused Hobbes to fear man's anarchical and evil nature more than did Locke. In any case, Hobbes took the very worst view of man in a natural state. He assumed that there would be a war of "every man, against every man," each distrusting the other and all desiring power; that there would be no industry or culture in such conditions; and, that, in his famous words, "the life of man, [would be] solitary, poor, nasty, brutish, and short."[12]

Hobbes did not pretend that such a condition of affairs ever actually existed. He was not trying to depict a historical development but to construct a logical one. He did comment, however, that this was the state of affairs during a civil war.

> It may peradventure be thought, there was never such a time, nor condition of war as this; and I believe it was never generally so, over all the world. . . . Howsoever, it may be perceived what manner of life there would be, where there were no common power to fear, by the manner of life, which men that have formerly lived under a peaceful government, use to degenerate into, in a civil war.[13]

What Hobbes really did, in constructing Leviathan, was to dissolve mentally the bonds of civil society into a natural state, that is, into its constituent atomistic parts, so as then to be able to achieve a new synthesis. It was the method *par excellence* of seventeenth-century experimental science. He was able to do this because he conceived of the commonwealth as an artificial body, or animal, Leviathan, created by the art of man.

> Nature, the art whereby God hath made and governs the world, is by the art of man, as in many other things, so in this also imitated, that it can

[12] *Leviathan,* p. 82. Later, Rousseau made a point of disagreeing with Hobbes on this view of human nature. In *A Discourse on the Origin of Inequality,* Rousseau declared, "Above all, let us not conclude with Hobbes, that because man has no idea of goodness, he must be naturally wicked." Rousseau went on to assert that natural man possesses the sentiment of compassion rather than viciousness. To Hobbes' contention that man is naturally intrepid, Rousseau opposed anthropological evidence to show that savage man is really timid and afraid. It was, in short, essential to Rousseau's whole political philosophy, as we shall show, to refute Hobbes' pessimistic view of man in a state of nature.

[13] *Leviathan,* p. 83. Beside *Leviathan,* Hobbes also wrote a history of the Civil War, i.e., of an actual time of the original "state of nature." He called his history *Behemoth.*

make an artificial animal. . . . For by art is created that great LEVIATHAN called a COMMONWEALTH, or STATE, in Latin CIVITAS, which is but an artificial man; though of greater stature and strength than the natural, for whose protection and defence it was intended.

Later, as we shall see, a conservative like Burke was to deny the artificial nature of the state and to assert that it was a natural, spontaneous growth. But Hobbes, in agreement with Machiavelli's idea of the state as a "work of art," left himself free to construct it *ab initio*, from first principles, as he willed.

Hobbes took it for granted, as we have seen, that human beings, in the absence of a state, can only behave toward one another with the greatest ferocity. He believed that, all being of equal ability to kill one another and, therefore, in a condition of constant insecurity, they would arrive at a desire for law and order. This could only be achieved by a mutual covenant. They would all agree to place someone over them, with absolute authority to tell them what to do.

Hobbes put forward this explanation in all his writings, but chiefly in the *Leviathan*. He suggested that a collection of human beings elect or appoint a king (or, it may be, an assembly, although Hobbes frowned upon this) for the sole purpose of his giving them orders to prevent them from being constantly at one another's throats. "The only way to erect such a common power, as may be able to defend them from . . . the injuries of one another . . . is, to confer all their power and strength upon one man, or upon one assembly of men, that may reduce all their wills, by plurality of voices, unto one will."

It followed, in Hobbes' view, that such a king has an absolute right to lay down whatever laws he chooses. He owes no responsibility to the individuals who chose him, except to keep the peace. There exists no contract between him and them. The only contract which exists is among the individuals themselves. It is simply an agreement to appoint somebody whom they will obey.

One consequence of Hobbes' view was to deny the Aristotelian belief that man is a "social animal." Instead, Hobbes asserted that no society or community exists before the covenant of submission. More importantly, however, Hobbes was setting up an absolute authority, unfettered by any contractual or "natural law" restraints (except for the individual's right of self-preservation, which Hobbes treated as a law of nature).[14] True, he did provide for artificial chains, called "civil laws," to bind the sovereign, but these, as Hobbes himself admitted, were "weak." Thus, in contrast to Coke and the constitutional lawyers who wished to limit the power of the king, Hobbes entrusted all power to the ruler to enforce unity and obedience.

The strange thing is that, by this reasoning, Hobbes both alienated the Parliamentarians and the Royalists in England. He alienated the former because of his theory of servile absolutism. Thus, after the circulation in manuscript, around 1640, of his *Elements of Law, Natural and Politic,* Hobbes felt himself in sufficient danger from the Puritans to leave the country and flee to France.

He alienated the Royalists because, though he believed in absolute monarchy, he did not base it on divine right. The Royalists felt intuitively that Hobbes' notion, that the monarch owes his position to those who elected him even though he has no further responsibility to them, was dangerous. They sensed, too, that Hobbes' doctrines justified the rule of an usurper as well as a legitimate king; that they opened the way for revolt by the assertion that the subjects entered into no promise or contract with the ruler but only with one another. We see the strength of the Royalists' alarm in the fact that Clarendon, the eminent historian of the Royalist cause, wrote a refutation of Hobbes.

The wrath of the clerics also fell upon Hobbes. For one thing, Hobbes was in favor of an Erastian church, that is, one subservient

[14] See on this whole subject Chap. 21 of *Leviathan.* Hobbes' denial of the existence of natural rights was consistent with his view of ethical values, for he denied the existence of absolute good and evil. For Hobbes, "whatsoever is the object of any man's appetite or desire, that it is which he for his part calleth good; and the object of his hate and aversion, evil." Thus, according to Hobbes, there are no known "natural" rules or laws in ethics. Later, we shall see that Rousseau followed Hobbes in having no natural-law restraints upon the sovereign power. Of course, in Rousseau, this sovereign power is the whole people and their "general will"; but we must remember that Hobbes, too, spoke of the king as reducing the people's wills "unto one will." Thus, both Hobbes and Rousseau, with totally different views of human nature, could arrive at the same totalitarian conclusion, if the line of reasoning used here is followed.

On the entire question of natural right, see Leo Strauss, *Natural Right and History* (Chicago, 1953); Chap. 5 is especially concerned with Hobbes and Locke. See, too, Strauss' *The Political Philosophy of Hobbes* (Oxford, 1934). Strauss believes that Hobbes' "new moral attitude" was anterior to his involvement with the new science; he also stresses Hobbes' "bourgeois" values and, thus, Hobbes' position as the real founder of liberalism, although himself not a liberal.

to the state. Even more an anathema to the clerics, however, were Hobbes' general religious views. For example, God's power was restricted and man's free will denied by Hobbes' materialistic philosophy. Indeed, so damaging was Hobbes' reputation as an irreligious thinker that he was forced to leave the exiled royal court in 1651, upon the publication of *Leviathan,* and to flee Paris. Later, around 1666, when Parliament began to hunt down atheism, a committee was appointed to examine the *Leviathan,* and it was only the intervention of Charles II which saved Hobbes from possible persecution.

Hobbes' doctrines pleased only those who wanted an argument for absolute power based on a secular justification. His book was, in this sense, a new version of Machiavelli's *Prince.* The really appreciative readers of Hobbes (although they would not admit it) would be the free-thinking "enlightened despots" of the next century.

Hobbes followed his own doctrines by accepting whatever authority was in power at the time. Thus, after fleeing Paris in 1651, he returned to England and made his submission to Cromwell's government. Then, at the restoration of Charles II in 1660, Hobbes was in the welcoming crowd. His former mathematics pupil rewarded him by receiving him at court and, later, with a pension. In short, Hobbes' life seemed to show that the other side of the coin of absolute sovereignty was absolute obedience.

VII

Despotic as were the implications of Hobbes' doctrines, they provided, in at least one essential way, the starting point for Locke's development of liberalism. Hobbes' great accomplishment in the *Leviathan* was to make government into an object of rational analysis rather than a veiled and divine institution above and beyond examination. Although he used scriptural arguments to convince his opponents, his was a completely secular statement of political power. As a result, Hobbes duplicated Machiavelli's feat of freeing political thought from theological tutelage; a feat which required duplication because the Reformation had obscured the Renaissance attitudes and reinstated a religious dominance over political theory.

Hobbes, however, not only rejected theology as a basis for political theory but proposed a new basis. He wished to create a political *science.* To do this, as we have seen, he based politics on the findings and method of seventeenth-century natural science. He did not stop there, however, but coupled to his physics a "scientific" theory of human nature.[15] Thus, with Hobbes, much of political theory began to revolve around the "science" of psychology.

The physics of Galileo and Newton had supplanted the doctrine of Christianity as the basis of political theory. It had supplied the model of society as an artificial construction, made up of individual atoms of humanity. But it had left unanswered the basic question: the nature of the fundamental building block—man. To try to answer that question became the task of philosophy and political science. Thus Hobbes, himself, wrote not only a *Leviathan* but a treatise on *Human Nature;* and Locke wrote an *Essay Concerning Human Understanding* as well as the *Two Treatises of Government*. Hobbes had opened the new avenues of approach; it was John Locke who went farthest, in the seventeenth century, along them.

VIII

In terms of practical experience, if not of reflection, John Locke was better equipped than Hobbes to deal with political theory. Locke's patron was Lord Shaftesbury, the leader of the Whig faction and, for a time, chancellor of the government. Under Shaftesbury, Locke, who was both his personal physician and his special adviser, held various minor official posts. For example, he was made secretary of presentations, and he helped draft the Fundamental Constitutions of Carolina. He was also financially interested in the Bahamas trade.

The connection with Shaftesbury, however, involved Locke in more than academic help to the Whig, mercantile interest in the colonies. It involved him with a revolution—1688—and the men who plotted it. Indeed, six years before the successful issue of the Glorious Revolution in 1689, Locke was forced to follow his patron into exile in Holland (where his views on politics and religion were confirmed by the Dutch example). After the Revolution, Locke became the intellectual leader and spokesman of the Whigs in their efforts to set up and defend new institutions of government. The publication of his *Two Treatises* served as both the "manifesto" (after the fact) and the justification of the 1688 upheaval.

[15] It may be that Hobbes' theory of human nature was derived also from theology, with its insistence on man's evil nature and original sin.

What had prepared Locke for his important role? He was born in 1632, and had been studying at Oxford during the Civil War. His father, a lawyer, was a captain in the Parliamentary army, and all of Locke's sympathies were on the Puritan side; in fact, he wrote verses praising Cromwell. His Puritan sympathies undoubtedly directed Locke to write his first important exposition, the *Essay Concerning Toleration* (1667); all in all, he wrote four "letters" on toleration. These make clear that his political views were most strongly affected by his religious beliefs.[16]

While at Oxford, Locke had been much impressed by Descartes. Later this influence was supplemented by Newton's, and it may have been Newton who fostered the empirical tendency in Locke. Although he never took a doctor's degree, Locke also studied medicine and, as we have seen, became Shaftesbury's personal physician. In that capacity, Locke performed a difficult operation on Shaftesbury. It involved lancing an abscess and inserting a silver tube to drain it, and it probably saved the life of his patron.

In medicine, Locke, who was a friend and co-worker of the great Sydenham, was a confirmed empiricist. He rejected the "theories" of Galen and other ancients and looked at disease with his own eyes. This independence was strengthened by his membership in the Royal Society. The Society's early stress on inductive science and the promotion of useful knowledge fitted perfectly with Locke's outlook. Thus we see him writing not only on medical subjects but on such practical matters as the consequences of lowering the rate of interest and the problems involved in education.

Locke's fame, however, derives chiefly from his *Essay Concerning Human Understanding* and his *Two Treatises of Government.* His problem in the *Two Treatises,* as we see it, was to continue Hobbes' emancipation of political theory from theology while dissociating a reasoned approach to politics from Hobbes' conclusions. More nar-

[16] We ought also to mention that Cudworth, one of the Cambridge Platonists who were fighting for latitudinarianism in the Anglican Church, was also Locke's close friend and strengthened his Puritan-derived desire for toleration. See further C. H. Driver, "John Locke," Chap. 4 in *The Social and Political Ideas of Some English Thinkers of the Augustan Age,* ed. by F. J. C. Hearnshaw. Locke's relations with deism are dealt with in S. G. Hefelbower, *The Relation of John Locke to English Deism* (Chicago, 1918), an unexciting book which concludes that Locke cannot be accounted "the father of the deistic movement"; it is a book to be used with caution.

rowly, of course, he had also to justify the coming of William to the throne of England; and, thus, indirectly, the execution of Charles I. Locke accomplished all his tasks brilliantly. As a result, he laid the permanent foundations of the liberal movement.

The *First Treatise,* too often neglected, roundly attacked Sir Robert Filmer's *Patriarcha* and the "False Principles" of divine-right monarchy which it contained.[17] It was, also, in a subtle way, an attack on Hobbes and tried to separate his rational politics from his absolutist conclusions.

Locke assailed Filmer and the religious justification of power by declaring that it was foolish to say that God gave power to Adam and through Adam to Adam's eldest and thence down to his present-day royal descendants. We are all descendants of Adam, Locke gravely pointed out, and it is impossible to know who is, today, the eldest son. Then, in an involved philological and semantic argument, Locke dissolved in ridicule the argument for the divine right of kings.

In this way, Locke directed his rebuttal of Filmer, not only on the question of theology, but also on the question of tradition. He rejected mere tradition as a justification of the exercise of power. In the eleventh chapter of the *First Treatise,* he stated the important question: "Who Heir?" "The great question which in all ages has disturbed mankind, and brought on them the greatest part of those mischiefs which have ruined cities, depopulated countries, and disordered the peace of the world, has been, not whether there be power in the world, nor whence it came, but who should have it." Then, with what might be a slight tilt at Hobbes, Locke added, "And the skill used in dressing up power with all the splendour and temptation absoluteness can add to it, without showing who has a right to have

[17] Filmer's *Patriarcha* was written before 1653, but it was not published until 1680, when the Tories brought it to light as a defense of their position. In an article, "A Second Thought on Locke's 'First Treatise,' *Journal of the History of Ideas* (January, 1956), Herbert H. Rowen expresses the view that Locke's main attack is on Filmer's acceptance of the state as a piece of dynastic property. There is evidence that Locke originally wrote the *Two Treatises* about 1681 to promote Shaftesbury's revolutionary movement for a Protestant successor to Charles II and then modified the book to suit the needs of 1689. Cf. Cranston, *op. cit.,* Chap. 15 and Peter Laslett's edition of the *Two Treatises.* Whatever the original date of composition, however, it does not affect the fact that the *publication* of the *Two Treatises* served to justify the coming of William to the throne of England.

it, will serve only to give a greater edge to man's natural ambition, which of itself is but too keen."

Against tradition, against the existence of "what has been," Locke employed the acid of reason and natural rights. He declared emphatically that "an argument from what has been, to what should of right be, has no great force." He commented satirically, "But if the example of what hath been done be the rule of what ought to be, history would have furnished our author [Filmer] with instances of this absolute fatherly power in its height and perfection." And thereupon Locke retold the story, from Garcilaso de la Vega's *History of the Incas of Peru,* of fathers begetting children on purpose to fatten and eat them at the age of 13. It is a delightful story.[18]

The *First Treatise* is long and, for our modern taste, often boring. Nevertheless, it effectively cleared the way for Locke to offer an argument in political science based on reason rather than on theology or tradition. Thus the *Second Treatise* could be a work of construction now that the *First Treatise* had done the preparatory work of demolition.

We have seen that in the *Essay* Locke analyzed human nature as it is before experience and without innate ideas. In the *Second Treatise* he abstracted man from society and placed him in a state of nature. The method in Locke's psychology and political science was the same —reduction to the simplest parts and conditions. In Locke's own words, to understand political power, we must "derive it from its original, we must consider what state all men are naturally in."

Locke's "original," his state of nature, differed radically from Hobbes'. With Locke, the state of nature itself was characterized by reason. It was not merely the scene of man's wild, irrational, and ungovernable passions. In Locke's description: "The state of nature has a law of nature to govern it which obliges every one; and reason, which is that law, teaches all mankind who will but consult it that, being all equal and independent, no one ought to harm another in his life, health, liberty, or possessions."

Locke insisted, and it is an important point, that man has no right

[18] *Two Treatises,* p. 44. The New World was not only a source of investment for Locke but also one of inspiration. A perusal of his *Two Treatises* will show how frequently he quoted from writers like Garcilaso de la Vega and from travelers' descriptions of the life and customs of the Indians. Thus Locke's picture of the state of nature must be seen in the context of his views on the New World.

to the exercise of his freedom until he attains the use of his reason; that is, until the child grows into the adult. Freedom without reason is mere license. As Locke put it, "lunatics and idiots are never set free from the government of their parents." Fortunately, most humans are not lunatics or idiots, and "thus we are born free as we are born rational, not that we have actually the exercise of either; age, that brings one, brings with it the other, too."

As a result of this infusion of reasonableness into the state of nature, men, from the outset, behave toward one another in such a way that society is possible. In this society, men are free and equal by the law of nature. They have inalienable rights to their life, liberty, and property. They also have the natural right to punish any who offend against them or their possessions.

Unfortunately, according to Locke, some men, not under the common law of reason, attempt to put other men under their power and to deprive them of their natural rights. This is to institute a state of war. At this point, Locke claimed, man resumes his natural rights and appeals to the God of battle. Here, succinctly, we have Locke's justification of revolution. He has fulfilled his immediate purpose, of justifying the Whigs in calling William to the throne. He has defended the reasonableness of the events of 1688–1689.

IX

We said earlier that one aim of the *Two Treatises* was to attack Hobbes' conclusions indirectly while yet retaining his support as an ally against theology and tradition. On the very first page of the *Second Treatise* Locke both made his thrust and indicated what he would offer in the place of his opponent's system:

> . . . He that will not give just occasion to think that all government in the world is the product only of force and violence, and that men live together by no other rules than that of beasts, where the strongest carries it, and so lay a foundation for perpetual disorder and mischief, tumult, sedition, and rebellion—things that the followers of that hypothesis so loudly cry out against—must of necessity find out another rise of government, another original of political power, and another way of designating and knowing the persons that have it than what Sir Robert Filmer hath taught us.

For Filmer, let us substitute the name Hobbes in this passage: it was Hobbes who had declared that the state of nature was a state of war, in which men behaved without reason and needed an absolute sovereign to control them. Locke cleverly turned this argument around. He asserted, as we have noted, that reason ruled in the original state of nature.

To avoid the disturbances and to curb the violators of the natural state, Locke declared, men enter "political or civil society." They leave the state of nature, band together in commonwealths, and appoint a government to act as a common judge over them and to protect their rights of life, liberty, and property. Thus government is freely created by the people to protect already existing rights. It derives its power from "the consent of the governed."

We shall avoid going into details of the particular arrangement Locke employed. It is important to say only that for him supreme power rests with the people. They exercise it through a parliament which can neither pass its legislative power to somebody else, such as an absolute monarch, nor levy taxes without the consent of the taxed. There exists, also, a separation of executive and legislative functions.

It is implicit in Locke that the true function of government is not to impose laws on the people but to discover what the laws of nature are. Locke seems to have believed that there are laws of nature which govern human society exactly as there are laws of nature which govern the speed of falling bodies. The government, in Locke's view, ought to be (as it were) a research establishment run by political scientists, whose business it is to discover what these laws are and to govern accordingly. Thus, if Newton saw God as a regulator of clockwork, Locke saw the state as a judge maintaining the social workings; both had as their ideal a mechanical stability, based on laws of nature.

A theory of natural law prevails, also, when we turn to Locke's treatment of property. Locke held that all men have a natural right to property (defined in this context as that with which a man mixes his labor). This right is restricted only by the fact that no man may claim more from the common treasure house of nature than he can use personally; any more would simply spoil and waste away. One of Locke's assumptions was that there would always be more than enough free land, as in the New World, for anyone who wanted it.

However, this natural right to the property one can work with one's own hands loses it force when, almost at the end of his discussion on property, Locke slipped in the use of money. Money, he claimed, does not spoil. And, as mankind has agreed to its use, it means that a man may acquire more property than he himself needs for the sustaining of life. Thus ends the idyllic existence of Locke's state of nature![19]

Nevertheless, the "equalitarian" implications of Locke's theory remain. Locke explicitly pointed out that "the preservation of property being the end of government and that for which men enter into society, it necessarily supposes and requires that the people should have property, without which they must be supposed to lose that, by entering into society, which was the end for which they entered into it—too gross an absurdity for any man to own." In the eighteenth century and after, supposed followers of Locke often embraced the "absurdity." Locke himself was too rational to deny the logical consequence of his thought. He insisted, therefore, that the test for participation in government was the possession, not of property, which he assumed all men should have, but of reason. It is this element in Locke's thought which made his work a seed bed for the democratic as well as the liberal movement of the times to come. The result of a belief in "common sense" was a belief in government by the common man.

X

Locke's point of view is usually called liberal while that of Hobbes has been labeled authoritarian or conservative.[20] Locke's view of human nature is optimistic while Hobbes' is pessimistic. In the analogy we used before—of society as a concourse of atoms—in Hobbes' picture the atoms are always in collision and liable to knock

[19] For a treatment of Locke's views on property, see Paschal Larkin, *Property in the Eighteenth Century* (Dublin and Cork, 1931).

[20] For the question as to whether Hobbes really was a conservative, see the chapter on Hobbes in the forthcoming book by Bruce Mazlish, *The Nature and Origins of Conservative Thought*. For further considerations of Locke's political philosophy, the following can be consulted with profit: J. W. Gough, *John Locke's Political Philosophy: Eight Studies* (Oxford, 1950); Willmoore Kendall, *John Locke and the Doctrine of Majority Rule,* Illinois Studies in the Social Sciences, Vol. XXVI, No. 2 (Urbana, Illinois, 1941), which emphasizes the democratic and egalitarian implications of Locke's doctrines; and Sterling Power Lamprecht, *The Moral and Political Philosophy of John Locke* (New York, 1918).

one another to pieces. In Locke's conception, the atoms settle down to a balanced and self-adjusting state, and all that is needed is for government to see that any intruders are properly dealt with. Government merely regulates or makes stable the state of nature.

These two views of society—the authoritarian and the liberal—were given classic form in Hobbes' *Leviathan* and Locke's *Two Treatises.* Conceived in the stress and strain of the Puritan Revolt and its aftermath, the two books contained, nevertheless, elements and directions of thought which transcended the particular events which called them into existence.

Hobbes and Locke, partisan in part, attempted to free political thought from supposed extraneous considerations so that it might become scientific. It may be that, in so doing, they dealt with man in too mechanical and atomistic a fashion. In this, they shared both the virtues and the limitations of the physical sciences, whose dominance over the thought of the time stood so high and whose shadow we have seen fall so sharply over their respective philosophies. But their attempt was noble. As a result, the political and philosophical thought of Hobbes and Locke has remained vitally important from the time they wrote until today.

CHAPTER 12

The Method of Descartes

FRANCE IN the seventeenth century moved in a strange contradiction: her government grew more absolute while the thought of her best minds grew bolder and freer. A flourishing bureaucracy tended to centralize authority more and more: even such a scholarly institution as the Académie Française, which was to give order and uniformity to the language, was in large measure an expression of authority. Yet, at the same time there were created, formally, the Académie des Sciences and, informally, the salons, which expressed the intellectual vigor of France during this time. These two currents of life—an awakened public opinion and an absolute government—had sooner or later to clash head-on in France, as they had in England in 1640.

The clash in France did not come until 1789. French thought took more than a century to reach the independence of mind of the English and the Dutch. When French thinkers did learn this freedom, they became the intellectual leaders of Europe. Our plan, therefore, in the next chapters is to go back in French history and discuss her leading thinkers in the seventeenth century. In this way, we shall form the background for the great intellectual movement from 1730 onward, when France caught up with the political ideas of Locke and the scientific ideas of Newton, and carried them beyond their development in England. We are making a small detour out of the regular sequence of history, as it were, in order to set the scene for the great explosion of liberty in the eighteenth century.

I

We begin with René Descartes, who lived from 1596 to 1650. He was born, therefore, roughly when Shakespeare began to write; in his early manhood, Galileo had the greatest influence in Italy; and he died a few years after Newton was born. The great struggle within France betweeen Catholics and Huguenots was just over when Descartes was born and had been healed, so far as it could ever be healed, in the single person of Henry IV. Descartes was brought up, from 1606 to 1614, at the Jesuit College of La Flèche, which was under the particular protection of Henry IV. As we have already explained, he willed that on his death his heart should repose in the college chapel, and Descartes was at the ceremony in 1610 which honored this wish.

The Jesuit teaching stayed with Descartes all his life; he remained a devout Catholic, even when later he found it more prudent to retire to Holland to write and print his books. From the age of 10 to the age of 18, he learned the classics, some science, and scholastic philosophy at La Flèche. He went on to the University of Poitiers, apparently to study law and probably some medicine. He then, as he tells us, at the age of 20 decided to read in the "book of the world" and for this purpose enlisted as a voluntary gentleman in the Dutch army under the command of Maurice of Nassau.

Holland was at this time in a ferment of intellectual and scientific development, and Descartes was strongly influenced by a Dutch thinker Isaac Beeckman, who gave him an interest in mathematics and physics.[1] This stimulus was vivid in his mind when he returned to France in 1619, by way of Germany, and went through a mystical experience which shaped his life.

The nature of this experience is not very clear: perhaps all mystical experiences are at bottom incommunicable. But the fact of the experience is not in doubt, nor is the message which Descartes drew from it. On the night of November 10, 1619, there was revealed to Descartes at the age of 23, in some overwhelming way, the fact that

[1] For the significance of Dutch scientists and their influence on French thought, see Pierre Brunet, *Les Physiciens hollandais et la méthode expérimentale en France aux XVIIIe siècle* (Paris, 1926). Also, Ernst Cassirer, *The Philosophy of the Enlightenment*, Chap. 2.

the structure of the universe is mathematical and logical. So profound was the sense of revelation which came to Descartes that night that he treasured the memory all his life, and made a pilgrimage to the Lady of Loreto in Italy to give thanks for it.[2]

The times were "jointed" right for Descartes' work. It was roughly at this time that Galileo wrote the famous dialogues in which he expounded the astronomical system of Copernicus in popular language, and for which the Inquisition at last broke him. Ten years later, in 1629, Thomas Hobbes fell in love with geometry almost as suddenly as did Descartes, by picking up a book of Euclid's. Perhaps it was Descartes' education which caused his conversion to mathematics to take a more dramatic form than those of Galileo and of Hobbes. But the drama is not out of place: these men made a revolution in science when they suddenly grasped the reach of the mathematical method in the sensual world.

II

Mathematics is no more than a symbolism. But it is the only symbolism invented by the human mind which steadfastly resists the constant attempts of the mind to shift and smudge the meaning. It is the only exact symbolism and, by being exact, it is self-correcting.

This is why, to this day, our confidence in any science is roughly proportional to the amount of mathematics it employs—that is, to its ability to formulate its concepts with enough precision to allow them to be handled mathematically. We feel that physics is truly a science, but that there somehow clings to chemistry the less formal odor (and odium) of the cook book. And as we proceed first to biology, then to economics, and last to social studies, we know that we are fast slipping down a slope away from science. We feel this because more and more

[2] For an account of Descartes' dream, see the translation from Baillet's *Vie de Descartes* (1691) in Norman Kemp Smith, *New Studies in the Philosophy of Descartes* (New York, 1952), pp. 33–39. The best life is *Vie et oeuvres de Descartes* (Paris, 1910), by Charles Adam, which forms Vol. XII of the *Oeuvres* cited in fn. 4 below. A very reduced version of this is the same author's *Descartes: Sa vie et son oeuvre* (Paris, 1937). S. V. Keeling, *Descartes* (London, 1934) is a serviceable English survey of life and works. See, too, A. Boyce Gibson, *The Philosophy of Descartes* (London, 1932). Gustave Cohen, *Ecrivains français en Hollande dans la première moitié du XVIIe siècle* (Paris, 1920), deals extensively with Descartes' sojourn in Holland. The first important biography was Adrien Baillet, *La vie de M. Descartes* (Paris, 1691); although not always trustworthy, it is the source from which all other biographies of Descartes have started.

of the argument in these is carried by words, and words are symbols which cannot be used without overtones and multiple meanings.

This may not in itself seem a great matter, but it has turned out to be critical to the development of science. When Aristotle established the four categories of fire and air, earth and water, he may well have intended to have them used quite precisely. But in the nature of language, these symbols were subject to constant shifts of meaning; and when the symbols are uncertain, the thought equally shifts its ground. It may be, for example, that science in China failed to build on the important discoveries made there, because the language was unsuitable for exact expression. Certainly in Europe an important step was taken in chemistry in Descartes' lifetime when a chemist grew tired of the multiple uses of the word "air" and coined the word "gas."

It was, of course, not simply a matter of deciding that mathematics should be applied to nature. The revelation made to Descartes went deeper than this. For the Aristotelian categories carried with them, and were part of, an assumption that the natural world can only be described in those general terms for which ordinary language is adequate. Aristotelian physics and biology always carried the sense that nature is more subtle than any exact description, and that the categories were to give only the rough skeleton on which her infinitely graded varieties take form. This was the sense that the vision of Descartes overthrew. In its place it put the conception of nature as a mechanism, every part of which follows exact and logical formulas.

The new conception was shared by Galileo, by Hobbes, and by others. Their interests, however, were both different from Descartes'. Galileo's major interest was in empirical science. Hobbes' interest was in the Euclidean procedure—that is, in the method of deriving results from a set of axioms by a sequence which, in the physical world, goes from cause to effect. Descartes' interest was simpler than either of these: he wanted to describe the universe, and mathematics was the unequivocal language which alone could form an intellectually satisfying description.

Accordingly, the work for which Descartes is best known is that which describes geometrical figures by the formulas of algebra. Descartes was, in fact, the first man who consistently thought of a straight line as an equation of the first degree, and of a conic section as an equation of the second degree, and who looked for their points of intersection in the common solutions of these equations.[3]

Descartes did work in astronomy, in the theory of light, and in general physics. He formulated a theory that all matter consists of vortices, a theory which began by being a stimulus and ended, however, by being an incubus of science. It was, of course, natural that his love for mathematics, and especially for geometry, should tempt him to look for such a priori foundations in nature; and that his later critics, and especially Isaac Newton, should find these a hindrance. All intellectual progress is ambivalent, and we cannot have Descartes' superb geometrical vision without some of the prejudices of the geometer.

III

Descartes wrote many philosophical works of which, characteristically, the most searching was a preface to a book on meteors and geometry. This preface makes a small book under the title *Discourse on Method,* and became (and remains) one of the most influential books in all philosophy, and particularly in the philosophy of science.

The title of Descartes' book reveals his interest, and the important interest of his time, in the method of reasoning about the universe. The Aristotelian method, because it was verbal, implied that the final detail of all structure always escapes analysis: it is somehow too supple, too imponderable for exact description. Descartes' vision broke sharply with this view: he had seen that the key to the universe was its mathematical structure, and from that moment nothing less could content science. Therefore, Descartes' method is designed ruthlessly to unmask the imponderables, and to find in everything the lucid and exact structure. Here are his four rules of logic.

My first rule was to accept nothing as true which I did not clearly recognize to be so; to accept nothing more than what was presented to my mind so clearly and distinctly that I could have no occasion to doubt it.

The second rule was to divide each problem or difficulty into as many parts as possible.

3 See J. F. Scott, *The Scientific Works of René Descartes (1596–1650)* (London, 1952), p. 86, for a discussion of Fermat's claims to priority in this matter.

The third rule was to commence my reflections with objects which were the simplest and easiest to understand, and rise thence, little by little, to knowledge of the most complex.

The fourth rule was to make enumerations so complete, and reviews so general, that I should be certain to have omitted nothing.

Let us single out from these rules first those which insist on clarity and simplicity—the first rule and the third. These present, and hinge on, the view that the world is built up in an orderly and understandable way from precise and definable entities. This is both the strength and the weakness of Descartes' outlook. It is his strength, because it makes him insist that the world can be explored and mapped with logical tools. It is his weakness because, in aping the geometry which had been traditional since Euclid, it implied that the fundamental entities of the universe and the laws which connect them should be self-evident.

To show the weakness, let us quote one of Descartes' arguments. He argues in several places for the existence of God, and in one place he says that once we understand the nature of God his existence becomes self-evident—just as, he says, once we understand the nature of a triangle it is evident that the sum of its three angles is two right angles. Unhappily, time has shown that nothing of the kind is evident about triangles. Even in the geometry of Euclid there is needed an additional axiom, which Euclid did not state, before it can be proved that the angles of a triangle make up two right angles. The additional axiom which is needed states that, through any point, there can be drawn one and only one line parallel to a given line; this axiom was supplied by the Englishman Playfair about 1795. There is an infinity of geometries unknown to Euclid, in which the angles of a triangle never make two right angles. It is by no means self-evident which of these geometries fits the universe at large distances from us.

In the same way, the physical sciences have found that their fundamental entities and axioms are a great deal less obvious than was hoped. Particularly as we penetrate into the world of small-scale phenomena, we find, like Gulliver, that we are no longer at home. To Descartes and to scientists for 300 years after his birth, it seemed natural to suppose that the small units from which all matter is built must behave more simply than does matter in the large. But it is not so, and the foundations of science have proved to be more elusive and less predictable than he supposed.

We must not think Descartes short-sighted in this. The task he had in mind was to construct the machinery of the world rigorously by deductive methods. He therefore needed to operate on a fundamental set of axioms. The Jesuits who taught him had found the sanction for their axioms in authority: the authority of Aristotle, St. Thomas Aquinas, and the Ecclesiatical texts. What sanction could Descartes propose? He proposed the sanction of the intellect. The axioms of natural science, like (he supposed) the axioms of mathematics, must spring of themselves to the human mind. They must be so evident that they cannot be doubted. We reach the rock of thought when we can no longer doubt; and therefore we reach it only by doubting.

IV

This radical doubt is the crux of Descartes' method. As his second and fourth rules show, his method is analytic: it works by taking things and thoughts to pieces. And the tool with which it takes them to pieces is given in the first rule: "to accept nothing as true" until the mind step by step reaches those pieces, those foundations "that I could have no occasion to doubt."

The crux of the Cartesian method is expressed in a phrase which he often used, *de omnibus dubitandum* (we must doubt everything). This may seem strangely cynical advice from a religious man; and, indeed, it did not make him popular with religious men. Yet, other religious men found their salvation by this method: Blaise Pascal, as we shall see, did so. When Descartes used the method of doubt, he reached a religious position somewhat by paradox. But his aim in using the method was always clear. Unlike Michel de Montaigne and other skeptics against whom he wrote, Descartes had no interest in a modish attitude of doubt for the sake of doubting. His aim was by way of doubt to reach down to what can be shown with certainty. This was essentially the scientific procedure of seventeenth-century physics, in which doubt played a constant part: "to accept nothing as true" until it was established, as far as possible, beyond doubt.

And this is the way that Descartes behaved in his own life. He tells us in the *Discourse on Method* that he came to reject the bookish authorities quoted to him at La Flèche, and resolved, as we have already mentioned, to read in "the book of the world" for himself. There he learned that the ways of men are different in different places, and that a custom which is thought ridiculous in one place is held to be sacred in another. This is the form of comparison which Charles Louis de Secondat Montesquieu used a century later in the *Persian Letters,* and Descartes anticipates him when he writes, "it is well to know something of the manners of various peoples, in order more sanely to judge our own, and that we do not think that everything against our modes is ridiculous, and against reason, as those who have seen nothing are accustomed to think."[4]

The books of scholars had shown Descartes that there is no certain truth in authority; the book of the world had shown him that there is none in custom. He had now to consider himself as a book—the book which was the evidence of his senses. Suppose then that this book is also written to deceive him, and that God is an evil genius who wants to keep the truth from him.

> I shall consider that the heavens, the earth, colors, shapes, sounds, and all other external things, are nothing but illusions and dreams.
>
> I shall consider myself as having no hands, no eyes, no flesh, no blood, nor any senses; yet falsely believing myself to possess all these things.
>
> If, by this means, it is not in my power to arrive at the knowledge of any truth, I may at least do what is in my power, namely, suspend judgment, and thus avoid belief in anything false and avoid being imposed upon by this arch deceiver, however powerful and deceptive He may be.[5]

By these drastic means, Descartes uprooted all his accepted ideas, whether they reached him from authority, from custom, or through his senses. He had left his mind free: not indeed blank, in the sense of John Locke, but free of all ideas except those innate to the mind. Like Socrates, he had arrived at the point where he knew only that

[4] *Discourse on Method,* Part I. *Oeuvres de Descartes,* publiées par Charles Adam et Paul Tannery, 12 vols. (Paris, 1897–1910), is the standard Works. *Philosophical Works of Descartes,* tr. by E. S. Haldane and G. R. T. Ross, 2 vols. (Cambridge, 1911), is a handy English edition of some of the works. *The Correspondence of Descartes and Constantyn Huygens, 1635–1647,* ed. by Leon Roth (Oxford, 1926), supplements the *Oeuvres.* Etienne Gilson, *Discours de la méthode, texte et commentaire* (Paris, 1925), is the fundamental edition of the *Discourse.* Leon Roth, *Descartes' Discourse on Method* (Oxford, 1937), is a most suggestive discussion of the importance of the *Discourse.*

[5] *Meditations,* First Meditation.

he did not know; and from here he wanted to construct himself and the world anew. He would have to do this by pure reason, and he would have to persuade others by the same means. He did not doubt that the means of reason were within reach of all: "good sense," he writes at the beginning of the *Discourse,* "is the most equitably divided thing in the world." He writes in French, not in Latin, because it is this "common sense" in everybody to which he must appeal.

V

If you doubt everything, what remains beyond doubt? Only the fact that you are doubting. This may seem a philosophical trick: yet at heart, surely, it is sound. You cannot doubt without being a doubter, that is, a person who is something more than a bundle of sense impressions which he is doubting. Because you doubt, you do not accept the sense impressions as they come; the universe is no longer something outside yourself, impersonally making its marks on you. Your relation to the universe has become personal; by the process of doubt, you explore and so rebuild it. And you explore yourself; you prove and make yourself. You alone resist your doubt, and your doubt creates your self.

This is the thought which Descartes put into one of the most famous sentences in philosophy: *cogito ergo sum* ("I think, therefore I am"). He might have put it more correctly by writing *dubito ergo sum* ("I doubt, therefore I am"). The change from doubting to thinking may have been only a verbal one to him, since his method of thinking was always to doubt. Yet such verbal shifts are dangerous, even to the clearest thinker. It is illuminating to follow one of the more remote consequences into which it led Descartes, perhaps not without some relief on his part.

When Descartes elsewhere moved from the proof of his own existence to that of God, he used the same form of argument: the ontological argument first made explicit by St. Anselm about 1070. In effect, Descartes says that he can conceive a being who is perfect; that the ability to conceive this implies the existence of perfection; and therefore "we must conclude that God necessarily exists as the origin of this idea I have of Him."[6] Existence is, as it were by definition, a necessary part of perfection. This is to say that "I think of God, there-

[6] See *Discourse,* Part IV, and *Meditations,* Third Meditation.

fore he exists." The argument would sound less convincing if it were phrased as "I doubt God, therefore he exists."

Today, we are no longer sympathetic to such linguistic arguments. Indeed, it is a result of Descartes' mathematical vision that we find them unattractive, and that we ask for definitions which compel a more rigorous use of words. We are inclined, and rightly inclined, to abandon Descartes at the moment that he shifts from doubt into thinking. For the method of doubt is the analytic method, and there Descartes made a basic contribution to science. Thinking, in the sense that he uses it in the ontological argument, is a method of synthesis, and this is an inductive process which Descartes did not reach; it was clearly formulated only by his Dutch junior, Christian Huygens.

And though science and indeed our daily life could not go forward without the inductive method, there never has been a proof of it. We cannot formally vindicate any method that goes beyond Descartes': we cannot justify rigorously and logically our inductive belief in the regular behavior of things, any more than we can justify our belief in their existence. Whatever argument we can produce to support these beliefs does not derive from the mind but from our own necessary behavior in the physical world. We are today in the state in which James Boswell was when he came out of Harwich Church with Dr. Samuel Johnson on August 6, 1763.

> We stood talking for some time together of Bishop Berkeley's ingenious sophistry to prove the non-existence of matter, and that everything in the universe is merely ideal. I observed, that though we are satisfied his doctrine is not true, it is impossible to refute it. I never shall forget the alacrity with which Johnson answered, striking his foot with mighty force against a large stone, till he re-bounded from it. "I refute it *thus*."

Many a man has felt this kind of irritation with doubting philosophers; and many a man has vented it, as Dr. Johnson did, by kicking a stone when he would have liked to kick the doubter.

VI

Faced with the physical universe, Descartes evolved a theory which had some of the qualities of success which make general theories so plausible and so dangerous. It is now only a scientific curiosity. But it is a curiosity of the kind which nonscientists find pleasing, and in one form or another it goes on being invented over again by many minds. We shall sketch it briefly.

In Descartes' scheme, the material universe does not consist of separate pieces: he was opposed to any atomic theory. For him, matter is continuous, has unlimited divisibility, and fills everything. There are two essential forms of this fundamental matter: a fine "first matter" which fills all space as a kind of infinite ether; and a denser matter which makes up the solid bodies with which we are familiar.

These solid bodies are carried about by the all-pervading "first matter," because the "first matter" has been divinely shaped into innumerable vortices or whirlpools. Thus each solid body, such as the sun or a planet, rides in its own vortex: the sun in a large vortex, a planet in a smaller vortex. There are no forces between the bodies except those caused by the spinning motion of the vortices in which they ride; when a coin drops to the ground, for example, it does so because the small vortex round the coin is drawn toward the larger vortex round the earth.

This theory of the behavior of material things was a major break with the theory which the church had accepted from Aristotle. With his vortex universe, Descartes became the first man in modern times to present a fundamental and comprehensive view of the universe different from the Aristotelian-Christian one; thus, his new cosmology was world-shaking as well as world-shaping. His theory was a mechanical theory, which broke once for all with the tradition that nature had feelings of love, of striving, and of abhorrence. To Descartes, nature was a machine in perpetual motion, whose movements were predictable and caused of themselves by the mutual attractions and repulsions of its spinning vortices.

There are some attractive features in Descartes' theory; for example, it has some of the geometrical merits of the general theory of relativity, in which also a material body is pictured as a hollow in a universal space-time. The vortex theory had enormous influence, which was overthrown only by Newton; for Newton showed that vortices could not do forever what Descartes asked of them, even qualitatively. And of course, the quantitative accuracy of Newton's system was altogether greater than anything that the vortex theory could give. Yet, Descartes did a service even to empirical science in advocating and for a time imposing the vortex theory, because he thereby opened the way for acceptance of more precise mechanical theories. He put an end, here as elsewhere, to the view that nature has motives, and in its place put the view that she obeys mathematical rules.

And the vortex theory elegantly saved Descartes and others from heresy. Galileo had been brought before the Inquisition in 1633 because his *Dialogues* implied that Copernicus was right and that the earth moves round the sun. The church now insisted that there was only one orthodox opinion: the earth does not move. Of course it does not, said Descartes; it is the vortex round the earth which moves. The earth is a solid body which is stationary in its own vortex and the sun is another solid body which is stationary in its vortex; and all that moves is the earth's vortex which goes round the sun. Since neither the Scriptures nor the commentators had mentioned vortices, there was no heresy in this. At one stroke, Descartes had made Copernicus respectable. In his own tortuous words:

> As to the censure of Rome, concerning the motion of the earth, I do not in this see any resemblance. For I emphatically deny this motion. I do indeed feel that it might be thought at first that I deny it so as to avoid censure for retaining the system of Copernicus. But when my reasons are examined I claim that they will be found to be weighty and well founded.[7]

VII

Was man, too, only extended matter? In the fifth part of his *Discourse on Method,* Descartes explains that man is distinguished from other animals by possessing an immortal soul. Like God, this soul or mind exists only for (that is, is only perceived by) the understanding. Unlike matter, it is not perceived by the senses; man is, therefore, distinguished from extended matter. But except for this soul or mind, man, like other animals, is a machine.

Descartes asks the reader, unversed in anatomy, to dissect an animal in order to see its heart. He contradicts the view of Harvey that the heart is an active muscle, and asserts that it is a "warm organ." At great length, Descartes explains how, from this "warm organ," all the motions and passions of the body can be explained by mechanics. And he declares that, from our knowledge of "automata," we should not be surprised at the human machine, especially when we consider that it has been constructed by God, the master craftsman. Machines lack speech and language; man, however, possesses un-

[7] Quoted in Scott, *op. cit.*, p. 169.

extended mind of which language is the proof and, since it is not part of his body and does not die with him, immortality. This is what makes man different from animal "machines."

Thus, Descartes introduces a deep cleavage between the inward look into mind and the outward look into matter. He offers no solution to the epistemological problem of how the mind knows, other than the vague notion of God as the connecting medium. Instead, he establishes a sharp dualism between mind and matter, in which self-consciousness is the property of mind and extension of matter. Man's passions are rejected as irrational intrusions, and the imagination is distrusted as a source of delusions.[8] It is ironic that the man who had an ecstatic experience, revealing to him the nature of the universe, should introduce into modern philosophy the dualism of mind and body which has plagued thought ever since.

Descartes' outstanding contribution to thought is that from which we began: the introduction into science of the ruthless methods of doubt and of mathematics. His second contribution, however, was a kind of rationalism—of logically deriving effects from causes—which has had great influence and has remained essentially the French way of thinking on all problems.[9] For example, he accepted as fact, without any empirical testing, the notion that the clouds could rain blood, and simply went on to explain this supposed phenomenon.[10] So, too, his theory of the physical universe was a speculative theory in the old and grand manner. It had little contact with experiment, and belongs still to the Greek tradition of what things are, rather than of how things work.[11]

[8] Nevertheless, with all his distrust of the imagination, Descartes had a high regard for poetry. For a discussion of René Descartes' view of man and machines, see Leonora Cohen Rosenfield. *From Beast-Machine to Man-Machine: Animal Soul in French Letters from Descartes to LaMettrie* (New York, 1941).

[9] Cf. F. S. C. Northrop, *The Meeting of East and West* (New York, 1947), pp. 72-73.

[10] See Butterfield, *The Origins of Modern Science*, p. 86.

[11] For the ties still holding Descartes to the scholastic way of thinking, see Etienne Gilson, *Etudes sur le rôle de la pensée médiévale dans la formation du système cartésien* (Paris, 1930). Norman Smith, *Studies in the Cartesian Philosophy* (London, 1902), concerns itself with Descartes metaphysics as distinct from his philosophy of nature, and regards the former as "in essentials scholastic."

Other studies worth consulting are: Henri Gouhier, *Essais sur Descartes* (Paris, 1937), and *La pensée religieuse de Descartes* (Paris, 1924); Albert G. A. Balz, *Des-*

Descartes' rationalism conquered France, but not Germany or England. In Germany, his ideas and method were strongly and successfully opposed by Leibnitz. In England, they were opposed by Henry More (a friend of Newton) and other Cambridge Platonists. By the time that Descartes died, in 1650, the Invisible College was already meeting in England; and its outlook derived from men like Francis Bacon, whose stress had been on experimental science. Descartes was a rationalist whose mind was most powerful in analysis, and alone; it is reported that he did his best thinking by spending his days in bed. By contrast, Francis Bacon was a man of action in public life, with a passion for the factual and the empirical method. Perhaps these two approaches still reflect a basic difference in attitude between English and French scientists. And science cannot develop without the constant combination of the two methods. Bacon's notion that experiment will give results of itself is obviously ill founded. So, equally, is Descrates' notion that the universe can be constructed by thinking alone. The empirical method and the rational method have to go hand in hand to make a science which is both realistic and orderly.

---

cartes and the Modern Mind (New Haven, 1952); James Iverach, *Descartes, Spinoza and the New Philosophy* (New York, 1904); Alexandre Koyré, *Entretiens sur Descartes* (New York, 1944); Ernst Cassirer, *Descartes, Corneille, Christine de Sùede*, traduit (from the German) par Madeleine Francès et Paul Schrecker (Paris, 1942); and Pierre Mesnard, *Essai sur la morale de Descartes* (Paris, 1936), which has an excursus on "Morale et politique: Le Prétendu machiavélisme de Descartes."

CHAPTER 13

The Contribution of Pascal and Bayle

PASCAL

I

THE NEXT great figure in French intellectual history whom we shall discuss is Blaise Pascal (1623–1662). He was a prodigy. Like most prodigies, he exercised his talents in a subject which lends itself to youthful intuition: mathematics. Philosophical systems are almost never the work of young men because philosophy is a subject in which maturity of ideas comes with experience. But there have always been prodigies in subjects like chess, mathematics, and music, in which the structural relations are of a kind with which the mind does well even before exposed to the ordinary experiences of life.

Pascal's natural genius was further heightened by his early training. He was educated by his father, who was himself a gifted mathematician and a friend of the outstanding scientists of the period. At one time president of the *Cour des Aides,* at another, assistant intendant of Normandy, Pascal's father had independent, though not great, wealth and an assured social standing as a member of the *petite noblesse.* A cultured man and a great reader, he determined to bring up his children along the lines indicated by Montaigne.[1]

Originally intending to educate his son in the classics and literature and to reserve the study of mathematics for later, Pascal senior

[1] See, for example, Montaigne's essay "Of the Education of Children."

was turned from his purpose by the discovery that Blaise, at the age of 12, had worked out Euclid's 32nd theorem—that the sum of the angles of a triangle is equal to two right angles—on his own. Henceforth, free rein was given to Pascal's mathematical interests. In accord with this, while still 13 or 14 years old, he was taken to the weekly conferences on science held at the home of Father Mersenne.[2]

Under such a stimulus, Pascal wrote a book on conics before he was 16. Unfortunately, the book as a whole was never published, and we have only fragments from it. However, it was seen in its original form by Descartes and Leibnitz. The former refused to believe that it was Pascal's own work, while the latter hailed it with great admiration.

It was only natural that Pascal's work was in geometry, for this was the mathematical approach appropriate to his time. Thus, one of his discoveries was the famous theorem, named after him, which relates to hexagons, and from which he drew 400 corollaries.[3] We have already referred to Descartes' work in analytic geometry. In an equally important sense, we may say that Pascal provided the foundation for all subsequent projective geometry (i.e., that branch of geometry which deals with the properties in space of configurations which remain invariant under projection).

Pascal was not only a geometer. He made important contributions to physics as well. For example, in his work of 1647, *Expériences nouvelles touchant le vide,* he attacked the Aristotelian notion that nature abhors a vacuum. Shortly thereafter, he proved, by performing the experiment, that the height of mercury in a tube was different at the top and the bottom of a mountain; at the bottom, it was higher. Thereby, he completed the experiment of Torricelli, and showed that the effects attributed to Nature's supposed horror of a vacuum were due instead solely to the weight and pressure of the air.

[2] These meetings served as the origin of the Académie des Sciences. The group of mathematicians and scientists—men like Roberval, Desargues, and Mydorge—meeting at Mersenne's, around 1635, were more or less critical of Descartes' use of deduction rather than demonstration. They wished him to supply empirical verification of his theories. Thus, the same stress on experimental science, as we see, appeared at one time in France and England.

[3] Pascal's famous theorem is the one which asserts that a hexagon inscribed in a conic section enjoys the property that the three meeting points of the opposed sides are always in a straight line.

II

Meantime, while engaged in all this scientific work, Pascal was converted to a mystical religious movement called Jansenism. The movement took its name from Cornelius Jansen, a professor of theology at the University of Louvain, who was a Catholic of deep but anti-Jesuit convictions. He claimed to derive his doctrines from Saint Augustine. Basically, however, the Jansenist movement was an expression of the need for a religion of the "interior soul."

The details of Jansenism are difficult to explain. It was, however, an extremely important movement in France.[4] It resembled the Puritan and Quaker religions which were its contemporaries. Indeed, it might have turned into a form of Protestantism if the field had not already been usurped by Huguenot Protestantism; but it did not. It remained a Catholic revivalist movement, until banned by Louis XIV and the pope in 1713.

As such, and like Savonarola's revival, it demanded the rejection of the world. Naturally, an intense interest in science was difficult to reconcile with Jansenism. Thus, Pascal was torn and divided between the two sides of his nature; later, as we shall see, he was able to reconcile the method of science with the content of religion. At first, however, he had grave difficulties; and at no time was he able to give up completely his scientific work.

Pascal's conversion to Jansenism took place around 1646. This was about the time at which he entered upon a period of very poor health. The lower part of his body was paralyzed and he experienced severe headaches and pains in his stomach. His illnesses, during which he was generally forbidden to study or work, were frequent episodes in his short life, and must always be remembered as background to Pascal's achievements.

Gradually, Pascal recovered his health again. His religious fervor slackened for a time and he turned, partly on the advice of his doctors, toward social life. At this time, around 1649 to 1653, he seems to have gambled, frequented the salons, written verses, and spent his time with libertines and *précieuses.* He read Epicurus and Montaigne and meditated on Stoicism. These were all very important experiences for Pascal. Later, they were to affect both his

[4] For an account of the movement and its doctrines, see the article "Jansenism," in the *Encyclopaedia Britannica.*

work in mathematics (especially in probability theory) and in philosophy.

III

When Pascal was 31, however, an important event occurred which turned him away from the world and its pleasures. He experienced a second conversion to Jansenism. This took dramatic form, on the night of November 23, 1654, in a mystical experience. It was like the experience of Descartes, which we have described; but, unlike Descartes' revelation, Pascal's was purely religious.

Previously, Pascal had embraced Jansenism for intellectual reasons. Now, during an ecstatic period of almost two hours, Pascal's heart *felt* God. He immediately wrote down the details of his illumination in what has come to be known as Pascal's *Memorial.* For the rest of his life he carried the report of his experience, written on a piece of parchment, sewn into the lining of his clothing. A brief glimpse of the *Memorial* gives its exultant flavor:

L'AN DE GRACE 1654

Lundi 23 novembre jour de saint Clément pape et martyr
et autres au martyrologe.
Veille de saint Chrysogone martyr et autres.
Depuis environ dix heures et demie du soir jusques
environ minuit et demi.

FEU

Dieu d'Abraham, Dieu d'Isaac, Dieu de Jacob,
non des Philosophes et des savants.
Certitude. Certitude. Sentiment, Joie, Paix.
Dieu Jèsus-Christ.

. . . .

Père juste, le monde ne t'a point connu, mais je t'ai connu.
Joie, Joie, Joie, pleurs de joie.

His memorable revelation made Pascal decide to abandon the world. He put this decision into effect by surrendering himself to the guidance of Monsieur Singlin, one of the Jansenist preachers, and by making short retreats at the Jansenist convent of Port-Royal.

This second and spectacular conversion of Pascal came at a very opportune time for the Jansenists. One of their professors, Antoine Arnauld, was in a serious fight with the Jesuits at the Sorbonne; indeed, a year or so before Pascal's adherence, five Jansenist propositions had been condemned by the pope. Thus the conversion of a famous mathematician like Pascal greatly aided the Jansenist cause in the public judgment.

The Jansenists, moreover, were aided by Pascal in another way. His retreat at Port-Royal brought him into contact with Arnauld, and led Pascal to write his famous *Provincial Letters*. He wrote eighteen letters and a fragment of another. In these, he defended the Jansenist position (akin to that of Saint Paul and Augustine) on the question of grace and free will. He attacked the Jesuits for their teaching that evil actions undertaken for a good end were permissible. Then, turning from the Jesuits' accommodation to this life (for example, the Jesuits defended, casuistically, the taking of usury), Pascal depicted a model of the simple, true Christian life.

Stylistically, the *Provincial Letters* were a model of ironic prose. Pascal utilized the form of a dialogue. Earlier, Galileo had employed the same form to throw into relief the validity of the Copernican and the weaknesses of the Ptolemaic system. Now, Pascal had his Jesuit character use actual quotations from Jesuit texts, thereby adding verisimilitude to his presumptuous statements. As one commentator on Pascal has remarked, "Demonstration has become drama."[5] Indeed, Pascal's use of the dialogue method of questioning was characteristic of the time. It is, essentially, the dialectic method of doubt.

IV

The *Provincial Letters* demonstrated Pascal's abilities as a polemicist. His fierce, ironic prose had proved itself against the Jesuits. But

[5] Jean Mesnard, *Pascal: L'Homme et l'oeuvre* (Paris, 1956), p. 79. This has a good bibliography; there is an English translation by G. S. Fraser (London, 1952). For an excellent edition of the letters, see *Les Lettres provinciales de Blaise Pascal*, ed. by H. F. Stewart (Manchester, 1920). Other lives of Pascal are: Léon Brunschvicg, *Blaise Pascal* (Paris, 1953); F. Strowski, *Pascal et son temps*, 3 vols. (Paris, 1907–1913); Jean Steinmann, *Pascal* (Paris, 1954); Jacques Chevalier, *Pascal* (Paris, 1922), with an English translation by Lilian A. Clare (London, 1930); Emile Boutroux, *Pascal*, tr. by Ellen Margaret Creak (Manchester, 1902); Roger H. Soltau, *Pascal: The Man and the Message* (London, 1927), which has a short bibliography, especially of works in English; and Léon Brunschvicg, *Pascal* (Paris, 1932), which has sixty wonderful illustrations. For the specifically religious side of Pascal, see F. T. H. Fletcher, *Pascal and the Mystical Tradition* (Oxford, 1954); and H. F. Stewart, *The Holiness of Pascal* (Cambridge, 1915) and *The Secret of Pascal* (Cambridge, 1941).

there was other religious work to be done. Pascal conceived the idea of writing against his libertine and atheistic friends of former times, not only to confound them, but to convert them.

We ought to mention at this point that Pascal was still doing work on problems of probability and gambling. In fact, the correspondence between Pascal and Fermat is the basis for modern probability theory. This work of Pascal was connected with his libertine (i.e., skeptical) friends. One of them, the Chevalier de Méré, had asked him why, in playing at dice, some frequencies come up more often than others. Thus it was Méré who influenced Pascal to work on probability theory; and it was Méré and people like him whom Pascal wished to convert, using, as we shall see, the very idea of probability theory itself.

Pascal, however, had more personal reasons for writing than the desire to convert his friends. In a sense, he was also trying to convert himself. Increasingly in bad health, from around 1658 onward, his tortured mind turned toward theodicy—the vindication of God's justice in allowing evil to exist—and he wrote a *Prière pour demander à Dieu le bon usage des maladies.* In this work he tried to convince himself of God's righteousness in visiting him with disease. In an effort to reach God, Pascal indulged in fleshly chastisement and asceticism. At one point, he believed he had succeeded in talking with Christ.

Tormented with doubt and anguish of soul, wishing to convert himself and others, Pascal wrote, before his death in 1662, at the age of 39, the extraordinary set of thoughts which was published under that name—the *Pensées.* This was not the title which Pascal himself gave to the work; he left it, at his death, unpublished and in rather disorderly fashion. It was first published by his nephew, in 1670, under the title *Pensées de M. Pascal.*[6]

[6] Pascal's thoughts were published after his death in various editions by people who had very sharp axes to grind. For example, the Jansenists, in an early edition, left out everything that Pascal said for which they did not care, and put in a great many things that he never said about which they did care. About a century later, a set of antireligious people, by simply reversing this procedure, were able to produce an equally extraordinary account of what Pascal was thought to have written. Today, however, there are several good editions of the *Pensées.* Fortunat Strowski, *Les Pensées de Pascal* (Paris, 1930), is a study and an analysis. Strowski has also edited Pascal's complete works (Paris, 1923–1931). The other edition of the complete works is by Brunschvicg, Boutroux, and Gazier (Paris,

Clearly, in the *Pensées,* Pascal was greatly influenced by Descartes' belief that one cannot obtain conviction until one has challenged the accepted doctrine at every stage. As a result, Pascal tried to set up a religious theory without using outside authority, and to reach religious certitude through total doubt.

We are accustomed, today, to associate the method of doubt with skeptical conclusions. One doubts the existence of God or the nature of the universe, and one finishes by being an unbeliever, a skeptic. However, this is not the only way of regarding doubt, and it was not the way Descartes or Pascal used the method of doubt.

Let us return to Descartes for a moment. Descartes doubted the accepted traditions and the accepted evidence about the world. Nevertheless, on the basis of his doubt he concluded that both mind and matter exist, and that a good deal could be formulated about them by simply asking questions. In other words, the process of doubting led him to assert the triumph of reason. One could get to almost any depth of a problem by simply thinking and reasoning long enough.

Pascal came to very different conclusions from the same point of departure. Like Descartes, he asked us to look into our own minds, and out at the world, and to ascertain what we can understand. While Descartes came to the conclusion that, as a result of doing this, we can understand everything, Pascal concluded that we discover that we can understand nothing. We can understand neither the nature of the universe nor our own personality. As Pascal said: "Man is altogether incomprehensible by man."

A deep spirit of *Angst* or anguish permeates Pascal's thoughts. Words like "terror' 'and "alarm" abound in his writings. For example, one of the most famous of the *Pensées* is the one in which he confesses to feeling "engulfed in the infinite immensity of spaces whereof I know nothing, and which know nothing of me, I am terrified. . . . The eternal silence of those infinite spaces alarms me."

Throughout the *Pensées,* it is often a puzzle to know whether Pascal is speaking in his own person or in the person of the doubter. The reason for this is that, in reality, they were one and the same

---

1904–1914); Brunschvicg's edition of the *Pensées* is fundamental. An excellent and handy edition of Pascal's works is *Oeuvres complètes,* texte établi et annoté par Jacques Chevalier (Paris, 1954). *Pascal's Pensées* is a French-English edition by H. F. Stewart (London, 1950).

person. They were Pascal in the mood of despair and abandonment and Pascal in the mood of pious resignation. And the mood of resignation to the will of God is, in our judgment, except at rare moments, just as depressing as the mood of more obvious despair.

Pascal was a completely divided and unhappy personality who, the more he discovered, felt the more that he was unable to discover anything. The conclusion he came to was that man really knows nothing. He is powerless. In these circumstances, his only resource is to accept, by faith, the external universe, to believe in the mediating powers of Christ and to embrace the notion of an all-powerful God who controls us completely and into whose hands we must give ourselves.

Pascal defended, and perhaps arrived at, his belief in God by resorting to a notion derived from his gambling friends and his mathematical work on probability. This is Pascal's famous "wager." " 'Either God is, or is not,' we can say. But to which side shall we incline? Reason cannot help us. . . . What will you wager? . . . There is no reason for backing either the one possibility or the other. You cannot argue reasonably in favor of either." The only solution, Pascal argued, was to bet on God. If He exists, you win all; and if He does not, you lose nothing.

In this strange way, Pascal carried on the method of Descartes and solidified it so that in the next century it came to be used by the great antireligious French skeptics as a matter of course. In fact, as we commented earlier, the method of doubt has really been one of the fundamental methods of French thinking ever since. Almost nothing has been accepted on simple authority. This attitude has given a great deal of vitality to French life. It has also led to much trouble in forming stable governments, founding taxation systems to which the people would pay attention, and arranging other such mundane affairs of life.

V

Pascal and Descartes were united on the method of doubt. They differed seriously as to the nature and extent of reason. Pascal really made the distinction himself when he declared that there were two spirits—the "geometric spirit" and the "intuitive spirit." Unlike Descartes, Pascal had learned, from his sojourn in social life, that there was a realm to which geometry did not apply. In fact, it was his

libertine friend Méré who wrote Pascal, pointing out that there were things of the spirit differing from geometry and beyond it.

Pascal reinforced this idea in his own mind from his work in mathematics. He came to realize from his study of the mathematics of infinity that there was a superior logic to the logic of "understanding"; that there were limits to geometry and geometrical reasoning. Thus, he insisted that free will and grace (i.e., necessity) could coexist. It was only our logic of understanding, the "geometric spirit," which separated them. The "intuitive spirit" (*esprit de finesse*), Pascal insisted, allowed for the paradox.[7]

Descartes had tended to scorn the emotions and to dismiss them as a hindrance to reason. Pascal asserted the contrary. Reason is below passions, he said (thus anticipating Rousseau), and he coined the famous phrase: "The heart has its reasons that reason knows not." In his *Discourse on the Passion of Love,* Pascal tried to explain that the logic of the heart was different from the logic of reason, but no less logical.[8] He carried this idea into the *Pensées,* where he declared that there was an art of convincing the understanding (geometry) and an art also of persuading the will (reasons of the heart). Declining to demonstrate the truth of religion as if it were geometry, Pascal tried to convert his reader by making him "feel" religion. In Pascal's view, passions only gave way to passions.[9] It is why "Demonstration" must become "Drama."

[7] This was really an anticipation of Kant. In order to save God, free will, and the immortality of the soul, Kant was forced to set up two realms of knowledge: phenomena (i.e. appearance) and noumena (i.e., things-in-themselves, or reality as it really is). If things only existed as they appear to us, freedom and necessity, for example, could not coexist. Fortunately for Kant's beliefs, however, "the doctrine of nature and necessity and the doctrine of morality and freedom may each be true in its own sphere." This is very close to a rephrasing of Pascal's argument.

[8] There is some question as to whether Pascal actually wrote *The Discourse on the Passions of Love*. For a discussion of this question, in which the evidence seems to show that Pascal did write it, see further Mesnard, *op. cit.,* pp. 57–60.

[9] Pascal's style of writing was based on this view, as well as on the Jansenist belief in clarity and purity rather than ornamentation and formalism of style. Thus, Pascal employed, in turn, the simple and the paradoxical, the ironic and the ridiculous. We should point out here the relationship between the Jansenist emphasis on simplicity of style and the similar emphasis of the English Royal Society; this is pertinent when we remember that the Jansenist influence was, like the Puritan influence in the "Invisible College," very strong in the group meeting around Father Mersenne.

Pascal's views on the art of persuading were linked with his view of man. Pascal wished to deal with the real man, as found in experience, rather than some "ideal" man, set up for us by reason. Experience showed man as a mass of contradictions, as "man the unknown." "A cesspool of uncertainty and error; the glory and the scrapings of the universe—such is man." His reason cannot help him, said Pascal. His only solution is to listen to the promptings of his heart and embrace God. The heart, Pascal informs us, is the best proof of God's existence, for it tells us that man needs God.

What we have just said may give the impression that Pascal completely scorned reason. This is not true. What he did was to erect a dualism of his own in which two realms existed: one of the heart and one of the mind. In religion, unlike Descartes, he applied the logic of the heart. In mathematics and physics, however, Pascal used the same geometry as did Descartes.

VI

Yet there is something in Pascal's science which is different from Descartes'. There is something in Pascal's thought which is essentially modern in its conception of the working of the physical world, whereas Descartes' is essentially scholastic. For example, Descartes' doctrine of the vortices and, in a sense, even the distinction between mind and matter, are very scholastic doctrines. Pascal, on the other hand, was much more aware than Descartes of the fact that, in a nicely organized and mechanical world, it is difficult to know what is the place of thought. In short, Pascal was more aware than Descartes of the epistemological problem which has characterized modern times.

At the age of 19, Pascal had constructed a calculating machine which thrust the whole problem of "what is thought" before him. He tells us that he constructed the machine to help his father, who was obliged in one of his government posts to make a great many calculations. He also hoped to exploit his machine commercially and to this end obtained a royal patent.[10]

After making about fifty models, Pascal finally arrived at a satisfac-

[10] Pascal was by no means averse to engaging in commercial enterprises. He participated in a scheme to drain the marshes of Poitou, and he invested in an omnibus company in Paris. Thus his withdrawal from the world was punctuated by commercial as well as by scientific interests; in this respect, his Jansenism was again much like Puritanism.

tory machine. However, the price of it was too high for its manufacture to be a commercial success. Nevertheless, he built quite a number of machines; six or seven still remain in existence, some of them in working order. A detailed sketch of the machine can be found in Diderot's *Encyclopedia,* and an examination of this, or of one of the existing machines, shows that several arithmetical operations could be performed on it.[11] In principle, it anticipated the modern electronic computator, except that it used gears instead of tubes.

The interesting thing is that, having made the machine, Pascal remarked, "The arithmetical machine produces effects which approach nearer to thought than all the actions of animals." In effect, he was saying that a logical machine is as good a logical reasoner as any human being could conceivably be. At one stroke, this placed the entire Cartesian position in jeopardy, for Descartes had asserted that the only thing differentiating man from the animals and automata was reason.

Even Pascal trembled at his own audacity. He wavered in the face of such a drastic conclusion. Thus, in one of his famous phrases, he referred, in Cartesian terms, to man as the "thinking reed."

> Man is but a reed, the weakest in nature; but he is a reed which thinks, a thinking reed. It is not necessary for the whole universe to arm itself to crush him. A vapor, a drop of water, is enough to kill him. But, were the universe to crush him, man would still be more noble than what kills him, because he knows that he dies, and he knows the advantage that the universe has over him; the universe knows nothing about it.

But, in other moods, Pascal admonished man that he is a paradox; and he commanded, "On thy knees, powerless reason."

Pascal, starting from the method of Descartes, gave it quite suddenly an essentially modern twist. He asked: If machines can duplicate logical thought, if a collection of atoms and cogwheels can "reason," what is it that distinguishes the human mind? Pascal's answer was that the machine "does nothing which can allow us to say that it has will, like the animals." It is not reason, Pascal was saying, but will and self-consciousness which distinguish us from brute matter in the form of a machine.

[11] There is a good sketch and photograph of Pascal's machine in *Oeuvres complètes,* pp. 350–351.

This question, that Pascal asked of himself, has never been satisfactorily answered, even today. Pascal therefore concluded that, the question being unanswerable, we are forced into the realm of faith. This was the essential step of Pascal: that doubt leads to faith, because doubt makes it certain that there is no answer to the question of self-consciousness. The view of Pascal was that the answer to the epistemological question is—that there is no rational answer. We must simply place our bet on faith.

Bayle

I

We have said of Descartes and Pascal that both belonged to the French tradition of the method of doubt. There was a third member of the trinity of doubt, who showed clearly the irreligious implications which the church felt in Descartes and Pascal, in spite of their professed piety. This was Pierre Bayle, who lived from 1647 to 1706, and whom we shall discuss only briefly.

Bayle was one of those people who are important pioneers without real originality or genius. In his own work, for example, he was unable, or unwilling, to distinguish the significant from the insignificant fact, and often gave more time and thorough research to the latter. The originality of Descartes and Pascal was not his; he merely applied their method, doggedly and with courage, to fields they had refused to cultivate.

He also went farther along the road than they did. Thus, the skepticism which led Descartes and Pascal to religious belief was pursued by Bayle with other results. Born the son of a Huguenot minister, Bayle attended a Jesuit college, became a Catholic, and then renounced his new faith. The end of his religious quest was, therefore, skepticism of all organized religions.

From religious skepticism he came to advocate religious toleration. Thus, on the occasion of Louis XIV's revocation of the Edict of Nantes, he wrote demanding freedom of conscience, not only for Protestants (who had enjoyed the Edict), but for all religious believers. Even atheists were partially defended by Bayle, who asserted that superstition was a greater danger to true religion and morality than was atheism.[12]

II

Bayle's religious skepticism found ample expression in his *Dictionary* (*Dictionnaire historique et critique*), which appeared in 1697. This was an extraordinarily popular book with the *philosophes* of the eighteenth century. Indeed, it became one of the models for Diderot's *Encyclopedia,* and, in fact, some of the articles in Diderot's work were simply "lifted" from Bayle's *Dictionary*.

It was not only Bayle's material which was copied; more important, his method was imitated. This was like the Socratic method of Galileo's dialogues or the dialogue form Pascal used, in which people were persuaded of their errors through a subtle and disguised dialectic of doubt. What Bayle did, his peculiar development of the method, was to cross-refer all of his articles, so that the reader was constantly being shuttled back and forth among articles with different headings but the same basic viewpoint. Thus, cunningly, a mosaic of new belief was built up in the often unsuspecting mind of the reader. In this rather gentle way Bayle, for example, disseminated agnosticism in the *Dictionary*.

Bayle was one of the first thinkers openly to criticize, in the name of reason, the ethics to be found in the Bible. Bayle lured his reader into agreement with him by first asking whether a particular, contemporary action was immoral; obviously, it was. Then, Bayle demonstrated that this same action is recorded in the Bible, where it is commended or left uncensored. Rhetorically, he asked whether because David committed the act, rather than, for example, Cesare Borgia, it was any more moral. The seeds of doubt were planted in the reader's mind; henceforth, he would never rest easy or read his Bible in total faith.

It is worth listening to the gentle irony with which, for example, Bayle discusses David.

> David, having dwelt some time in the chief City of King Achish, with his little Band of Six hundred bold Adventurers, was afraid of putting this

[12] Bayle had a personal motive for feeling strongly about religious intolerance in France. His brother died as a result of punishment in a dungeon, where he had been imprisoned for religious reasons.

Prince to Charges, and begged he would assign him some other place of Abode. Achish gave him the Town of Ziklag. David removed thither with his Adventurers, and suffered not their Swords to rust in their Scabbards. He often led them out in Parties, and killed, without Mercy, both Men and Women. He left nothing alive but Cattle, which was the only Booty he returned with. He was afraid, lest the Prisoners should discover the whole Mystery to Achish, and therefore he carried none of them away, but put all to the Sword, both Male and Female. The Secret, he would not have revealed, was, that the Ravages were committed not upon the Country of the Israelites, as he made the King of Gath believe, but upon the Lands of the ancient Inhabitants of Palestine. To speak plainly, this Conduct was very unjustifiable: in order to conceal one Fault, he committed a greater. He imposed upon a King, to whom he had Obligations, and exercised great Cruelty, to cover the Imposition. If any one had asked David, by what authority dost thou these things? what could he have answered? Can a private person, as he was, a Fugitive, who finds shelter in the Dominions of a neighbouring Prince, have a Right to commit Hostilities for his own Advantage, and without a Commission from the Sovereign of the Country? Had David any such Commission? on the contrary, did he not act in Opposition to the Intentions and Interest of the King of Gath? If a private Man, how great soever by Birth, should behave himself now-a-days, as David did on this Occasion, he would, undoubtedly, be called by Names of little Honour. I know the most renowned Heroes, the most famous Prophets of the Old Testament, have sometimes approved the putting to the Sword every thing that had Life; and therefore I should be very far from calling that Cruelty, which David did, had he been warranted by the Order of any Prophet, or if God, by Inspiration, had commanded him to do so: But it evidently appears, from the silence of the Scripture, that he did all this of his own accord.[13]

III

Bayle's major achievement, as great as his impact on religious belief, was in history. He applied the Cartesian method to a field that Descartes had explicitly rejected as the "merely factual." Descartes' use of doubt led him simply to eliminate the entire realm of history on the grounds that it was unsusceptible to certainty. Bayle, however,

[13] *Selections from Bayle's Dictionary,* ed. by E. A. Beller and M. duP. Lee, Jr. (Princeton, 1952), pp. 101–102. This is a very serviceable edition in English; it has an introduction.

did what Pascal had done; he denied that mathematical knowledge was the only kind of knowledge attainable by the method of doubt. Pascal removed from the aegis of Cartesian reason the things of the spirit. Bayle claimed a similar exemption for history, asserting that historical certainty was of another kind from mathematical certainty. Facts of history, such as that the Roman Empire had existed, he pointed out, were as certain as anything in mathematics.

Bayle may have taken his cue from Descartes' comment in the *Discourse:* "Even the most faithful histories, if they do not change or augment the value of things, to render them more worthy to be read, at the least omit almost all the baser and less illustrious circumstances."[14] From this quality of factual distortion in history, Descartes asserted, people drew both misleading conclusions and false principles. Bayle's response to this Cartesian position was to concentrate on the "facts." He did not try to "render them more worthy," but showed instead that the reality of history consisted mainly of crimes and errors. From Bayle's "realistic" response to Descartes, eighteenth-century thinkers derived the notion of history as a record of man's evilness to man, a record enlightened only by the gradual progress of reason.

Descartes had said, too, that we must uproot from the mind all past errors and start as if the mind were void of all previous knowledge. Therefore, following Descartes, Bayle took as his motto the following: "I do not know whether one could not say that the obstacles to a good examination do not come so much from the fact that the mind is void of knowledge as that it is full of prejudice." The historical task was to doubt every fact, to view it as a product of prejudice and superstition and to subject it to the test of reason.

The great enemy was religious dogmatism, which claimed that the true sources of historical fact were the Bible and tradition. Bayle expended great pains in his *Dictionary* to show that the two sources mentioned were empirically not trustworthy. In article after article, he proved concretely the falseness of historical "facts" so derived.

The result was that he accomplished a revolution in historical science. As Machiavelli had freed politics, and Galileo physics, so Bayle freed history from the shackles of theology. In the phrase of Howard Robinson, he achieved the "profanation of sacred history," elimi-

[14] Descartes, *Discourse on Method,* Part I.

nating the artificial division between sacred and profane history.[15] Bayle did this by demanding that reason refuse to accept so-called historical fact until it had first been examined. The historian, further, must serve no prejudice of country or creed but take only truth as his mistress. "He must," said Bayle, "as far as possible, adopt the state of mind of the Stoic who is moved by no passion."

IV

What, however, was the historian to do with his facts once he had ascertained their truthfulness? Bayle did not answer this fundamental question. Like his master, Descartes, who did not say how mind was to know matter, Bayle did not explain how "order" was to be placed upon the facts. His subtle and pervading doubt did not permit him to see any regulating plan in history. The epistemological and philosophical difficulties of history Bayle left to his eighteenth-century followers, like Montesquieu and Voltaire, to solve.

Bayle, instead, conceived his immediate task to be the exposure and uprooting of error rather than the construction of a rational explanation of history. In this, the dictionary form of organization played a significant role, for it permitted the coördination of facts rather than a system of deductive knowledge.

Bayle turned the dictionary form into a delicate means of persuasion. As a result, the *Dictionary* demonstrated that, in spite of Descartes' religious affirmations and in spite of Pascal's mysticism, the view which a skeptical inquiry was generally (and perhaps ultimately) bound to produce was Bayle's civilized and discreet agnosticism. In Bayle's hands, the dialectic method of doubt served a twofold purpose. It served the Cartesian purpose of criticizing tradition. It also served the non-Cartesian purpose of clearing the ground for a "science" of history and of leveling the structure of revealed religion.

15 See Howard Robinson, *Bayle the Sceptic* (New York, 1931). This is the best work in English; it contains a bibliography. J. Delvolvé, *Religion critique et philosophie positive chez Pierre Bayle* (Paris, 1906), is a broad treatment. The relevant pages in Ernst Cassirer, *The Philosophy of the Enlightenment,* should also be consulted.

CHAPTER 14

Voltaire: Science and Satire

We come now to the moment when the French rationalist movement made its juncture with the English empirical-analytic movement. It did this primarily through Voltaire and at about 1730. Voltaire's ancestors in French rational thought bequeathed to him the method of doubt, and Voltaire proved by the body of his work that Descartes had been right in believing that the method which he introduced into philosophical and scientific thinking was far more important than the results he achieved. Voltaire's personal achievement was to link Descartes' method with the ideas of Newton and Locke. Thus, a confluence of British and French thinking (usually labeled the French Enlightenment) was achieved.[1]

I

We do not propose to treat Voltaire in a comprehensive way, but to pick out of his life and work certain highlights. We shall focus on him as the man in whom, in addition to the English ideas, there was canalized the current which flowed through Descartes and Pascal directly into the powerful stream which flooded over into the French Revolution. Although Voltaire died eleven years before the Revolution, it was the effect of his ideas on men like Diderot and Beaumarchais that helped to form the intellectual background for the events of 1789.

Perhaps the most powerful influence on Voltaire's life was the visit

[1] Of course, Voltaire was aided in this task by others, principally Maupertuis, who visited England in 1728 and introduced into France the Newtonian explanation of the solar system, and by d'Alembert.

of almost three years (1726–1729) that he paid to England just before the death of Newton. Newton, an old man at the time, was president of the Royal Society. His death at the age of 84, during Voltaire's stay in England, helped to stimulate Voltaire's interest in Newton's work, in his philosophic background and, particularly, in the universal reverence in which Newton was held in England. From Voltaire's letters, one receives the strong notion that, of all the things which impressed him in England, the most prominent was that a man like Newton enjoyed such widespread popularity and respect. Coming from a France just emerging from the shadow of Louis XIV, Voltaire was painfully conscious of the fact that intellectual distinction did not ensure acceptance or even recognition in French society.

To be in England in the late 1720's, when George II had just come to the throne, was a large and liberal education for a Frenchman. The very methods by which the Georges had become kings must have seemed extraordinary. In France, the Estates General, last convened in 1614, did not meet again until 1789. In England, between 1614 and 1728, one king had been beheaded, another exiled, and two men brought from outside to become kings of England. On all this, we have laid stress in previous chapters.

Voltaire was much taken by the status of the English middle classes, whose ascendancy expressed itself in the Whig Party and in Parliamentary government. He was greatly impressed by the Royal Society and the esteem in which it was held. He noted especially the literary freedom granted Englishmen, and the fact that freedom from prejudice was one of the major characteristics of English society.

Voltaire could appreciate the English social system with particular feeling. He himself was a person of middle-class origins. The main reason for his going to England was a characteristic quarrel with a French noble. One night, at the Paris Opera, the Chevalier de Rohan had tried to insult Voltaire by drawing attention to his family name. "M. de Voltaire, M. Arouet [Voltaire's real name], what's your name?" he taunted. The proud answer, "The name I bear is not a great one, but I at least know how to bring it honor," almost led to a fight then and there. The chevalier, however, bided his time, and a few nights later had Voltaire cudgeled, in a surprise attack, by six of his men. Voltaire's audacious response—a challenge to a duel—outraged the Rohan family, and they had him arrested. Voltaire was able to get out of the Bastille only by agreeing to go to England, i.e., into exile.

It was natural that by the time Voltaire returned to France he was completely converted to what might be called the English way of life. There is a charming, though untrue, story that he thought the first important action on meeting his mistress was to inculcate in her the principles of Newton. His *Philosophical Letters Concerning the English* (*Lettres philosophiques sur les Anglais*) was hailed by Condorcet with extravagant praise, although the French government had it publicly burned as a "scandalous work, contrary to religion and morals and to the respect due to the established powers." All in all, Voltaire was a magnificent propagandist for the English attitude to society and government and for the Newtonian system of science and thought.[2]

How much Voltaire changed France is shown by a remark of the Marquis d'Argenson who, on the eve of dying, in 1757, commented: "Fifty years ago the public was wholly uninterested in the news. Now everybody reads the *Gazette de France*. Much nonsense is talked about politics, but they are no longer ignored. England has conquered us."[3]

II

What, essentially, was the nature of this "Anglicizing" of French thought? We have already indicated, in discussing the Royal Society,

[2] Voltaire backed Newton in opposition to Fontenelle, who, in his *Conversations on the Plurality of Worlds*, still popularized the Cartesian system. For Voltaire's relation to Newton and science, see Ira O. Wade, *Voltaire and Madame du Châtelet: An Essay on the Intellectual Activity at Cirey* (Princeton, 1941) and *Studies on Voltaire* (Princeton, 1947), as well as Margaret S. Libby, *The Attitude of Voltaire to Magic and the Sciences* (New York, 1935). For Voltaire in general, perhaps the best short survey in English is by Henry Noel Brailsford (London, 1935), although its author's dislike of organized religion is given more space than would seem necessary. S. G. Tallentyre, *The Life of Voltaire*, 3rd ed. (New York, 1910), is frequently recommended, but it is written like an historical novel and is often untrustworthy. John Morley, *Voltaire* (London, 1923), may also be mentioned. Outstanding are Gustave Lanson, *Voltaire*, 2nd ed. (Paris, 1910), and Norman L. Torrey, *The Spirit of Voltaire* (New York, 1938) and *Voltaire and the English Deists* (New Haven, 1930).

[3] Quoted in G. P. Gooch, *Louis XV* (London, 1956), p. 252. Voltaire's influence took other people, such as Buffon, the great naturalist, to England to learn at the same fount. In return, Buffon aided Voltaire's popularization of Newton by translating, in 1740, some of the latter's work. Buffon, who visited England shortly after Voltaire, always modeled himself on the English thereafter; that is, instead of talking in the excitable French way, Buffon gravely and slowly nodded his head. He created a great impression in French circles by this means.

that its accomplishment, under Newton, was to give definitive form to the method of analysis, a method which consisted in combining rational deduction and empirical induction. A "systematic spirit" was substituted for the seventeenth-century systems deduced from first principles.[4] Until Voltaire and his friends helped propagate the English analytic and empirical traditions, France had been largely under the spell of Descartes' rationalism. Although the basis of French thought has always remained Cartesian, the Newtonian influence was a welcome leavening.

In expounding the elements of Newton's philosophy, Voltaire extrapolated from the master's work and proposed the application of the Newtonian method of analysis to all knowledge. He advocated a concern with how things work and not with their "essence." For Voltaire, as for Newton, the fact came before the principle.

Naturally, Voltaire's attitude to Newton implied the acceptance of Locke's empirical psychology. Locke was a "true philosopher," and Voltaire eulogized the author of the *Essay* by saying: "After so many speculative gentlemen had formed this romance of the soul, one truly wise man appeared, who has, in the most modest manner imaginable, given us its real history. Mr. Locke has laid open to man the anatomy of his own soul, just as some learned anatomist would have done that of the body."[5] Voltaire's encomium of Locke established the English philosopher as the guiding light of French thought on psychology. Others, it is true, such as Condillac, attempted to go beyond Locke and to dismiss not only innate ideas but also the innate operations of the mind which Locke had permitted to remain; but they did this in Lockean terms.

The result of Voltaire's work, along with that of d'Alembert and Maupertuis, in introducing the ideas of Newton and Locke was to change substantially the direction of French thought. The French, after Voltaire, wished to deal with "practical" matters. Such problems as the freedom of the will and the nature of grace were dismissed as meaningless. The desire of large segments of the French population

[4] Cf. Ernst Cassirer, *The Philosophy of the Enlightenment,* pp. 6–7 and our Chap. 10. Of course, even before the work of Voltaire there was the Gassendi-Mersenne tradition of empiricism; as always, it is a question of degree of influence rather than of a total novelty.

[5] Voltaire, "Locke," in the *English Letters.* For Voltaire's corpus of writings, see *Oeuvres complètes,* ed. Louis Moland (Paris, 1877–1885).

to reform the old regime had been given a solid, philosophical foundation.

III

The desire for reform was a response to the absolutism of Louis XIV, which had left a France morally and physically weakened. A few voices, at the time, had been boldly raised in protest. Fénelon wrote his *Examination of Conscience for a King* in order to point out to Louis himself the abuses of his absolutist reign. In a *Letter to Louis XIV*, Fénelon drew a harsh, and even exaggerated, picture of France in 1691. "Your peoples are dying of hunger. Agriculture is almost at a standstill, all the industries languish, all commerce is destroyed. France is a vast hospital. The magistrates are degraded and worn out. It is you who have caused all these troubles." Fénelon was not alone in his criticisms. Practical administrators, such as Vauban (the military engineer), Boulainvilliers, and Boisguillebert also spoke forth in the seventeenth century. All their words, however, had little effect and reached few people.

By the eighteenth century the atmosphere had changed. The philosophers had begun to do their work. The protests of a d'Argenson, who exposed the internal rottenness of an externally splendid France, in his *Considerations on the Past and Present Government of France* (written and circulated around 1737 but not published until 1764), created, and expressed, public discontent.[6] Thinkers, like Voltaire and d'Alembert, provided the intellectual and metaphysical support behind the desire for economic and political reform entertained by practical men like d'Argenson and, in general, by the middle class.

The link between the empirical attitude derived from England and the French desire for reform was further developed in the *Encyclopedia*. This venture was directly prompted by an English example—Chambers' *Cyclopaedia*—published in 1728, which attempted to summarize, in the main, English technical achievement at

[6] D'Argenson, and others, also prepared the ground for the development of Physiocratic doctrines. The seeds of this school were sown by Vincent de Gournay, who served as director of the Bureau of Commerce around 1751 and took decisive steps toward greater freedom of trade and manufacturing in the practical as well as in the theoretical sphere. We shall deal more fully with the Physiocrats when we come to Chap. 19 on Adam Smith.

the time. Indeed, when Diderot started the French *Encyclopedia* with d'Alembert (who later withdrew), he started with the idea that he would simply translate Chambers' volume into French.

However, the idea grew until it embraced not only the presentation of technological achievements but the general state of contemporary culture. The latter was to be handled along the lines of Bayle's *Dictionary,* which we mentioned earlier as being the real model and inspiration for the method of the *Encyclopedia.* In fact, Diderot declared expressedly that the purpose of the *Encyclopedia* was not only to communicate a definite body of information but to produce a change in the way of thinking: "*pour changer la façon commune de penser.*" Thus, strikingly, we have the connection between the English empirical interests and the French method of doubt consciously present in Diderot's mind as the purpose behind the *Encyclopedia.*

The confused state of French public opinion at the time is exhibited by the details of the *Encyclopedia*'s publication. Starting in 1751, it took almost twenty years to come out and it was alternately welcomed and suppressed by the censors.[7] On one hand, the sale of the *Encyclopedia* made a fortune for its publisher. On the other, it involved Diderot in several imprisonments and occasional confiscations of the plates and volumes.

The amusing story of his having to hide the papers of one of the volumes in the boudoir of Madame de Pompadour, the king's mistress, who was favorable to the *philosophes,* in order to save them from confiscation, illustrates nicely the direct involvement of society in the philosophical conflict of the time. Whereas in England the *Encyclopedia* could have been printed openly and with far more outspoken matter, in France it was hampered and suppressed as a

[7] The director of publications, Malesherbes, was mildly favorable to the enlightened ideas. However, even his approval could be annulled by Parlement or a royal council; thus, if possible, suspect books were printed abroad, in Amsterdam or Geneva. However, many books, like d'Argenson's, were simply copied by scribes in longhand and circulated in manuscript among the members of the fashionable salons. For details of the *Encyclopedia* see the relevant article in the *Encyclopaedia Britannica;* also Louis Ducros, *Les Encyclopédistes* (Paris, 1900). Raymond Naves, *Voltaire et l'Encyclopédie* (Paris, 1938), argues for an unusual interpretation of the relationship between Voltaire and the encyclopedists; he finds "a serious lack of understanding between them, and a parallelism rather than a convergence of efforts."

source of disturbing ideas. In short, philosophy, in England a rather speculative pastime for a few minds, was a revolutionary force in France.

Voltaire was only one of the many contributors to the *Encyclopedia*. The essential creator of the volumes was Diderot. Yet, although Voltaire did not write many articles for the *Encyclopedia*, we stress his role. It was the inspiration which he brought from England which made distinguished Frenchmen feel that the real charge which would transform the French social system was the type of free intellectual approach which the *Encyclopedia* fostered. They felt that the English public mind had been created by the generation of the Royal Society; and they hoped that their writings and the *Encyclopedia* might do the same for France.

IV

The task of the *philosophes* was really twofold: they had to demolish the old system and to construct a new one. In order to accomplish the first purpose—of destroying an old set of beliefs—they metamorphosed the method of doubt into the method of satire. Previously, we have talked about the development of scientific method. More recently, we dealt with Descartes' method of doubt. Now, we wish to concentrate at length on the method of satire used so superlatively by Voltaire and other philosophers.

Satire is a mode of challenging accepted notions by making them seem ridiculous. It usually occurs only in an age of crisis, when there exists no absolute uniformity but rather two sets of beliefs. Of the two sets of beliefs, one holds sufficient power to suppress open attacks on the established order, but not enough to suppress a veiled attack. Further, satire is intimately connected with urbanity and cosmopolitanism, and assumes a civilized opponent who is sufficiently sensitive to feel the barbs of wit leveled at him. To hold something up to ridicule presupposes a certain respect for reason, on both sides, to which one can appeal. An Age of Reason, in which everyone accepts the notion that conduct must be reasonable, is, therefore, a general prerequisite for satire.

We have mentioned satirical works before—notably Erasmus *Praise of Folly*—when discussing the age of the Reformation. Erasmus could write satirically because Lorenzo Valla's critical pen had

cut through the first layers of unthinking faith and convention. The eighteenth-century efforts in this field went even farther than those of the earlier period. Philosophers, writing during an Enlightenment which was more powerful than the Renaissance movement of rational thought because it could base itself on the latter's accomplishments, had an advantage. Thus, they could finish the task of demolishing, by means of satire, the old foundations of state, society, and religion. Descartes continued what Valla had started; and Voltaire took the place of Erasmus.

Earlier, Galileo and Pascal had used the dialogue form to cast ridicule upon their opponents; and, later, Beaumarchais will employ this method in his plays. Now, starting as it were from Descartes' comment, that travel among varied peoples will unburden our minds of the notion "that everything against our modes is ridiculous, and against reason," Montesquieu and Voltaire inverted his suggestion. Travel, they claimed, showed that not other people's customs but our own are ridiculous. Thus, the imaginary travel story became the favorite form of satire for the eighteenth century.

V

Let us take our initial example from Voltaire's predecessor in this field, Montesquieu. We shall deal with Montesquieu's political and historical ideas in the next chapter, but here we shall concentrate on his work in satire. He wrote the first really successful work in the genre of the satiric travel story. In fact, Montesquieu made his great reputation as a very young man, when, in 1721, he published, anonymously and from outside France, the book called the *Persian Letters*. The book soon had many imitators. For example, Goldsmith made money and a reputation in England sometime afterward by publishing a book obviously modeled on Montesquieu's, but which purported to be written by some Chinese visitors.

Montesquieu's method is simple. His book of travel professes to be written by two Persians, Rica and Uzbek, who are visiting France. They write home and, in the course of their letters, manage to depict as extraordinary all the customs which seem quite natural to Frenchmen. Thus, with straight faces, they make fun of the conventional aspects of social life at the time—the habits of dress, the ways of behavior and the formalities of salon life.

It is because the *Persian Letters* start, not by talking of the church and the king, but by asking why it seems so obvious to Europeans for men to wear trousers when in Persia they are worn only by women, that it is so effective. Only after the reader has been jollied along, by progressively making him challenge all the accepted things of the time—how people have their hair cut and why they wear wigs—can he be made to ask himself more pertinent questions. Only then can he be drawn to ask whether, for example, the Persian is right in describing the pope as a conjuror; for the Persian "naïvely" states that: "There is another magician [The first one is the King of France, who, "If he has only a million crowns in his treasury and needs two, . . . has only to say that one crown equals two and they believe him"], . . . who is called the Pope, who makes people believe that three are only one, that the bread one eats is not bread or that the wine one drinks is not wine, and a thousand other things of the same sort."[8] By degrees, in this way, the large doctrinal issues of the Roman Catholic Church are then exposed to the same type of questionings. These, if put on the first page, would have been sure to make the unsympathetic reader close the book.

Instead, the *Persian Letters* enjoyed a great popularity, especially as Montesquieu did not hesitate to add to his satiric and ironic gifts an appeal to the salacious interests of his readers; thus there are slightly scandalous passages dealing with life and gossip in a Persian harem. As a result of his skillful blending of styles, Montesquieu's work was perfectly calculated to appeal to the polite, urbane, and sophisticated members of French salon society.

VI

At about the same time that the *Persian Letters* were published in France, a number of outstanding satires were also published in England. For example, Swift published *Gulliver's Travels* in 1726, and Pope's *Dunciad* and Gay's *Beggar's Opera* came soon after.

We have pointed out that Voltaire visited England during this period. We know, too, that Voltaire met Gay and saw his *Beggar's Opera* before it appeared on the stage, and that he enjoyed the society of Pope and Swift.[9] He even encouraged his friend Thieriot to translate *Gulliver's Travels* into French.

[8] Letter 24.

[9] Cf. J. Churton Collins, *Voltaire, Montesquieu and Rousseau in England* (London, 1908), p. 36.

Thus there is reason to believe that Voltaire was influenced, not only by Montesquieu and the French tradition of doubt and irony, but also by the powerful social satire which these defeated Tory writers of England were letting loose on the public. May we, then, attribute not only much of Voltaire's science but, in some measure, his satire to his trip to England in 1726–1729? If so, the confluence of English and French thought which took place at this time would include style as well as content.

VII

We have stressed, until now, only the destructive aspect of the French satirical "travel" literature. There was also a constructive purpose behind these works. Comparative travel literature led to the assertion of a natural religion or morality, which existed above and beyond the particular habits and customs of a particular faith. The *philosophes* made the claim that the world of men and their ways could be subjected to Cartesian reason and be made to produce "laws" as clear as those of geometry. These would be "natural laws," evident to all men of common sense. As Voltaire said: "Even though that which in one religion is called virtue, is precisely that which in another is called vice, even though most rules regarding good and bad are as variable as the languages one speaks and the clothing one wears; nevertheless it seems to me certain that there are natural laws with respect to which human beings in all parts of the world must agree."

It was on this positive assertion of the existence of natural laws and, therefore, of natural rights amid the welter and confusion of varied customs, that revolutionary doctrine could be built. Voltaire and Montesquieu, as we shall show, wanted reform, and would have been aghast at the notion of revolution. Yet, the assertion that men enjoyed certain "natural rights" which were above local custom and tradition contained the seeds of universal revolution. Thus, the American Revolution was marked by the Virginia Declaration of Rights and the Bill of Rights, and the French Revolution by a Declaration of the Rights of Man and the Citizen.

Little of this, however, was consciously in Voltaire's mind when he wrote. He had in mind the church. He wished to show that a natural religion and morality existed, common to all men everywhere, of which the Catholic Church was only a particular and, at that time, corrupted embodiment. His famous war cry was "*Ecrasez l'infâme.*" He meant by this, war against the infamy of church superstition, not against natural religion.

The church, which claimed to be catholic and universal, tended to be intolerant and parochial. Against this, Voltaire, following Pierre Bayle, took up the cause of tolerance. Like Erasmus, he aimed his shafts, tipped with wit, against the foibles and stupidities of the Catholic priesthood. To clear the memory of Jean Calas, a Protestant, who had been barbarously executed in 1762 on the false charge that he had murdered his son to prevent his conversion to Catholicism, Voltaire wrote his famous *Treatise on Toleration* (1763). He succeeded, after three years, in having the decision in the Calas case reversed by the king's court of appeals.

Voltaire's own religion was what is called Deism. He believed in a personal God, but a God who did not interfere, after the initial creation, in the world of men with miracles and in response to prayers. The world created by Voltaire's God operated according to natural laws, such as those discovered by Newton. In this sense, then, Voltaire's "natural" religion made God, for all practical purposes, a natural rather than a supernatural Being.

VIII

In a supposedly harmonious and lawful universe, how do we explain the existence of evil? On this profound question of theodicy, Voltaire found himself in opposition to the facile optimism of men like Leibnitz. Against the view that "all is well in this best of all possible worlds," Voltaire placed the "real" world of his hero, Candide. The book is too famous to need paraphrasing. It is enough to say that wandering from Westphalia, through Paraguay and El Dorado, and finally ending in Turkey, Candide experienced disasters almost all the way. *Candide* is, like Montesquieu's *Persian Letters,* a magnificent satire in the form of a travel story.

In a brilliant analysis, the critic Erich Auerbach has laid bare the elements of its style. He points out that Voltaire employed a blinding mixture of the comic and the tragic. He coupled two dissimilar ideas in an unexpected way, simplified complex problems into a black-and-white contrast, and stated truths which are only half-truths.

Above all, it is the tempo in which Voltaire manipulated all these elements that explains his success. In Auerbach's words: "Especially his own in his tempo. His rapid, keen summary of the development, his quick shifting of scenes, his surprisingly sudden confronting of things which are not usually seen together—in all this he comes close to being unique and incomparable; and it is in this tempo that a good part of his wit lies."[10]

Candide was prompted by the kind of disaster of which it is full: the seemingly inexplicable Lisbon earthquake of 1755 which killed over 20,000 apparently innocent people. How was this to be reconciled with a good God? In a poem written shortly after the event, "The Lisbon Earthquake," Voltaire took up the notion of Pascal that man, a "thinking reed," can comprehend nothing. But Voltaire gave a materialist twist to Pascal's formula, by calling man a "thinking atom," and made man's resignation a stoical rather than a pious religious one. In his rather straightforward poetry, Voltaire told of

> Atoms tormented on this ball of clay,
> The sport of death, of hazard's strokes the prey,
> Yet thinking atoms, atoms whose clear eyes
> Guided by thought have measured out the skies,
> Into the infinite we fling our gaze
> Yet cannot see ourselves, nor count our days. . .
>
> Humbly I sigh, submissive I await
> Without a challenge the decrees of fate.
>
> . . . .
>
> Groping in darkness for a guiding light
> No murmur shall escape me in the night.[11]

In *Candide* itself Voltaire took up the same attitude. He held that the riddle of God's purpose is unanswerable. " 'But to what end was this world formed?' said Candide. 'To infuriate us,' replied Martin." Even when Candide finds "the country where everything is for the

[10] Erich Auerbach, *Mimesis* (New York, 1957), p. 356; see further pp. 353–364. The work by William Bottiglia, *Voltaire's 'Candide': Analysis of a Classic* (Geneva, 1959), is definitive on Voltaire's prose style.

[11] We have used the translation in Brailsford, *Voltaire*, p. 154. A full, and different, translation of the poem can be found conveniently in *The Viking Portable*, ed. by Ben Ray Redman (New York, 1949); Redman's introduction is also very interesting.

best"—El Dorado—he becomes bored rapidly with its equality and equanimity and leaves for the outside world. The world is a mixture of good and evil, Voltaire informs us. Three great evils, boredom, vice, and need, confront us. The solution to them is to "cultivate our garden," for through work the three evils are kept at bay. Voltaire's final advice reflects the influence of English empiricism: "Let us work without theorizing, 'tis the only way to make life endurable."

Voltaire's balanced, common-sense, middle-class view of things is also put forth in another, earlier tale, *Babouc; or, The Way of the World* (1746). In this book, Voltaire stressed the fact that from man's evil passions come his creative impulses; the moral weaknesses of man are the necessary concomitants of his refined culture.[12] Rather than denying Pascal's anguished cry that man is a mass of unknowable contradictions, Voltaire accepted the notion but made of it man's highest glory. It is in man's contradictory nature—good and evil—Voltaire suggested, that we must seek the source of the manifold human achievement.

Voltaire, however, rejected Pascal's belief that evil was a result of man's original sin. On this basis, Pascal had believed that there could be no change in man's unhappy status as long as he remained man. But Voltaire, with his age, believed in progress. Scientific ideas had replaced religious ones. No longer, it was thought, need man atone for his original sin here on earth. Instead, man might, and should, attempt to improve his temporal existence by reforming his institutions and manners. Work and projects were to take the place of ascetic resignation and facile optimism. In Voltaire's mind, the existing world was by no means the "best of all possible worlds"; but, at least, it could be made better.

IX

The fact that progress was possible and that, in reality, a betterment of the world had occurred could be demonstrated by a study of history. Such a study would illustrate the triumphant march of man's reason; that is, if it were properly handled. Thus, Voltaire turned to

[12] We shall see later that Rousseau dealt with exactly this point, and concluded that the price is too high. Earlier, Mandeville, with his *Fable of the Bees*, had shocked his contemporaries by claiming that economic well-being depended on the "evil" and self-interested part of man's nature.

history. He did so in order to study, not the innumerable incidents, but the "spirit of the age."

His first history, written in 1728, dealt with Charles XII. The subject of Voltaire's second history was not a single man but an entire age: *The Century of Louis XIV* (1754). His third major historical work, the *Essay on Customs* (*Essai sur les moeurs et l'ésprit des nations*), is his most important, and admirably shows the broadening of Voltaire's vision. It aimed at depicting and explaining the causes for "the extinction, revival, and progress of the human mind" from Charlemagne to Louis XIII.

The *Essay on Customs,* begun for Madame du Châtelet in 1740 and published finally in 1756, evolved an original methodology of history by concentrating on the cultural rather than the political achievements of man. Voltaire set himself the task of showing "by what stages mankind, from the barbaric rusticity of former days, attained the politeness of our own." He wrote, therefore, about opinions and the enlightenment of the human mind which alone made "this chaos of events, factions, revolutions, and crimes worthy the attention of wise men."

Equally original in the *Essay* is Voltaire's empirical attitude. That is, Voltaire tried to explain events in terms of man, not Providence; in terms, therefore, of natural or secondary rather than so-called divine or primary causes.[13] Thus he applied to history the Newtonian concern with how things work rather than with their essence.

In the *Essay on Customs,* Voltaire also introduced the phrase "philosophy of history." By this he meant that history was to be looked at as by a *philosophe.* He wished the critical apparatus of a Bayle to be utilized for the study of society, thus duplicating in that field the work of a Newton in natural science. This, an application and extension to the study of modern times of Bayle's similar critical treatment of more ancient history, was Voltaire's third contribution to the development of the human sciences.

Voltaire did not try to show that there existed "laws" in history. He did, of course, make generalizations. But even his generalization to the effect that the human mind was progressing did not rest on

[13] See Robert Flint, *History of the Philosophy of History* (New York, 1894), pp. 289–304, for a full discussion of this point and of Voltaire's historical work in general.

any law that Voltaire perceived in human affairs. Rather, Voltaire believed human progress to be the result of a happy accident; and this fitted in with his view that God's purpose was inaccessible to man's knowledge.

In his histories, then, Voltaire employed a secular, critical, and empirical approach to a subject matter which he took to be man's total culture and civilization.[14] Nevertheless, in spite of his "scientific" proclivities he did not neglect colorful narrative in his histories. He did not bog down in a collection of dry facts but made history an amusing, lively, living thing. When his work was criticised by another historian, Voltaire commented: "A historian has many duties. Allow me to remind you here of two which are of some importance. The first is not to slander; the second is not to bore. I can excuse you for neglect of the first because few will read your work; I cannot, however, forgive you for neglecting the second, for I was forced to read you." Needless to say, Voltaire was never boring. His empiricism was not pedantic and it was always witty. Indeed, Voltaire's "science" of history was always as much satire as it was science.

X

History showed that men were capable of becoming enlightened. But what was the motor force behind this "progress of the human spirit"? How were the desired reforms to be brought about? By heroes or by the common people? Here, in the thought of Voltaire, we meet a curious contradiction.

From the time of Leonardo and Machiavelli, men had been fascinated by the condottiere, the brutal, roughshod, and yet heroic individual. Voltaire marks the temporary end of this tradition. In clear and determined words, he declared: "I do not like heroes; they make too much noise in the world." Here was the announcement that the link between the intellectual and the prince, between the man of thought and the man of action, had been broken.[15]

[14] It ought to be mentioned that, in the execution of this purpose, Voltaire often faltered badly; for example, he consistently and seriously underestimated the positive role of religion in shaping culture.

[15] Cf. Paul Hazard, *La Crise de la conscience européenne* (1680–1715), 2 vols. (Paris, 1935), Vol. II, pp. 126–127, for the growing substitution, in the course of the seventeenth century, of the ideal of the sage or of the bourgeois for the ideal of the warrior or the hero. In this movement of ideas, Fénelon and Bayle anticipated Voltaire.

Yet, at the time, the divorce was not fully recognized by the eighteenth-century philosophers. This blindness led them into what amounts almost to a glaring contradiction. Thus, even though Voltaire had written his denunciation of heroes to Frederick the Great, in 1742, after the latter's victory in a battle, he persisted in believing that reform would come about under the hands of an enlightened despot.[16]

The idea of democracy in the modern sense was foreign to Voltaire, and his letters are filled with references to the "rabble." The *philosophes,* as the French historian A. A. Aulard says, "had no intention of confiding the government of the nation to what we call universal suffrage: a thing so strange to the thinkers of the eighteenth century that it had not, so far, a name."[17] Further, Aulard points out that the illiteracy rate, as shown by the 1788 *cahiers,* was enormously high; thus, the mass of the people would have been insensible to the philosophic propaganda.

In any case, the *philosophes* could not bring themselves to trust the people. They even felt that, to keep the masses from rebelling, religion was necessary. In a letter to Frederick the Great, Voltaire declared: "Your majesty will do the human race an eternal service in extirpating this infamous superstition [Christianity], I do not say among the rabble, who are not worthy of being enlightened and who are apt for every yoke; I say among the well-bred, among those who wish to think."

It may be that Voltaire was purposely appealing to the known opinions of Frederick, who despised the populace and called them "that cursed race."[18] In any case, there can be little question that

16 This was one of the decisive points on which Voltaire and most of the other *philosophes* broke with Montesquieu; they believed that absolute power, when enlightened, would secure to man all the necessary benefits of reform, whereas Montesquieu distrusted such an unchecked, centralized power. In a recent book by Peter Gay, *Voltaire's Politics* (Princeton, 1959), the view is stated that "Voltaire can be counted as at most an occasional and uncertain supporter" of enlightened despotism (p. 8, n. 15). But further statements, for example on p. 89, seem to support our position. Gay's is an important book on Voltaire's politics; its bibliography is very complete.

17 A. A. Aulard, *The French Revolution: A Political History,* tr. by Bernard Miall, 4 vols. (New York, 1910), Vol. I, p. 121.

18 See further W. F. Reddaway, *Frederick the Great and the Rise of Prussia* (New York, 1904), p. 161.

Voltaire and the majority of his fellow *philosophes* were not democratic revolutionaries. They favored reform imposed from above. In the phrases of de Tocqueville, "They wanted to employ the power of the central authority in order to destroy all existing institutions, and to reconstruct them according to some new plan of their own device; no other power appeared to them capable of accomplishing such a task. The power of the State ought, they said, to be as unlimited as its rights; all that was required was to force it to make a proper use of both."[19] The force which was to compel a right use of power was "enlightenment." Like Luther, turning to the princes; like Erasmus, believing in the princes' education (this time, however, not Christian), the *philosophes* placed their hopes in the despots.

At first, events seemed to bear out the *philosophes'* expectations. It became the fashion for kings and queens to pose as enlightened rulers; thus, Catherine of Russia, Joseph II of Austria, Charles III of Spain, and innumerable others held salons and patronized the philosophers. Frederick the Great, for example, expressed, as we have seen, "enlightened" views of religion, practiced toleration and considered himself to be the "first servant of the state." One might say that he collected great philosophers in the way his father before him collected outsized soldiers.

Gradually, however, it became obvious that the Enlightenment, as far as the rulers were concerned, had turned into another "court masquerade." In France, the failure of the Turgot-Necker program of reform from above discredited the advocates of enlightened despotism.

If the despots could not be properly enlightened, to whom should the *philosophes* turn? Why, if he believed that all men were possessed of natural common sense, did Voltaire not try to enlighten all men? Was it not a contradiction to assert that "there are natural laws with respect to which human beings in all parts of the world must agree," and then to call the people "rabble"? Fiercely to denounce heroes, and then to extol despots?

Part of the explanation may be that Voltaire was a "snob," who spurned those below him and wished to reach those above him. It may be, too, that Voltaire wrote for the salons, and in terms of the

[19] Alexis de Tocqueville, *The State of Society in France Before the Revolution of 1789*, tr. by Henry Reeve, 3rd ed. (London, 1888), pp. 59–60.

salons, because no other public was available to him.[20] In any case, Voltaire, with his practical common sense, cultivated the garden at hand. Unable to enlighten the people, he carried his light where he could.

But the light of reason was stronger than Voltaire had thought. It exposed the evils of state, society, and religion to the glare of publicity. It created a mood of reasoned dissent from traditional authority. When the suggested means of reform, enlightened despotism, failed to function, men turned to harsher means—revolution.

In his intentions Voltaire may not have been revolutionary. He was certainly revolutionary in his ideas. His combination of English empiricism with the French method of doubt was an explosive one. Thus, the same thoughts which formed, and were formed by, the English seventeenth-century revolutions, flowered anew into revolution (joined with other factors, of course) when carried by Voltaire to the soil of France. Voltaire, indeed, had cultivated his own garden, but with unexpected results.

20 As we shall see, however, Rousseau broke this set of circumstances by attacking the culture of the salons and appealing to those "common people" who could read

CHAPTER 15

Montesquieu

I

MONTESQUIEU PUBLISHED the *Persian Letters* when he was 32 years old. They were an instantaneous success. In the course of the year 1721 they went through eight editions. One reason for their success was that Montesquieu had hit exactly the right moment for publication: the Regency, which came into power after the death of Louis XIV, had reacted against the Sun King's reign of absolutism and religious hypocrisy and was prepared to laugh at the foibles of the old regime. Further, Jean Chardin had recently published his *Voyage to Persia,* and, in general, interest in the Orient was at its height in France.[1]

The *Persian Letters,* with their witty irreverence and their exotic flavor, were bound to appeal to a society in which boredom was accounted the worst evil. Montesquieu's writing was eminently readable. His use of short, aphoristic paragraphs (often consisting of no more than one sentence) and his trenchant, crisp language made for an immediate effect. Unfortunately, there was a danger hidden in such a style. It tended to become a series of pompous generalizations, and Montesquieu's later works, where he employed his aphoristic style in more serious vein, increasingly showed this vice. For the satiric purpose of the *Persian Letters,* however, Montesquieu's style was perfect.

[1] The first part of Jean Chardin's *Voyage en Perse et autres lieux de l'Orient* appeared in 1686, but the completed work, in ten volumes, was not published until 1711, in Amsterdam. Earlier, around 1701, a translation of *The Thousand and One Nights,* which Montesquieu read carefully, had also been published.

As early as the *Letters,* Montesquieu sounded the note which was to run constantly through all his works: a concern with virtue and liberty. We see the theme clearly in the unforgettable tale of the Troglodytes. A friend writes from Persia to one of the two travelers, Usbek, asking whether men are happier as a result of enjoying pleasures and the satisfaction of the senses or as a result of practicing virtue. Usbek responds, in Letter 11, that "There are certain truths which it is not sufficient to advocate [intellectually], but which it is also necessary to make one feel: such are the truths of morality. Perhaps this bit of history will affect you more vitally than a subtle philosophy."[2] Usbek then proceeds to tell the tale of the Troglodytes.

The Troglodytes are a people who have no principle of equity or justice. They kill their king and elect magistrates in his place; then they kill the magistrates and, in a state of natural savagery, think only of themselves. Their motto is: "I will live happy." On this basis, as Usbek shows, consulting only their particular interests rather than the general interest, the Troglodytes suffer economic ruin, the loss of their property, and the destruction of their families. Visited by a plague and cured by a foreign doctor, they do not pay him. When the plague recurs, the doctor returns to the land of the Troglodytes, this time, however, not to cure them but to tell them that they have no "humanity" and therefore deserve to die.

Usbek continues the story of the Troglodytes in Letter 12. Only two families, who, virtuous and honorable, had withdrawn to a remote spot to work and to raise their children, survived the plague and the end of the Troglodytes. These two families raise their children "to feel that the interest of individuals always finds itself in the common interest; to wish to separate oneself from it is to wish to destroy oneself." Intermarrying, the two families grow and prosper, living the happy, bucolic life and piously worshiping their gods. In lyric fashion Montesquieu, in the person of Usbek, describes the felicitous pastoral existence of the Troglodytes.

2 Montesquieu's attitude seems to indicate the influence of Pascal. It is also of interest, in this connection, that the *philosophes,* like Voltaire, considered history to be philosophy teaching by example. For Montesquieu's writings, see *Oeuvres complètes de Montesquieu,* ed. by André Masson (Paris, 1950–1955), and the *Pléiade* edition of the *Oeuvres complètes* (Paris, 1947), which has an excellent preface by Roger Caillois, who sees a "révolution sociologique" already prefigured in the *Persian Letters.*

By Letter 14, however, they appear to have become too happy. They decide to choose a king, and approach their wisest and most virtuous citizen, a venerated old man. To their surprise, he turns on them and cries, "Your virtue begins to weigh on you . . . this yoke [of virtue] appears too heavy to you; you prefer to be subject to a prince and to obey his laws, which are less demanding than your [virtuous] customs." And with that, the old man scorns the honor they wish to confer on him and weeps over the degeneration of his people.[3] Thus ends the tale of the Troglodytes.

In this tale, Montesquieu has presented his belief that the best life is not the hedonistic, libertine existence but the virtuous one. He has also stressed the moral that liberty and virtue are indissolubly tied. Later, in the *Spirit of Laws,* he will insist that virtue is the dominating principle in a republican form of government, and that, where virtue fails, freedom disappears into the hands of a monarch or a despot. Thus, in the very first and in the very last of his works, Montesquieu sounded the same dominant theme—the theme of virtue and liberty.

II

Montesquieu's ideas were presented more formally, and perhaps more seriously, in the two major works which followed the *Letters*—the *Considerations on the Grandeur and Decadence of the Romans* and the *Spirit of Laws.* To prepare for the writing of these works, Montesquieu acquainted himself at first hand with the political and social institutions of many European countries. He left Paris in 1728 and traveled through Germany, Austria, Hungary, Italy (especially Venice, Milan, Turin, Florence, and Rome), Switzerland, the Rhine country, the Low Countries, and finally, England. The greater part of Montesquieu's time was spent in England; he arrived there from Amsterdam in Lord Chesterfield's yacht in 1729, just after Voltaire had returned to France, and stayed until 1731.

Montesquieu's main interest was in the English political constitution. Although earlier concerned with science, he now paid little heed

[3] The contemporary psychologist Erich Fromm has written a book with the provocative title *Escape from Freedom* (New York, 1941), which sounds the same theme: that the burden of freedom and virtuous self-conduct is too much for modern man, who wishes to escape from it into a mass movement like fascism.

to Newton and Locke or to physics, psychology, or epistemology.[4] Unlike Voltaire, Montesquieu was impressed more by the position of the English aristocracy than by the status of scientists and businessmen. Himself a noble of the robe, former president of the Parlement of Bordeaux (a post he inherited from his uncle, who had it from Montesquieu's grandfather), Montesquieu hobnobbed primarily with his aristocratic compeers in England. He attended the sittings of both Houses of Parliament, was presented to the king, and was made a Fellow of the Royal Society. What he saw in England enchanced Montesquieu's prejudice for the nobility as a "balancing" power in the state. All in all, however, the impact of England on Montesquieu was by no means as forceful or important as it had been for Voltaire.

Returning to France in 1731, Montesquieu retired to his estate at Brède to write a study on the greatness and decline of Rome.[5] The subject he really had in mind was the larger question of what causes the decline and fall of *any* nation—although one feels that he was thinking especially of France. In fact, the *Considerations,* published in 1734, actually served the purpose of trying to awaken France to reform and recovery. What Montesquieu did was to employ the method of comparative history instead of comparative travels (as in the *Persian Letters*), but with the same references to contemporary affairs in his mind.

Montesquieu's *Considerations* is one of the earliest and finest examples of the "decline and fall" literature which seems to have so strongly attracted the eighteenth century.[6] His work, however, was

[4] In 1716, elected to the Academy of Bordeaux, Montesquieu changed its focus from an interest in music to a concern with science. He himself read papers on physics and zoology, especially the latter, at its meetings. He also advertised his intention of writing a history of the earth and actually sketched such a work in 1719. Apparently, he turned from the pursuit of science because of his weak eyesight; in any case, by the time of his English visit he seems to have lost his earlier interest in science. See further Désiré André, *Sur les écrits scientifiques de Montesquieu* (Paris, 1880). There is an excellent article by Henry Guerlac, "Three Eighteenth-Century Social Philosophers: Scientific Influences on Their Thought," *Daedalus* (Winter, 1958), which deals with the scientific interests of both Montesquieu and Voltaire.

[5] At Brède, Montesquieu also produced wine. The sale of his books, as he himself confessed, greatly aided the sale of his wine.

[6] Edward Gibbon's *The History of the Decline and Fall of the Roman Empire* is so well known that we need only mention it as a second example.

not a reflection of pessimism or despair. Although he was convinced of the mutability of all things (in the *Spirit of Laws,* he said about England, "As all human things have an end, the State of which we are speaking will lose its liberty. Rome, Lacedemonia, and Carthage have all perished"), Montesquieu believed that the decay of institutions can be checked if we know the causes. He attempted in his study of Roman institutions to discover these causes. He believed that the past can teach the present because men were the same then as they are now; that men always have the same passions and that, therefore, human institutions eternally function in the same way. In the *Considerations,* Montesquieu said: "Modern history offers us an example of what happened at that time in Rome, and this is very noteworthy: for, as men have always had the same passions in all times, although the occasions which produce great changes differ, the causes are always the same."

Montesquieu shared, it is clear, the eighteenth-century belief in strict causality, and extended it to the sphere of history. On the basis of such a faith, he could attempt to work out the laws of historical mutation. The stimulus to his work, as we have already indicated, was the desire for the reform of France and its political and social institutions. Thus his work was optimistic and forward-looking rather than, as in such twentieth-century examples as Spengler's *Decline of the West,* despairing. In the *Considerations,* Montesquieu wished to teach the French by the example of the Romans. Later, in the *Spirit of Laws,* he hoped that the French would learn from the English. Otherwise, he knew, the French would lose out to the English (for this, after all, was the main power conflict of the eighteenth century).

The *Considerations* is not, strictly speaking, a history of Rome but a series of considerations or comments on Roman history; in fact, if we may pun a bit, it assumes a considerable knowledge of Roman history.[7] Most of the salon members who would read Montesquieu's book had this classical background, and shared in what one recent

[7] For Montesquieu's sources of Roman history see A. J. Grant, "Montesquieu," in *The Social and Political Ideas of Some Great French Thinkers of the Age of Reason,* ed. by F. J. C. Hearnshaw (London, 1930), pp. 116–117; also, see Roger B. Oake, "Montesquieu's Analysis of Roman History," *Journal of the History of Ideas* (January, 1955).

author has called the "cult of antiquity."[8] The events of Caesar's life were as familiar (if not more so) to well-educated Frenchmen of the eighteenth century as those of Louis XIV, and references to the classical past which today mean nothing to us carried then tremendous overtones.[9]

In many ways, the *Considerations* may be thought of as a book of virtues and vices. Thus, during the rise of Rome to greatness, we are informed of the virtues which "caused" this political flowering, and, in the decadent period, of the causal vices. In general, Montesquieu focused on the virtues. Instead of Rousseau's "noble savage," he presented a picture of the "noble Roman." This noble Roman was virtuous, warlike, devoid of greed, avid of honor, and enamored of his country or city-state.

The virtues of the Roman institutions, however, were even more important than the virtues of the men, for the institutions were what guided the men. Thus, the sources of Roman greatness lay in the willingness to adopt other people's customs when they were better; in the constant exercise of the art of war (basing himself on Vegetius, as had Machiavelli, Montesquieu analyzed the "Art of War among the Romans"); and in the making of liberty the soul of the state (for this was what "moved" the state).

Of notable importance was the equitable division of land among the Romans. This meant that the farmer was a soldier and that the pursuit of luxuries did not exist. Like Rousseau after him, Montesquieu believed that the arts emerged upon the appearance of inequality: "When the laws [regarding equitable property] were not strictly complied with, the same occurred then as today among us; the avarice of some and the prodigality of others caused property to fall into a few hands, and so, at the start, arts were introduced through the mutual needs of rich and poor."

The decline of Rome occurred because its aggrandizements led inevitably to the corruption of the Roman virtues. Anticipating the *Spirit of Laws,* where he generalized the notion that a republic of vir-

[8] See further Harold T. Parker, *The Cult of Antiquity and the French Revolutionaries* (Chicago, 1937).

[9] We have mentioned the earlier use of the name Junius Brutus by the anonymous author of the *Vindication Against Tyrants*. See, on the entire subject, Gilbert Highet, *The Classical Tradition* (Oxford, 1949).

tue could only exist in a small territory, Montesquieu asserted that the Roman Republic was destined to perish once its principle (i.e., virtue) was violated. With the introduction of mercenary armies and landless masses, it was "only a question of how it was going to die and through whom it would be overturned." Sulla, Pompey, or Caesar, the Republic was doomed.

Montesquieu denied emphatically that class warfare was the cause of Rome's decadence. For him, the rivalry of patrician and plebeian was a sign of health. He insisted, in words which have importance for today, that

> Historians speak only of the divisions that destroyed Rome; but they do not see that these divisions were necessary, that they had always existed and would necessarily always exist. The expansion of the republic was what envenomed the situation and converted popular tumults into civil wars. . . . A general rule is that, always, in a state which calls itself republican, when absolute tranquility reigns, you can be assured that liberty does not exist there. What is called unity in a political body is a very ambiguous thing; true unity is a harmonious one, through which all the parts, opposed as they may appear to us, concur in the general good of the society; like dissonances in music, they concur in the total harmony.

To talk of unity in a despotism, Montesquieu declared, was to talk of a "dead" unity.

Rome declined, not because of political divisions, but because it lost its warlike and other virtues; because it violated its geographical principle and overextended itself (thus necessitating despotism, which meant a consequent decadence); because it allowed its money to be drained to the East, whereas in its ascent it had absorbed money from all its conquered provinces; and, finally, because the riches it did have corrupted it.

As a result, Rome degenerated into Epicureanism. It forgot the Stoicism which its best emperors (like Marcus Aurelius), and Montesquieu, so much admired. (Indeed, so enamored of Stoicism was Montesquieu that he risked severe church disapproval of his work by including in Chapter 13 a laudation of the Roman custom of suicide.) Epicurean (i.e., hedonistic) and without virtue, the Romans had set foot on the same path as the Troglodytes. Their doom and extinction was only a matter of time.

III

In the same way that Machiavelli's *Prince* was a fragment of a larger work, the *Discourses,* so, essentially, the *Considerations* was only a partial development of the *Spirit of Laws.* Published in 1748, the *Spirit of Laws* occupied Montesquieu, as he himself tells us, for over twenty years. It was intended as the copestone of his effort, setting out many of the ideas to be found in the *Letters* and *Considerations,* although in more generalized form, as well as many new ideas. Its popularity was immediate; it had twenty-one editions in eighteen months. Perhaps it has been allowed unduly to overshadow the *Considerations.* In any case, it exposes in heightened form, at one and the same time, the values and the weaknesses of Montesquieu's thought.

In his preface, Montesquieu asks us to judge the book as an entirety and not to approve or disapprove of some paragraphs. He is asking us, really, to pass decision on his method rather than on the correctness of particular facts that he may cite.[10] This method is stated clearly and unequivocally in the very first chapter of the *Spirit of Laws.* The statement is aimed especially against Hobbes and shows the strong influence on Montesquieu of Descartes.

> Laws are the necessary relations which derive from the nature of things; and in this sense, all beings have their laws: the divinity has its laws, the material world its laws . . . man has his laws. Those who have said that a blind fatality has produced all the effects that we see in the world have uttered a great absurdity; for what greater absurdity than a blind fatality which has produced intelligent beings. Therefore, there is an original reason; and laws are the relations which are found between it and different beings, and the relations of these beings among themselves.

Hobbes had considered laws as being of human institution, rather than deriving from the "nature of things," and had uttered the "great absurdity" about blind fatality in the course of developing his materialistic metaphysics.

From the "nature of things," Montesquieu set himself to derive the principles or laws which express necessary relations. Specifically, as he tells us in the full title of his book, he wished to derive "The Spirit

[10] For a discussion of the correctness of Montesquieu's facts, as they related to the field of travel literature, see Muriel Dodds, *Les Récits de voyages: Sources de l'Esprit des Lois de Montesquieu* (Paris, 1929).

of Laws, or of the Relation that the Laws should have with the constitution of each government, the customs, climate, religion, commerce, etc." Montesquieu had faith that, amid the "infinite diversity of laws and manners" to be found among men (discovered during his travels and his study of history), there were simple principles. "I have posed principles," Montesquieu reported, "and I have seen particular cases fall under there as of themselves, the history of all nations being only the consequences thereto."

Voltaire had been concerned with the "progress of the human mind." Montesquieu took legal and political institutions for his subject and focussed on the "spirit" which animates them. In his view, the major political institution is the government, and it is from the form of government that the "spirit of laws" derives.

Montesquieu singled out three forms of government: republican, monarchical, and despotic. The republican form, he postulated, is one in which the people, or a part thereof, govern; this definition permitted Montesquieu to classify aristocratic government as republican. Where one man governs according to fixed, established laws, we have monarchy. Where there is government by one man without laws, we have despotism.[11]

The form, or nature, of government is its particular structure; its principle is that which makes it move, or, more fully, the "human passions which make it move." In democracy, the principle is virtue (by this Montesquieu means political, not individual, virtue). In aristocracy, it is virtue to which is added moderation. In monarchy, laws take the place of virtue (as we have seen in the story of the Troglodytes), and honor is the moving principle. In despotism, fear is the moving principle. Montesquieu was careful to indicate that these are "ideal" types. It is not that a republic has virtue, he pointed out, but that it *should* have virtue, for otherwise "the government will be imperfect."

Each of the principles of the three governments has its corruption (Montesquieu persisted in talking about the "three governments" although he really treated aristocracy as a fourth form). Democracy

[11] Montesquieu deviated here from the traditional division of governments, based on the number of people governing, into monarchy, aristocracy, and democracy. When he came, as we shall see, to the principles animating these forms of government, he treated aristocracy as having a separate principle from the republican form

is corrupted by either the excess or the loss of the spirit of equality and virtue; aristocracy, when the nobles lose moderation and become arbitrary; and monarchy, when the prerogatives and privileges of corps and cities are removed, resulting in a despotism of all or of one. Despotism, itself, is corrupted, in principle, from the beginning. It can only pursue its corrupt path downward. Once the principle of government is corrupted, Montesquieu said, the best laws become bad.

In addition to his general theory of the forms and principles of government, Montesquieu presented a description and an analysis of the English constitution. What he had to say about the English form of government had a powerful influence on his contemporaries.

Montesquieu chose the English constitution as a case study because, as he explained, it had as its direct object political liberty. Liberty he defined as "the right to do that which the laws permit, and if a citizen can do that which they forbid, he would no longer have it, because the others also have this power." There is only one way to preserve liberty, Montesquieu claimed, and that is, as in England, by the balance and separation of power. Liberty can only, therefore, be "found in moderate [i.e., aristocratic] governments," where, "by the disposition of things, power stops power."

Montesquieu did not distinguish clearly enough between the mechanical division of power in the government and the real distribution of power in society. But the distinction is implicit in his work. Montesquieu divided power, mechanically, into executive, legislative, and judiciary power; all liberty, he claimed, is lost when "the same man, or the same body of princes, or of nobles, or of people, exercise these three powers." He also asserted that even the joint exercise of the executive and legislative powers by one man or one body would suffocate liberty; and he said this even though at the very time, in the England of 1730, Walpole and his cabinet were exercising both. We must remember, however, that Montesquieu was dealing with an "ideal" government.

Montesquieu's other division of power is into three social forces: the king, intermediaries, and the people. The intermediary power, in Montesquieu's theory, is the aristocracy of the "privileged." These men, distinguished by birth, riches or honors, should have, according to Montesquieu, a role in accordance with their value; to be judged by the people would enslave them. Thus, the nobles should be hereditary and in an upper house, where they might check the people's enterprises.

On the other hand, the people, in the lower house, would be able to check the deeds of the nobles. Montesquieu openly acknowledged the class rivalry between the two orders and admitted that they had "separate views and interests." But this, as we saw earlier in the *Considerations,* was what gave the state life and preserved it in a healthy "harmony" or balance of forces. Once again, then, Montesquieu has expressed his belief in the virtue and necessity of "political divisions," one of the key beliefs of modern liberalism.

IV

What specific political position did Montesquieu hold and what overall effect did his theories have? He was certainly not a democrat. The separation and balance of powers was opposed to democracy, and, in Montesquieu's scheme, social or political change could only come about through the concert and agreement of the three powers. He was, then, a liberal conservative or a conservative liberal, arguing in favor of a limited, balanced, constitutional government, i.e., government on the English model.

Did Montesquieu's theories go farther than this in either their intentions or their effects? He, himself, made it very clear in the preface to the *Spirit of Laws* that he had no revolutionary purpose in mind. "I do not write to censor that which is established in any country whatsoever. Each nation will find here the reason for its maxims." Montesquieu cautioned that if the people were enlightened they would realize that faults and virtues are tied together and they would proceed moderately. In fact, he added, his book would give those who obey a new pleasure in obeying. How rare it would be, he exclaimed, if the laws which suited one nation were to agree with another nation whose climate, terrain, and history are so different.

In spite of these obeisances in the direction of the *status quo,* Montesquieu was a dangerous innovator. It was all very well to state that countries must have different customs and laws; but to extol the English constitution for its "political liberty" seems to cast some doubt on the sincerity of Montesquieu's equal tolerance for all forms of government. In sum, his work showed clearly that he preferred a free, mixed government.[12] We see in this matter a conflict between Montesquieu the "scientist" and Montesquieu the "philosopher." The scientist must accept all forms as inevitable results of the "nature of things"; the philosopher sets up an "ideal type" to which he hopes reality can be made to correspond.

Basically, of course, Montesquieu's intentions, like Voltaire's, were reformatory rather than revolutionary. His ideas had, nevertheless, a great influence in both the political revolutions of the eighteenth century: the American and the French. In the framing of the American constitution, as well as in pre-Revolutionary propaganda, the theory of the separation and balance of powers was consciously invoked. The effect of the *Spirit of Laws* in France was to increase dissatisfaction with the old regime and to encourage men to seek liberty on the English example.

Voltaire had believed in reform by an enlightened despot; as we have seen, the despots betrayed their role and left the task to other hands. Montesquieu put his faith in the aristocracy and backed the Parlements against the monarchy.[13] But events were to betray his trust. It was the French aristocracy which first started the Revolution. In 1787, the Parlements opposed the monarchy's efforts to effect needed reforms. Instead, they insisted on the calling of an Estates General which they thought the aristocracy would control. When the Estates General met in 1789, it was the people, not the aristocracy, which came into power. Instead of balancing the state, the aristocracy had helped to overturn it.

Like the earlier humanists, Montesquieu had contributed, along with Voltaire and the other *philosophes,* to forming an atmosphere of reasoned dissent. He and the philosophers, however, were not in a position to control the desire for reform. It was Rousseau who, so to speak, "backed the right horse," and provided a philosophical justification for democracy which had been lacking in both Montesquieu and Voltaire.[14] For Montesquieu, the closest realization of his theory

12 Does this mean that Montesquieu thought that the principles of the four types of government could be mixed? The point is not made clear in his work.

13 Thus Montesquieu placed himself in the same camp as, for example, the Comte de Boulainvilliers, in favor of aristocratic control of the government, against, for example, the royalist Abbé Dubos and the Physiocrats.

14 Of course, without the preparation provided by the works of Montesquieu and Voltaire, Rousseau's efforts would have fallen on barren soil.

came, not in his own country, but in the United States of America, a country without an aristocracy. Such is the irony of the history of ideas!

V

Let us turn our attention, now, from the more immediate political consequences of Montesquieu's theories to their long-range methodological and philosophical importance. He himself thought that his work broke completely new ground. On the first page of the *Spirit of Laws,* Montesquieu placed the epigram *Prolem sine matre creatam* ("Child born without a mother").

It is doubtful, actually, whether Montesquieu's work was as original as he thought. Even if he had not borrowed from Vico, as some have claimed, the latter had preceded him in many places. And so, indeed, had Machiavelli.[15] Nevertheless, Montesquieu's pen did scratch new lines in the literature of human studies.

His achievement in history, all too often overlooked today, rests mainly on the *Considerations.* In that book, especially, he helped to introduce the historical method into modern thought. His special insight was that institutions undergo growth and decay and must be judged in reference to the period in which we observe them. Thus, he attempted to enter the "spirit of the age" with which he was dealing and not to judge it, as did Voltaire, on the standards of the eighteenth century.

When he came to write the *Spirit of Laws,* however, he lapsed. He forgot his own lessons from the *Considerations.* As one eminent critic has pointed out, he paid "little attention, generally none, to

[15] For a discussion of the relations of Montesquieu to Vico, see Flint, *op. cit.,* pp. 265–266. In a comparison of Montesquieu and Machiavelli, there are some striking parallels. Both men were engaged in writing commentaries on Roman history. They were both concerned with the art of war and used Vegetius as their basis of reference. They were both desirous of discovering "laws" in history, and they both believed that laws were possible because men everywhere had the same passions. And, in the broadest sense possible, their purpose was the same, inasmuch as they both intended to create a "science" of politics. (Montesquieu, like Machiavelli, also wrote lighter works. Thus, for the satisfaction of the more frivolous interests of the salons, he penned romances, such as *The Temple of Gnide* and *Arsace and Ismène,* noteworthy only because written by the author of the *Persian Letters* and the *Spirit of Laws.*)

the chronology of his facts, which is, however, the indispensable condition of their comparison."[16] Thus, Montesquieu's comparisons were "ideal," timeless comparisons; but this was to lose all the value of the historical method which he had so painstakingly developed in his work on the Romans. We see this clearly in his treatment, in the *Spirit of Laws,* of legal and political institutions as static things, mainly affected by conditions of size and climate, i.e., geography. The time factor is neglected, except in the notion that when the principle of a government is corrupted, decay follows. This, however, is merely an "ideal" decay in a timeless vacuum.

Montesquieu's betrayal of his own historical method, however, helped him to create another and newer field of study: sociology. In this sense, it can be said that his work was a "child born without a mother." As the modern French sociologist Emile Durkheim tells us, Montesquieu gave to the new field both its proper object of study and the path by which it should proceed in that study. He perceived that social facts were legitimate objects of science, and that they were subject, like any other facts, to general laws. He also saw that all social facts are related parts of a whole, and must be judged and evaluated only in their context. Finally, he laid down that the correct method to be pursued in the new science was comparison and classification; and, in relation to the latter, he set up the notion of social types.[17]

These, as we have seen, were "ideal types." From the point of view of history, they were faulty and represent a step backward in method. However, from another point of view, the introduction of "ideal types" marks an advance in the method of social science. The "ideal type" is a powerful tool of analysis, which permits us to seek the principle which underlies a mass of differing appearances.

[16] Flint, *op. cit.,* pp. 267–268.

[17] Durkheim's critique of Montesquieu is to be found in his doctoral thesis, reprinted, in a French translation from the original Latin, in *Montesquieu et Rousseau, précurseurs de la sociologie* (Paris, 1953). Other estimates and treatments of Montesquieu are to be found in Albert Sorel, *Montesquieu* (Paris, 1887), which, though frequently recommended, is an overrated book; Louis Vian, *Histoire de Montesquieu* (Paris, 1879), which should be used with caution, especially in relation to Montesquieu's trip to England; and F. T. H. Fletcher, *Montesquieu and English Politics, 1750–1800* (London, 1939), which covers a very broad range of material in very suggestive fashion and stresses less the influence of English thought on Montesquieu than the influence of Montesquieu on English writers, especially of the conservative persuasion.

Montesquieu was partially aware of the ambiguities involved in his method. He constantly stressed the fact that he was discussing not what is, but what is *ideally*.[18] Thus, he stated that he was not concerned with whether England *actually* enjoyed the liberty that he had described, but only that it was established by law; a glance at Montesquieu's *Notes on England* will show that he was keenly aware of how corrupt and unrepresentative eighteenth-century England was in reality.

He used "ideal types" because, with all its drawbacks and limitations, the "ideal type" permitted him to penetrate to the essential structure of political organisms and to observe clearly the interplay of the forces which animate them. We use it today for the same reason. It is, therefore, a tool which modern sociology has gratefully borrowed from Montesquieu.[19]

All in all, then, Montesquieu's prime contribution was to sociology rather than to the philosophy of history. His unique achievement was to "travel" in the spheres of custom and law, first for purposes of satire and then for purposes of science. In this, his mind reflected the cosmopolitan attitude, the international interests of the eighteenth century.[20] The result was that he could compare the institutions, not only of France or of Europe, but of all societies and, in his best moments, compare them impartially.

True, he was too Cartesian in his approach and, too often, used the facts to support a priori principles. This was his particular weakness. It did not mean that his general method was incorrect. Others could improve it by adding empiricism to his rationalism, and by working out a more scientific sociology. Perhaps motherless, Montesquieu's "child" was not to be without a progeny of its own.

[18] This is not quite the same thing as "what should be," and must be distinguished from the point we made earlier about the philosopher setting up an "ideal type" to which he hopes reality can be made to correspond; the latter is the Platonic aspect of Montesquieu's use of an "ideal type." For a further discussion of this whole aspect of Montesquieu's method, cf. Durkheim, *op. cit.*, pp. 91–93.

[19] "Ideal types" have been used especially by Max Weber, although in a way somewhat different from that of Montesquieu. See Max Weber, *The Theory of Social and Economic Organization,* tr. by A. M. Henderson and Talcott Parsons (New York, 1947), especially pp. 12 ff.

[20] See further Muriel Dodds, *op. cit.*, p. 174.

VI

In many ways, however, Montesquieu was a prophet without honor in his own time. He did not belong really to the circle of the *Encyclopedia.* He was unpopular, in spite of the rationalism of his *Spirit of Laws,* with the *philosophes* around Voltaire. They resented his notion that what suited man in one climate might not suit him in another; that there was no universally valid solution to the question as to how man should live in a political society. Of all the *philosophes,* only Diderot was in the funeral cortege which followed Montesquieu in 1755.

Montesquieu's own age had first hailed him as the writer of a brilliant satire, the *Persian Letters,* which questioned the established institutions and extolled liberty and virtue. When the *Spirit of Laws,* which questioned not only the established ideas of the old regime but the new, universal ideas of the *philosophes* themselves, came along, Montesquieu's contemporaries experienced a sense of apprehension. Montesquieu had seemed one of them; now they felt, somehow, that he had betrayed his own time.

They were right. Montesquieu was not totally of the "spirit of his age." Indeed it was this very fact which made him more attractive for the nineteenth century than almost any of his contemporaries. Thus, in the later period, Montesquieu's achievements could be valued anew for what they really were—not merely eulogies of political liberty and virtue, but solid stepping stones to a more adequate comprehension of history and to the new science of sociology.

CHAPTER 16

Rousseau

JEAN-JACQUES ROUSSEAU was the revolutionary, the impertinent, who, for the first time, directly and effectively challenged the accepted rationalist view held by the enlightened century in which he lived.[1] He made a real breach in that long tradition of reasonableness which, building up in North Italy before 1600, dominated the French and English academies in the seventeenth century and was carried on actively by Voltaire and the Encyclopedists in the eighteenth century. Partly under Rousseau's pounding, the formal structure of French salon life gave way to a more equalitarian society, and its belief in science and satire yielded to a view which seemed to glorify instinctive, irrational, and emotional behavior.

Rousseau was an odd and complex character, unable to come easily to terms with himself or his society, and his ideas reflect this fact. We shall picture him in three settings: as the moralist who attacked the notion that progress results from advances in science and technology; as the "outsider" who fathered the romantic sensibility and opposed it to the dominant rationalism; and as the revolutionary thinker who first inscribed on the political banners of modern times the opposing slogans both of democratic and of totalitarian government.

Rousseau's new set of ideas—for they were all related—unless worked into the rational system of society, threatened to blow it up. Thus, the second half of the eighteenth century was the scene of a precarious attempt by Western man to reconcile reason and emotions, science and morality, and to create for himself a society in

[1] Montesquieu also made the challenge, but his was not direct and it was by no means as effective, at least immediately, as Rousseau's.

which he could live the balanced life. This had been Rousseau's personal problem; he made it, irrevocably, that of modern man.

I

Rousseau was born in Geneva in 1712 and lived to the age of 66. The main source for the history of his life—private as well as public—is his *Confessions,* one of the world's great autobiographies. It is the candid, sometimes shocking, revelation of the mental and emotional processes of an exceptional person. He, himself, claimed to be unique, and on the first page of the *Confessions,* one reads: "I have resolved on an enterprise which has no precedent. . . . I am made unlike any one I have ever met; I will even venture to say that I am like no one in the whole world."

His mother died in childbirth, and Rousseau was raised by his father and an aunt. The psychological effect on Rousseau, of growing up without a mother's love, seems to have been very strong, and he suffered constantly from a feeling of not "belonging." Whatever the reason, his character exhibited a strange mixture of truculence and a slightly pathetic appeal for comfort from others. Later in life he got a good deal of comfort from women, whose maternal instincts he often aroused and to whom he turned constantly.

His father was an independent, almost ne'er-do-well character, who tried to be a dance master in a Geneva where dancing was forbidden, and who left his wife, shortly after marriage, to spend five years in Constantinople (it was a year after his return that Jean-Jacques was born). Professionally, Rousseau *père* was a watchmaker, and thus among the only high-precision technologists of the time. He was a well-read man as well, and Rousseau recalled his father "with the works of Tacitus, Plutarch, and Grotius lying before him in the midst of the tools of his trade." Then, thinking of himself, Rousseau continued, "At his side stands his dear son, receiving, alas with too little profit, the tender instructions of the best of fathers."[2]

[2] Jean-Jacques Rousseau, *The Social Contract and Discourses,* tr. with introduction by G. D. H. Cole, Everyman's Library ed. (London, 1947), p. 150. This is a very handy volume of some of Rousseau's work, including the Discourses on the *Arts and Sciences, Origin of Inequality,* and *Political Economy.* The standard edition of the political works is C. E. Vaughan, *The Political Writings of Jean-Jacques Rousseau,* 2 vols. (Cambridge, 1915), which supplies excellent notes to the French text; Vaughan's longish introduction emphasizes the collectivist aspect of Rousseau. For the rest of Rousseau's writings, see *Oeuvres complètes de J.-J. Rousseau,* 13 vols. (Paris, 1905).

The "best of fathers," wounding a man in a quarrel, fled Geneva. Rousseau was committed to the care of an uncle and then put out as an apprentice to an engraver. At 16, perhaps depressed by the discipline, Rousseau abandoned his job and the city of Geneva. He became a convert to Catholicism and began his relations with the woman, Mme. de Warens, whom he was henceforth to call "Mamma."

It is not our purpose to describe in any detail Rousseau's "picaresque" existence, so well related in the *Confessions*. We need point out merely the "unattached" quality of his early life. He went about (like Lazarillo de Tormes) with a holy beggar; he became a music master in Lausanne, and served as a footman, a surveyor, and a tutor. No wonder that in *Emile* he tried to devise an education that would permit a man to fill any post to which fortune called him![3]

Around 1742 he went to Paris to seek his fortune. His project for a new system of musical notation was heard by the Académie des Sciences but not considered of particular originality. Undeterred, Rousseau, after a short stay in Venice as secretary to the ambassador, settled in Paris with pretensions to be a philosopher and a man of letters. He had a small success in writing operas, attached himself to Diderot and became a habitué of the salons.

However, he was never really at home in the gay, sophisticated life of Paris. He was clumsy and boorish and unable to think of witty remarks until the occasion had passed. In a strange, complicated way, he was morally shocked at the lascivious and indolent tone of salon life, although he himself lived in sin with a former barmaid by whom he had five children, all sent off to the foundling hospital. In spite of his own moral lapses, Rousseau felt that he was by nature a good man; and he came to believe that it was only the warped, unnatural character of modern life which had corrupted him and all French society.

II

The real crisis in his life and beliefs occurred on a hot October afternoon in 1749. It was similar in its ecstatic quality to the episode of Descartes' suddenly realizing that the world was to be resolved by mathematics and to the experience of Hobbes in picking up the

[3] See Rousseau, *Emile ou de l'éducation*, nouvelle édition par François et Pierre Richard (Paris, 1951), p. 12.

copy of Euclid in the library of the Cavendishes. It gave to Rousseau the vision of a world which he was henceforth to try to bring into existence.

In 1749, the Academy of Dijon had advertised that a prize would be given for an essay on the following subject: "Has the progress of the arts and sciences tended to the purification or to the corruption of morality?"[4] Rousseau was on his way to visit his frend Diderot, then a prisoner at Vincennes for violating the censorship, when he read the advertisement in the *Mercure de France.* "The instant I read it, I saw another universe and I became another man," he confessed. Later in life, he wrote to M. de Malesherbes describing the incident:

> If anything was ever like a sudden inspiration it was the impulse that surged up in me as I read that. Suddenly I felt my mind dazzled by a thousand lights; crowds of lively ideas presented themselves at once, with a force and confusion that threw me into an inexpressible trouble; I felt my head seized with a vertigo like that of intoxication. A violent palpitation oppressed me, made me gasp for breath, and being unable any longer to breathe as I walked, I let myself drop under one of the trees of the wayside, and there I spent half an hour in such a state of agitation that when I got up I perceived the whole front of my vest moistened with my own tears which I had shed unawares. Oh, Sir, if ever I could have written even the quarter of what I saw and felt under that tree, with what clarity should I have revealed all the contradictions of the social system, with what force would I have exposed all the abuses of our institutions, in what simple terms would I have demonstrated that man is naturally good, and that it is through these institutions alone that men become bad.[5]

Inspired by such feelings, Rousseau decided to enter the Dijon competition. In a red-hot, gospel-like essay (one commentator has

[4] Often, in the eighteenth century, the word "arts" was used when referring to technical skills. The French used the expression *arts et métiers* when they wished to talk about the mechanical arts, and in England a Society of Arts and Manufactures was founded in 1754. With the same meaning, the American Academy of Arts and Sciences was founded in 1779. Rousseau, in the *Discourse on the Arts and Sciences,* appears to have used the word "arts" pretty much as we use it today but with a slight overtone of the technical aspect of the word clinging to his usage.

[5] Charles William Hendel, *Citizen of Geneva: Selections from the Letters of Jean-Jacques Rousseau* (New York, 1937), p. 208. Hendel's book has both a 126-page biographical sketch of Rousseau and about 250 pages of some of his best letters in translation. In French, see *Correspondance générale de J.-J. Rousseau,* publiée par Th. Dufour et P.-P. Plan, 20 vols. (Paris, 1924–1934).

perceived in it the tone of a Geneva preacher attacking the Whore of Babylon)[6] he indicted the arts and sciences for corrupting morality and all of life.

The Academy of Dijon, in all probability, had not expected this answer. Nevertheless, it awarded the prize to Rousseau; and it was awarded to him for what were, certainly in the eyes of *philosophes* like Voltaire and the Encyclopedists, as well as of the manufacturers in England, thoroughly outrageous sentiments.[7]

III

Rousseau's essay was the most important challenge to science since the Inquisition's sentence on Galileo in 1633. Science had had pretty much its own way for over a hundred years from that time. A few men, like Swift and Pope, had grumbled about science, but they had not done it with anything like Rousseau's profound conviction.

In itself, the *Discourse on the Moral Effects of the Arts and Sciences* is not one of Rousseau's best pieces, as he, himself, was aware: "Full of strength and fervor, it is completely lacking in logic and order; of all my works it is the weakest in argument and the least harmonious. But whatever gifts a man may be born with, he cannot learn the art of writing in a moment." Rousseau did the entire thing again, and better, later on. But 1749 was the critical moment.

Although Rousseau claimed, "It is not science, I said to myself, that I am attacking; it is virtue that I am defending, and that before virtuous men—and goodness is ever dearer to the good than learning to the learned," he really was attacking science. Even further, he was saying that civilization and cultured society, that is, a society based

[6] F. C. Green, *Jean-Jacques Rousseau* (Cambridge, 1955), p. 104. This is the most recent full-scale work on Rousseau and is a critical study of his life and writings. Although the author's distaste for his subject creeps into the book occasionally, it does not prevent Professor Green from giving a subtle and penetrating psychological analysis of Jean-Jacques. The result is a very valuable treatment of Rousseau which complements the *Confessions* by full use of the correspondence and all other sources. Arthur Lytton Sells, *The Early Life and Adventures of Jean Jacques Rousseau, 1712–1740* (Cambridge, 1929), must be read, of course, in conjunction with Rousseau's own accounts in the *Confessions, Rêveries du promeneur solitaire,* and *Dialogues: Rousseau juge de Jean-Jacques*. Another life in English is R. B. Mowat, *Jean Jacques Rousseau* (London, 1938).

[7] See Roger Tisserand, *Les Concurrents de J.-J. Rousseau* (Paris, 1936), and Marcel Bouchard, *L'Académie de Dijon et le premier discours de Rousseau* (Paris, 1950), for the other entries as well as for a discussion of the Dijon Academy.

on the arts and sciences, had produced the present evils of man.

Such a formulation permitted Rousseau, in his *Discourse*, and other works, to shift the theodicy problem from God to society.[8] "Everything is good as it comes from the hand of the Creator," Rousseau was later to declare in the opening lines of *Emile*. "Everything becomes evil in the hands of man." Further on, he added, "Man, seek no further the author of evil; the author is yourself."

By fixing the cause of evil in society rather than in man's original sin (as done, for example, by Pascal) or in God, Rousseau "secularized" the problem of evil. He accomplished for the theodicy problem what Machiavelli had done for politics and Galileo for science: freed it from theology. In so doing, he took the momentous step of disclosing a new vista wherein man, by reforming his society, could perfect himself.

In his own life, he set out to act upon this conviction. Placing before himself a new pattern of existence, he realized that he must withdraw from the world of society. Only in solitude, in isolation, would he find his true nature, uncorrupted by bad institutions. Thus Rousseau went into retreat at various idyllic spots. He gave away his fine clothing and dressed henceforth as a simple "citizen," without white stockings or sword. He even sold his watch, that fundamental instrument, practically and symbolically, of modern mechanical civilization.[9] Toward the end of his life, he commented of himself in the third person: "It was one of the happiest moments of his life when, renouncing all plans for the future in order to live from day to day, he got rid of his watch. 'Thank heavens!' he exclaimed in a transport of joy, 'I shall no longer need to know what time it is!' " Time had created all the institutions of corrupt society; Rousseau, in becoming an "anarchronist," had sponged it all away. Society could now be reconstructed, on a "natural" footing.

IV

Rousseau was very conscious of being "different" and out of harmony with the main tenets of his age. In *Emile*, he confessed: "I do

8 Cf. Ernst Cassirer, *The Philosophy of the Enlightenment*, pp. 154–156.

9 Was this Rousseau's symbolic rejection of his father, a watchmaker? Or a repetition of his father's rejection of the watchmaking trade?

not see as do other men." In the preface to his 1749 prize essay, he admitted that "Setting myself up against all that is nowadays most admired, I can expect no less than a universal outcry against me." He compared himself with Socrates and anticipated the same fate.

Like the Greek philosopher, too, Rousseau spoke in "praise of ignorance." In Rousseau's hands the praise of ignorance became a condemnation of the arts and sciences. He compared the *philosophes* to "mountebanks" and called printing a "pernicious art." Was Rousseau indulging in satire when he applauded the burning of the books at Alexandria? We doubt it. Rather, all of Rousseau's ideas seem to follow logically from the fact that he placed morality above knowledge and virtue above luxury and art.[10] In his prize essay on the arts and sciences, therefore, he was saying that knowledge and art are not to be free. They are to be subordinate to social needs and the moral life.

This particular aspect of Rousseau's thought did not horrify his contemporaries. Most of them agreed with him that increased knowledge ought to serve a moral purpose. The disagreement came when Rousseau rejected the elegant, cosmopolitan culture of his day as immoral and declared it the source of man's evilness rather than the crown of his achievement. For this declaration attacked the basic rationalist belief that the improvement in the arts and sciences meant moral as well as physical progress.

Instead, Rousseau suggested that the regeneration of man would come about, not by scientific and technological improvements, but by man's rediscovery of himself. As Cassirer points out, Rousseau demanded a radical political and ethical renewal instead of the steady "progress of the human spirit" which Voltaire and his friends favored.[11]

[10] There is much of Montesquieu's *Considerations*—with its stress on liberty and virtue—in Rousseau's *Discourse*. For example, Rousseau condemned the arts and sciences for bringing luxuries in their train and thus being the cause of the decline of nations, and he castigated money and commerce for weakening and destroying the warlike virtues.

[11] Ernst Cassirer, *The Question of Jean-Jacques Rousseau,* tr. and ed. by Peter Gay (New York, 1954), p. 67. Cf. Green, *op. cit.,* p. 240. The work by Cassirer, translated from an article which first appeared, in German, in 1932, attempts to synthesize all the aspects of Rousseau's thoughts and to show their essential unity. See, too, Cassirer's *Rousseau, Kant, Goethe,* tr. by Gutmann, Kristeller, and Randall, Jr. (Princeton, 1945). Charles William Hendel, *Jean-Jacques Rousseau,* 2nd ed. (London, 1934), is an account of the development of Rousseau's thought.

Rousseau's *Discourse* was, therefore, a direct stroke at the collective intellectual product of the time, the *Encyclopedia,* then in its first stages, to which, ironically, Rousseau was himself a contributor on the subject of music. D'Alembert recognized the seriousness of the attack. In his *Preliminary Discourse of the Encyclopedia,* published in 1751, he attempted to turn it aside with soft words.

Perhaps this would be the place to reject the shafts that an eloquent and philosophical writer recently directed against the arts and sciences, in accusing them of corrupting manners. It would ill befit us to be in agreement with him at the beginning of a work such as this; and the worthy man of whom we are speaking seems to have given his approval to our work by the zeal and success with which he has contributed to it. We shall not reproach him with having confused the culture of the mind and the abuse that one makes of it; he would doubtless reply that this abuse is inseparable from it: but we ask him to examine whether most of the evils which he attributes to the arts and sciences are not due to very different causes. . . . will it be necessary to proscribe the laws because their name serves to cloak a few crimes? . . . In short, were we here to make a confession to the disadvantage of human knowledge—which we are very far from doing—we are even farther from believing that one would gain by destroying it: the ills and evils would still be with us, and we should have ignorance in addition.[12]

Rousseau eventually accepted this contention. In 1755, he wrote to Voltaire indicating his modified position.

The love of letters and the arts arises in a people from an internal weakness which it only augments; and if it be true that all human advance is pernicious to the species, that of the mind and intelligence, which increases our pride and multiplies our errors, soon speeds the coming of our evil day. But a time arrives when the evil is so great that the very causes which gave rise to it are needful to prevent its becoming worse—it is a case of the weapon one must leave in the wound for fear the wounded man should expire if one were to draw it out.[13]

Such a statement was cold comfort for the men of the *Encyclopedia.* They realized that Rousseau was temporizing, and this was made abundantly clear in the books that followed the first *Discourse.* In

12 D'Alembert, *Discours préliminaire de l'Encyclopédie,* introd. par F. Picavet (Paris, 1929), pp. 124–125.

13 Hendel, *Citizen of Geneva,* p. 135. Cf. Cassirer, *The Question of Jean-Jacques Rousseau,* p. 54.

these, Rousseau pressed his point further and, indeed, broadened the area of attack.

V

In order to deal more adequately with Rousseau's work and writings, we must realize that he was a different kind of man from the other people whom we have discussed. In Rousseau, we have the eruption into political and philosophical writing of a man of petty-bourgeois—almost workingman—origin. Amid the elegance of French society, his attitude was naturally a truculent, chip-on-the-shoulder one. He was proud of his humble origin and declared, "My father, I own with pleasure, was in no way distinguished among his fellow-citizens."[14]

Previously, with thinkers like Locke or Voltaire, we have had rather prosperous people. Rousseau was a man who was poor all his life and died, as he had lived, in great poverty. There can be little question that Rousseau's social and economic status, while not explaining his views, certainly influenced them. He was always for the underdog and, as we shall see, believed in the natural goodness of the common people.

Undoubtedly, too, Rousseau's lower-class background gave him a greater insight into the hearts of simple people. His wife, Thérèse, for example, was illiterate, could not tell time, and could not remember the days of the week. Rousseau understood, however, what *moved* people like her. Accordingly, he stressed that the emotions and sentiments, not logic and reason, were common to men. Thus, he

[14] According to Rousseau, however, his father was virtuous. For him to have been "distinguished" in terms of eighteenth-century society would have meant participation and success in immorality. Rousseau was not preaching the overthrow of the existing order and the rise of the lower classes; he said, instead: "Nothing is more fatal to morality and to the Republic than the continual shifting of rank and fortune among the citizens: such changes are both the proof and the source of a thousand disorders, and overturn and confound everything; for those who were brought up to one thing find themselves destined for another; and neither those who rise nor those who fall are able to assume the rules of conduct, or to possess themselves of the qualifications requisite for their new condition, still less to discharge the duties it entails" (Everyman's Library ed., p. 255). What Rousseau *was* preaching was the glorification of the lower class *as the lower class*. If virtue, rather than wealth, glory, or position, was alone worthy of honor, then perhaps the poor might "inherit the earth" more readily than the rich and well-born.

played a major role in what is often called the "emotional revival," in the rise of sentimentality.

There is a parallel to this aspect of Rousseau in the religious history of the time which throws his work into perspective. Around the first half of the eighteenth century, a wide religious latitude was practiced in such countries as France, England, and Germany. Indeed, it was quite usual to call the English bishops at that time "latitudinarian bishops" because they took their doctrines so widely and so lightly.

Against this, there sprang up in several countries of Europe a fervent religious revival, of which Pietism and Methodism are perhaps the most interesting and important examples. In Germany, the Count von Zinzendorf led the Pietist movement. He attempted to spark a "religion of the heart" and, ignoring dogma to a certain extent, appealed to sentiment and a common morality.[15] In England, the Methodist movement, fired by the preaching of John Wesley and George Whitefield, had even more success and made the jump from a sect to a church.[16]

Perhaps a case could be made to show that Jansenism was the potential Pietism or Methodism of France; Pascal was certainly working in that direction. Jansenism, however, was beaten down by the Catholic Church and by the classically minded court of Louis XIV. Thus, it remained for French sentimentalism and emotionalism to be developed outside of an organized religious movement.

This was the task of Rousseau. He stressed the part played by the emotions, not only in religion, but in literature, politics, and philosophy. He offered a broad and original treatment of all fields on the basis of his new sensibility. The result, to a large extent of his work, was the so-called Romantic Movement. Thus, we have before

15 Pietism, by appealing to what was common in all men's hearts, tended to have a unifying effect on Germany and to lead to nationalism. See further K. Pinson, *Pietism as a Factor in the Rise of German Nationalism* (New York, 1934).

16 John Wesley, educated at Oxford, was calm and argumentative and, as a recent writer puts it, "His great psychological effect was due to the fact that he sought to produce in his hearers an experience similar to his own conversion" (Walter L. Dorn, *Competition for Empire, 1740–1763* New York, 1940, p. 241). What Wesley did was to apply the method of Pascal (that is, the method or art of persuading) but on a mass basis. Whitefield, who was uneducated and uncultured, was more the demagogue, passionately and almost hysterically appealing to the emotions of his audience. Between them, Wesley and Whitefield developed the tradition of revivalist preaching to the point where it became crowd oratory and the demagogic creation of mass opinion.

us the picture of Rousseau, the "outsider," who fathered the romantic sensibility and opposed it to the dominant rationalism of his time.

VI

The "romantic" aspect of Rousseau is fully displayed in his novel *La Nouvelle Héloïse* (1761) and in his treatise on education, in fictional form, called *Emile* (1762). Both books tell how, ideally, people ought to grow up and live naturally, that is, among the trees and in the country, and in response to the promptings of their heart and to the inner voice of conscience.

The *Nouvelle Héloïse,* like so much of Rousseau's work, was a projection of a personal episode in his own life. In love with a woman, Sophie d'Houdetot, who was both married and already provided with a lover, Rousseau idyllicized his hopeless passion in a written daydream—his novel. Other novelists had analyzed the passions; Rousseau depicted them, living and trembling under the reader's hand, and made the reader actually share them. As one commentator has pungently remarked: "Love borrows the ardent language of religious devotion."[17]

Rousseau was well aware that his passionate outpourings jeopardized his standing as a moralist. "And the grave citizen of Geneva, the austere Jean-Jacques, close on forty-five, suddenly became again a love-sick shepherd swain," he admitted in his *Confessions.* Indeed, he did not wish the novel to be included in his collected works.[18]

His next work, *Emile,* was a more solemn and astringent book. It attempted to advise how a boy should be educated so as to become a

[17] Green, *op. cit.,* p. 184.

[18] Actually, he need not have worried. The *Nouvelle Héloïse* combined morality and passion in a unique synthesis. Rousseau showed that the emotions, even sensuous ones, could not help being on the side of virtue. Extraordinary souls, like those of his two lovers, Julie (i.e., Sophie d'Houdetot) and Saint-Preux (i.e., Rousseau) "could not be judged by ordinary rules." They needed a "natural" morality to judge them. Such a "natural" morality would have the emotions, not the reason, pass judgment on actions and intentions.

In the end, however, Rousseau reversed his moral and had his heroine place family duty before her lover. Bourgeois virtue is triumphant. The passions run docilely into chats between the lovers on the duties of domestic life and on all the major problems of religion and morality.

The setting for these conversations is the rustic life. In the country, far from the salons and cities, man's nature is in harmony with his moral duties. That is the lesson Rousseau preached in his sentimental novel, the *Nouvelle Héloïse,* or, as it was first titled, *The Letters of Two Lovers* (*Les Lettres de deux amants*).

good man, uncorrupted by existing society. In a sense, Emile was the boy Rousseau wished he had been.

The starting point of *Emile* was Rousseau's basic belief that man is naturally good but is corrupted by society. Thus, a good education will "follow nature." It will be "negative," at least till the age of 10 or 12, in the sense that it will allow the child to progress in his own way and give him no positive injunctions. Instead of by abstractions, the child will learn by observation of facts. In place of precepts, he will see only examples. Only one book will be permitted him: Defoe's *Robinson Crusoe;* and this because its fictional isolation shows the true worth of things and not their social value.

Rousseau stated openly that he was not educating Emile for a particular social position, although the boy is described as being of good family. His is not Montaigne's education, for a gentleman. In a "century which overturns everything each generation," as Rousseau remarked, Emile had to be trained for any post fortune might give him.[19] Thus Rousseau prescribed risks and dangers, exposure to sun and wind to toughen his hero for life. Emile is even taught to be a carpenter, in case of need.[20]

An agonizing dilemma, however, is present in all of Rousseau's thoughts on education. He realized that the natural man stood in opposition to the civil man. "It is necessary to choose," he admitted, "between making a man or a citizen: for one cannot make both at the same time." Rousseau's choice was to create a citizen. Thus, he acknowledged the fact that all good social institutions, all correct education, must "denature" man, must "change the 'I' into the common unity."

How was this to be done? How were the man and the citizen to be reconciled? The only solution was to change man by working with

19 *Emile,* p. 13. Is the education of Emile the education of a Napoleon, a Julien Sorel, a man who seizes a "career open to talent"? It would seem so.

20 Emile is brought up till the age of 15 without knowing any religious dogmas. At that time, he is exposed to the famous "Profession of Faith of a Savoyard Vicar" (Part IV of *Emile*). In this "Profession," Rousseau outlined his own "religion of the heart." It was this part of *Emile,* especially, which raised a violent outcry against Rousseau. The Parlement of Paris condemned it, and Rousseau was forced to flee France. Geneva followed suit, and went so far as to burn the book. From this point on, Rousseau felt hounded and persecuted, a feeling which led him eventually to the brink of madness. For an exhaustive treatment of Rousseau's religion, see Pierre M. Masson, *La Religion de Jean-Jacques Rousseau,* 3 vols. (Paris, 1916).

and not against his natural bent. Vague as this may be, it was Rousseau's first, tentative solution to his problem.

In reality, Rousseau never lost his nostalgia for childhood. "Although born a man in some respects, I long remained a child and in many other respects I am one still," he confessed. The transition from child to adult, from natural man to social citizen, was one he never wished to make. Eventually, in the *Social Contract,* which we shall discuss shortly, he made the jump. But in *Emile* the ideal he had in mind was a "noble savage," who, while living in society, was still almost completely natural.

The idea of the "noble savage," stemming from the age of Thomas More and the New World discoveries, was an ideal current in Rousseau's time. According to the travel reports, the savages of North America, for instance, despised gold and silver. They lived freely and happily in the forests, dwelling harmoniously in small groups. They needed no laws or social restraints, and ordered their lives by instinct rather than by reason; for example, they found their way home, instinctively the eighteenth century thought, by following the direction of the sun at different times of day and not by reading a compass. As interpreted by Rousseau and others, the noble savage was the man who, with a wonderful set of emotions, and rejecting entirely his intellectual gifts, achieved for himself the simple and perfect life which civilization has since destroyed.[21]

There are some obvious flaws in this picture. For example, Rousseau, like many people before and since, was deceived by the fact that savages wear very few clothes into supposing that savages have very few laws. In fact, the written and unwritten laws of savage societies are many and complex.[22] However, in Rousseau's day, the South Seas were virtually unexplored territory, and a word like "taboo,"

[21] See Daniel Mornet, *Rousseau: L'Homme et l'oeuvre* (Paris, 1950), p. 39, for the travel books read or consulted by Rousseau. Paul Nourrisson, *Jean-Jacques Rousseau et Robinson Crusoé* (Paris, 1931), is critical of Rousseau and attributes much of his trouble to his "fixed notion of solitude" which he took as an ideal. Joseph Texte, *Jean-Jacques Rousseau and the Cosmopolitan Spirit in Literature: A Study of the Literary Relations Between France and England During the Eighteenth Century,* tr. by J. W. Matthews (London, 1899), throws light on Rousseau's attitude to the "natural" life by exploring the effect of English literature on the formation of Rousseau's romanticism. For a stimulating discussion, see Jacques Barzun, *Romanticism and the Modern Ego* (Boston, 1943).

[22] See, for example, Ruth Benedict, *Patterns of Culture* (Boston, 1934), and the books of Margaret Mead.

which comes from South Sea customs, had hardly been brought into the European languages. The "noble savage" was thought to live in a careless rapture. His life, at the same time that it was socialized, was supposed to be a "natural" life.

VII

What, however, is a "natural" life? How shall we—who are already overcivilized and therefore corrupted—know what is man's real "nature"? There were two ways, Rousseau suggested, of discovering an answer. We can look into our own hearts. And we can look at nature herself; not the nature of the scientist but the nature of the romantic solitary.

Rousseau, once having borrowed the inspiration of the "noble savage," broke with the notion that the real nature of man could be found through the method of comparative travels. He rejected anthropology in favor of psychological introspection. Look inside man, not at man, was Rousseau's advice. He declared that "A retired and solitary life, a love of reverie and contemplation, the habit of introspection and of seeking within himself, in the calm of the passions, these primitive [*premiers*] traits which have disappeared in the multitude, could alone enable him to recover them. In a word, it was necessary that one man should depict himself in order thus to show us primitive [*primitif*] man."[23]

Furthermore, in his *Discourse on Inequality,* Rousseau attempted to show that the savage man looked within himself while the modern man looked without. "The savage lives within himself," he said, "while social man lives constantly outside himself, and only knows how to live in the opinion of others, so that he seems to receive the consciousness of his own existence merely from the judgment of others concerning him."

Rousseau passed a moral judgment on the "other-directed" man.[24]

23 *Rousseau juge de Jean-Jacques,* 3rd Dialogue, quoted in Green, *op. cit.,* p. 259.

24 Rousseau's phrasing of the problem anticipates the work of the modern sociologist David Riesman. In his book *The Lonely Crowd* (New Haven, 1950), Riesman makes a differentiation between the Puritans and modern man. The Puritans were "inner-directed," that is, they took their moral directions from their own consciences. Modern man, on the other hand, is "other-directed," that is, he swings around like a radar set in order to pick up cues from his fellow man as to how to behave. Riesman, unlike Rousseau, does not censure the "other-directed" man.

He declared that "Always asking others what we are, and never daring to ask ourselves, in the midst of so much philosophy, humanity, and civilization, and of such sublime codes of morality, we have nothing to show for ourselves but a frivolous and deceitful appearance, honour without virtue, reason without wisdom, and pleasure without happiness."

Having now asked ourselves "what we are," the other way of discovering our true nature is to interrogate external nature. In this case anticipating poetic seers like Wordsworth, Rousseau urged man to "retire to the woods" to consult nature in order to discover himself. "O man, of whatever country you are, and whatever your opinions may be, behold your history, such as I have thought to read it, not in books written by your fellow-creatures, who are liars, but in nature, which never lies. All that comes from her will be true." In this bold expostulation Rousseau took the same step which had been essential to the development of natural science in the seventeenth century. He asked man to turn from books, from written authority, and to consult the facts. In the case of human nature, he believed, these facts were to be found either within our own hearts or in the "impulse from a vernal wood." Thus, his suggested method may be termed a sort of romantic empiricism.

By making this suggestion, Rousseau was explicitly rejecting the society of man as it existed, because that existence was based on the wrong facts and therefore on a distortion of man's real nature. This was, of course, merely another variation on the theme he had broached in the *Discourse on the Arts and Sciences*. However, what we have called Rousseau's romantic empiricism went far beyond that in its importance. In our judgment, it was a prerequisite—with its insight that man's nature is not what it is but what it is potentially—for the development of a science of man, and a necessary complement to Montesquieu's emphasis on societies as they exist.

VIII

Can we, however, throw off the mental shackles of society and see man in a state of nature? Rousseau tackled the question directly:

> It is by no means a light undertaking to distinguish properly between what is original and what is artificial in the actual nature of man, or to form a

true idea of a state which no longer exists, perhaps never did exist, and probably never will exist; and of which it is, nevertheless, necessary to have true ideas, in order to form a proper judgment of our present state. It requires, indeed, more philosophy than can be imagined to enable any one to determine exactly what precautions he ought to take in order to make solid observations on this subject; and it appears to me that a good solution of the following problem would not be unworthy of the Aristotles and Plinys of the present age. What experiments would have to be made, to discover the natural man? And how are these experiments to be made in a state of society?

To this profound question, Rousseau devoted much of his *Discourse on the Origin of Inequality*.

What he did was to make a "thought-experiment." We have already seen an example of a "thought-experiment" in our account of the reasoning which made it unnecessary to drop weights from the Tower of Pisa. Now we see another example, but in an attempt at a human science. To discover man in the pure state of nature, Rousseau, mentally, stripped him of all the attributes of society. Then, he reconstructed the evolution of man "experimentally," in his mind. Rousseau was here very conscious of his methodology, as we see when he said: "I confess that, as the events I am going to describe might have happened in various ways, I have nothing to determine my choice but conjectures: but such conjectures become reasons, when they are the most probable that can be drawn from the nature of things, and the only means of discovering the truth."

In this manner, Rousseau introduced a new and difficult tool of social science. It was a further development of what, earlier in our work, we have seen Cassirer call a "causal definition," and, in a certain sense, of Montesquieu's "ideal type." It differs from both of these, however, and deserves a separate title, which we have given by calling it a "thought-experiment."

Perhaps Rousseau's theory of language led him to this methodological discovery. As he stated:

Every general idea is purely intellectual; if the imagination meddles with it ever so little, the idea immediately becomes particular. If you endeavour to trace in your mind the image of a tree in general, you never attain to your end. . . . The definition of a triangle alone gives you a true idea of it:

the moment you imagine a triangle in your mind, it is some particular triangle, and not another, and you cannot avoid giving it sensible lines and a coloured area.

Rousseau was suggesting that the only true idea of man's nature could be reached by means of definition, by what we have called a "thought-experiment," for all other images of man's nature were particular ones, formed by, and mirroring, particular societies. He was saying that the state of nature has never really existed but it is a pure "idea of reason." Yet, it is only by elaborating the imaginary construct of a state of nature that we can discover man's *real* nature; for "nature" actually meant to Rousseau the full development of man's capacity, of which contemporary society was only one partial and incomplete exposition.

IX

The next step was for Rousseau to "define" correctly the right state of nature for man and to construct, by a "thought-experiment," the correct society for man's newly discovered true nature. Once this was done, Rousseau had then to take *man* from his state of isolation and bring him into society as a *citizen*. All of this he attempted to do in the *Social Contract* (1762).

A wide separation is often felt between Rousseau the "wild man," who wrote *Emile* and started the glorification of the child and the cult of the "noble savage," and Rousseau the sage and balanced philosopher of the *Social Contract*.[25] Rousseau himself denied this. Furthermore, Book V of *Emile* is a preview of the *Social Contract* itself. To miss the connection between Rousseau's attack on the arts and sciences, his romanticism, and his democratic politics, is to miss the heart of his thought. The *Discourses*, the *Nouvelle Héloïse*, and *Emile* give us Rousseau's view of man's psychology; the *Social Contract* builds on that basis a politics for man.

The *Social Contract* is preoccupied with one central question: What is the basis of society if it is not the consent of its members, past or present, implicitly or explicitly expressed? Thus, starting with man in a state of nature—which in Hobbes is a state of war, in Locke a state of peace, and in Rousseau a state of isolation—we must bring

[25] In a play by Palissot, *Les Philosophes* (1760), a professed disciple of Rousseau is depicted as entering on all fours, munching a lettuce.

him into society by his own consent, i.e., by way of a social contract.

There was of course nothing new in Rousseau's use of the idea of a social contract. Hobbes and Locke had based their governments on social contracts. The novelty in Rousseau's formulation was that the people, who entered into the general contract, remained sovereign throughout the development of their society and, in fact, had to exercise their own rule. The elements of this thought were, of course, in Locke, but Rousseau stated the thesis of democratic sovereignty clearly and unequivocally, so that even illiterates might understand it when they heard it.

Once stated flatly, however, the principle of democratic sovereignty raised questions. What do we mean by the "will of the people" and how are we to discover it? This was the profound inquiry which Rousseau treated, with tremendous repercussions, more deeply and at greater length than any previous writer. His perceptive answer to the first half of the question, coming from his own introspective psychology, was that the basis of any state is a collective consciousness. He suggested, further, that a state which enlisted only man's interests and not his passions would not endure for long.

Rousseau's position was that the laws of society, whether about property, crime and punishment, or anything else, are neither God-given, as implied by Luther, nor arbitrarily imposed by a tyrant, as in Hobbes, nor natural laws which one simply has to discover, as in Locke. Rousseau's claim was that the laws can and do operate only by the consent of the whole population. They represent the *way of life* which the society has adopted for itself.

Although the fullest treatment of the general will is in the *Social Contract,* Rousseau had begun the discussion in his *Discourse on Political Economy* (1755). In that work he stressed two points about the general will. The first was that the general will can always be identified with the welfare of the whole. The second was that the general will is the standard of morality; in fact, Rousseau contended that before man's entrance into the social contract and, thus, the rise of a collective consciousness, morality did not exist. In his own words, "The body politic is also a moral being, possessed of a will, and this general will, which tends always to the preservation and welfare of the whole and of every part, and is the source of the laws, constitutes for all the members of the State, in their relations to one another and to it, the rule of what is just or unjust."

The momentous consequences of these two points—that the general will is the welfare of the whole, and is also a valuation of what is just and unjust—did not really emerge until the *Social Contract*, where Rousseau labored to distinguish the general will from the will of all. He was not clear and consistent on the matter. However, the gist of his views seems to be that the general will is an "ideal construct," representing what is best for the state, and that the will of all, or the will of the majority, is not necessarily what is best for the state.

Yet, practically speaking, how can the general will be found? Rousseau did not want to give his preliminary answer, but he was forced to do so: by counting heads. Let us follow his tergiversations. His first statement in the *Social Contract* is:

> There is often a great deal of difference between the will of all and the general will; the latter considers only the common interest, while the former takes a private interest into account, and is no more than a sum of particular wills: but take away from these same wills the pluses and minuses that cancel one another, and the general will remains as the sum of the differences.

This appears to favor the discovery of the general will by counting votes. However, a little farther on in the book, Rousseau declares:

> If, to save words, I borrow for a moment the terms of geometry, I am not the less well aware that moral qualities do not allow of geometrical accuracy.

Thus, if the general will is the source of our knowledge of what is "just and unjust," it is unquestionably a moral quality and cannot be determined mathematically. Finally, however, Rousseau seems to waver again and to stress the mathematical nature of the general will, but now coupling it implicitly with our interpretation of it as expressing the way of life of the people:

> The constant will of all the members of the State is the general will; by virtue of it they are citizens and free. When in the popular assembly a law is proposed, what the people is asked is not exactly whether it approves or rejects the proposal, but whether it is in conformity with the general will, which is their will. Each man, in giving his vote, states his opinion on that point [Rousseau should have added, "ideally"]; and the general will is

found by counting votes. When therefore the opinion that is contrary to my own prevails, this proves neither more nor less than that I was mistaken, and that what I thought to be the general will was not so. If my particular opinion had carried the day I should have achieved the opposite of what was my will; and it is in that case that I should not have been free.

This may sound like double talk. The reason is that Rousseau had so loaded his concept of the general will with moral significance that he thrashed about on the question of how it is determined. It is a question of enormous significance. If the general will is determined by the counting of votes, we have a democratic interpretation. If it exists as a Platonic idea, independent of the will of all, it is only accessible to philosopher-kings; and this is the totalitarian interpretation. Let us examine these two possible developments of Rousseau by exploring more fully his total position.

X

Rousseau's major interest was in the problem of political obligation. He wished to provide a rational sanction for society. Historically, man had entered society by accident or by necessity but not out of a moral inclination. Now, by imaginatively returning to the original state of nature, he must build the state again, but this time choosing his form of society freely and rationally. Rousseau declared: "The problem is to find a form of association which will defend and protect with the whole common force the person and goods of each associate, and in which each, while uniting himself with all, may still obey himself alone, and remain as free as before. This is the fundamental problem of which the Social Contract provides the solution."

It follows, of course, that man will have to throw over or gradually reform his old society in order to enjoy the new model society; and, in this sense, Rousseau was revolutionary. But, unlike Locke, his basic problem was not to justify revolution. It was to justify restraint of the individual by the state. The full title of Rousseau's book is *The Social Contract or Principles of Political Right.* He was asking: What form of society will do away with the right to revolt by restraining men with perfect justice?

His answer was: Only that society in which sovereignty emanates from the people and is inalienable from them. Montesquieu had believed in the separation of the powers of government; Rousseau asserted the total supremacy of the legislative power. Montesquieu believed in a balance of social powers; Rousseau replaced this with the "general will." Montesquieu favored the aristocracy as a class; Rousseau, while admitting that elective aristocracy was the best form of administering government, lavished his love and affection on the people. "The voice of the people," he said, "is, in fact, the voice of God."

As an abstraction, Rousseau undoubtedly believed in the "people" and was not afraid to remove all barriers to their sole rule. In practice, however, he was more cautious and recognized "realities." Thus, when asked to draft a constitution for Corsica, he made an intensive study of the customs, history, and conditions of existence in the island; and his *Constitution of Poland* even allowed the continuance of serfdom as a temporary necessity.[26]

With equal realism, Rousseau foresaw the inevitable excesses of the French Revolution. He declared that he would not wish to live in a "republic of recent institution," for "Peoples once accustomed to masters are not in a condition to do without them. If they manage to shake off the yoke they still more estrange themselves from freedom, as, by mistaking for it an unbridled licence to which it is diametrically opposed, they nearly always manage, by their revolutions, to hand themselves over to seducers, who only make their chains heavier than before." Thus, Rousseau foresaw not only 1789 but the democratic movement lapsing into Robespierre and Napoleon as well. He did not realize, naturally, that these tyrants, just as had the democrats, would go forward under the banner of his doctrines.

Although Rousseau explicitly opposed despots and aggressive wars, his idea of the general will was particularly suited to be exploited by tyrants. For example, Germany in 1935 was filled with hordes of people shouting "*Sieg heil!*" in unison, in order to give

[26] Cf. Vaughan, *The Political Writings of Jean-Jacques Rousseau,* Vol. II, p. 78. On Rousseau's politics, Robert Derathé, *Jean-Jacques Rousseau et la science politique de son temps* (Paris, 1950), is fundamental. Derathé concerns himself with the antecedents, rather than the consequences, of Rousseau's political thought, and emphasizes Rousseau's indebtedness to the school of natural right—especially to Grotius, Pufendorf, Hobbes, and Locke—as well as indicating the steps he took beyond this school. There is also an excellent, semicritical bibliography. See, too, Derathé's *Le Rationalisme de Jean-Jacques Rousseau* (Paris, 1948). Anne Marie Osborn has an interesting book, *Rousseau and Burke* (London, 1940), which emphasizes the similarities rather than the obvious differences between the two men.

themselves the impression that, somehow, they were consenting to a general will which was interpreted for them by one man.

We can see the totalitarian seeds clearly if we continue with our analysis of the general will. In order to reach the general will, Rousseau decreed that there would be no partial societies, representing particular wills. "It is therefore essential," he proclaimed, "if the general will is to be able to express itself, that there should be no partial society within the State, and that each citizen should think only his own thoughts." This means, in effect, that there is to be no intermediary barrier between the citizen and the force of the state. The conception of "freely associated" groups, such as political parties and religious sects, held, for example, by the Puritans, is totally excluded.

Rousseau arrived at this one-party view because he believed sovereignty, which lies in the people, to be inalienable. Therefore, there can be no representative government—which Rousseau considered, disdainfully, as a feudal survival—and the people's sovereignty must be exercised directly.[27] For this to be effective, there must be no partial societies preventing the direct expression of the general will.

Essentially, Rousseau was saying that nothing must interfere with the unity of the state. For example, Rousseau firmly believed the general will to involve unanimity. How modern to our ears is his attack on parliamentary wrangling! "The more concert reigns in the assemblies, that is, the nearer opinion approaches unanimity, the greater is the dominance of the general will. On the other hand, long debates, dissensions, and tumult proclaim the ascendancy of particular interests and the decline of the State."[28] Montesquieu had talked rather of the harmony formed by dissonances in the body politic and called despotism a "dead" unity. Rousseau preferred the single note. As a result, the individual is isolated from his neighbors by the prohibition of political parties.

He is further isolated by Rousseau's refusal to allow separate loyalties to church and state. In the last chapter of the *Social Contract,* Rousseau called for a "civil religion" whose dogmas would be pro-

27 Rousseau may have been led to this belief because, at that time, there existed no *democratic,* representative government in Europe. England had a form of representative government, but it was corrupt and elected by a minority of the nation.

28 On the question of unanimity, see J. L. Talmon, *The Rise of Totalitarian Democracy* (Boston, 1952), and the article by David Thomson, "The Dream of Unanimity," *Fortnightly Review* (February, 1953).

vided by the state. Belief is compulsory, and the citizen who refuses to believe is to be banished from the state, not for impiety, but "as an antisocial being." And when belief has been accepted, the penalty for forsaking the dogmas is death.

Surely, we are in the presence of a doctrine of nationalism! By civil religion, Rousseau meant patriotism; by a profession of faith, he meant an oath of allegiance; and by the death penalty, he meant death for treason to the state, not to religion. In Rousseau's formulation, the state is all-powerful and must be given total allegiance. It is not the individual who is possessed of inalienable natural rights, but the sovereignty of the people which is inalienable. And this sovereignty is simply another word for the general will. Thus, all rights belong to the state, for it is the state which embodies the general will.

The totalitarian implication of this view is mitigated only if the general will is arrived at by counting votes. In that case, we need fear only a "tyranny of the majority." If, however, the general will is an abstract "justice," then the shadow of Robespierre and his "reign of virtue" looms upon the threshold of history. And this is exactly what happened.

In the *Discourse on Political Economy,* Rousseau asked the question: "In order to determine the general will, must the whole nation be assembled together at every unforeseen event?" "Certainly not," was his answer. "It ought the less to be assembled, because it is by no means certain that its decision would be the expression of the general will. . . . The rulers well know that the general will is always on the side which is most favorable to the public interest, that is to say, most equitable; so that it is needful only to act justly, to be certain of following the general will." This was no accidental statement. Rousseau repeated the notion a few paragraphs farther on. He commanded, "As virtue is nothing more than this conformity of the particular wills with the general will, establish the reign of virtue." What more could Robespierre need in order to believe that he was following Rousseau in setting up his "Republic of Virtue"? And, to confirm his view, Robespierre needed only to read further: "We ought not to confound negligence with moderation, or clemency with weakness. To be just, it is necessary to be severe; to permit vice, when one has the right and the power to suppress it, is to be oneself vicious."

In these words, we hear anew the voice of Calvin, whose city of Geneva and whose person Rousseau so much admired. And behind Calvin, we hear the equally ominous admonition of the Catholic Church: "Compel them to come in." The totalitarian message of Rousseau illustrates once again that the reign of pure virtue is hardly distinguishable from the "dead" hand of despotism.

XI

There is much that is sublime and much that is dangerous in the thought of Rousseau; the dangerous part is like reefs lying hidden under pure water. For both good and for bad, he has had an enormous and powerful influence on the development of thought and action in the modern Western world.

Above all his accomplishments, perhaps, Rousseau, like Hobbes, added to the powers of Leviathan. He provided the philosophical justification for the totalitarian as well as the democratic national state. He planted seeds which flourished in the revolution of 1789, the reign of Robespierre, and the dictatorship of Napoleon. This he did when he wrote the *Social Contract* and his other works on politics.

Rousseau was also a powerful forerunner of the great romantic reaction against rationalism which arose after the French Revolution. He stressed the natural and the spontaneous, the emotional and the unlettered parts of man's make-up and wished to turn him back to his childhood. Abandoning, in this romantic mood, the attempt to make man a citizen of a national state, Rousseau tried to convert him into a "noble savage," although he knew that this was only an anachronistic notion. This aspect of his thought he presented in such books as the *Nouvelle Héloïse* and *Emile.*

Rousseau's starting point had been an attack on the belief that man's progress depended on the advancement of the arts and sciences. His claim was, in fact, that their advancement meant man's increasing moral degradation. He made his assault primarily in his *Discourses,* and he did it just on the eve of the Industrial Revolution.

In the perspective of present history, that tremendous technological and organizational upheaval, bringing in its wake the emergence of the masses to power, seems to have fulfilled Rousseau's Cassandra-like warnings. It has left man more adequately attired but still unredeemed. But so, it seems, did Rousseau's "Republic of Virtue." These two judgments will be seen in action in the development of the Industrial Revolution and the French Revolution, as we enter the period of the "Great Revolutions."

PART III

THE GREAT REVOLUTIONS:

From Smith to Hegel,
1760-1830

CHAPTER 17

The Industrial Revolution

WE HAVE given the title *The Great Revolutions* to the period which we are now to discuss, because it includes three revolutions: the Industrial Revolution, the American Revolution, and the French Revolution. To use the word "revolution" to describe all three of these events is not wholly an artifice of language, because these revolutions form an historical unity. About 1760 there occurred the explosive moment in European and world history in which the central features of contemporary life were created in a remarkably short time. One essential feature of contemporary life is large-scale, mechanized industry: this was created in the Industrial Revolution in England. A second feature of contemporary life is that the center of gravity of the Western world today lies somewhere between Europe and North America; and the shift westward to this imaginary point in the Atlantic begins with the American Revolution. And third, the substitution of democratic, elected governments for the traditional and absolute monarchies of Europe took its momentum, outside of England, from the French Revolution, which set off a chain of revolutions of the nineteenth and into the twentieth century.

I

The Industrial Revolution took place in England and not on the continent of Europe, in part because, even when it began, England was already a society with a relatively advanced industry, by the standards of the time. Hitherto, this had been a village or cottage industry; and for the first half of the eighteenth century England basked in a last Indian summer of village industry and pleasant overseas trade. The summer faded, trade grew more competitive and the needs of industry harsher and more pressing about 1760.

It might be thought that the sharp upswing of industry at that time derived from the work of the great scientists of Newton's age. And it is indeed true that the new industries could not have gone forward without the tradition of experiment and invention which had grown in England. But this tradition was not carried on, and the industrial inventions were not made, by the established scientists in such professional bodies as the Royal Society. For the established scientists and the professional bodies had already become arid and academic. They had been browbeaten by the success of Newton's system, and, as a result, science had come to mean to them something abstract and speculative—a set of theories which must be universal and mathematical.

Because the Royal Society was now dominated by this intellectual snobbery, the fine technical work of the eighteenth century was done outside it. It was done by rather unorthodox men, many of them nonconformists, who had therefore been excluded from the universities and had learned their applied science at first hand, in their own trades. And they came to form their own societies, which we shall discuss separately.

Men such as these made the practical inventions on which the Industrial Revolution depended; and yet, even their inventions were not the core and crux of that revolution. For the essential change which the Industrial Revolution brought was not in machines but in method. The Industrial Revolution was only incidentally a change in industrial techniques; it was more profoundly a change in industrial organization.[1]

To appreciate this change, we must look first at the organization of village industries as they existed in England before 1760. Some of these industries, such as the woolen industry, went back to the Middle Ages; others, such as the making of needles, had been advanced by technical innovations made in the inventive Elizabethan age. Their organization had one thing in common: it was divided into small units.

[1] Cf. T. S. Ashton, *An Economic History of England: The 18th Century* (London, 1955), pp. 105–117. This is a fundamental work, to be constantly consulted; it includes a number of useful statistical tables.

It is, of course, plain that only those industrial processes which produce raw materials—such processes as the mining of coal, the smelting of iron, and the making of glass—are necessarily large-scale processes. When these raw materials are turned into goods, i.e., secondary products like nails, the scale of the processes needed is only as large as the goods themselves. There is nothing inherent in the production of nails or of hats or of woolen cloth which requires it to be done in a factory in the presence of several hundred people. And in fact, nails and hats and woolen cloth were well made in the homes of villagers for centuries. They were made as toys are still made in Germany and Japan and as the parts of watches are made in Switzerland. A Swiss watch is an instrument of great precision, and we therefore think of it as made in a modern factory, all glass and chromium. In fact, however, what is important in the manufacture of the watch is that it is made of many parts, and these parts lend themselves to being made separately by people in their own homes. Watchmaking was run in this way, as a highly subdivided home industry, in Islington in London in the eighteenth century.[2]

The cardinal change which the Industrial Revolution brought was to move many of these industries from the home into the factory. Within two generations, roughly between 1760 and 1820, the customary way of running industry changed. Before 1760, it was standard to take the work to villagers in their own homes. By 1820, it was standard to bring everyone into a factory and have them work there.

This change was not brought about by any single factor: for example, it was not simply brought about by new inventions. Like all the great movements of history, it has no single explanation. It was the result of the interplay of many factors, some small in themselves, whose cumulative weight combined to overbalance the traditional way of making things so that it became modern industry.

II

It will be well to look at the influences at work in one industry; and we choose one of the key industries in the world before 1760, the making of woolen cloth. Characteristically, this was carried out in many steps. Sheep were reared and then shorn; the wool was cleaned and combed. It was then spun into thread, and the thread went to

[2] Cf. *ibid.*, p. 103.

the individual weaver, who had a weaving frame in his own home and wove the cloth on it. In principle, the weaver was a private manufacturer; that is, in principle he bought the thread, he owned his frame, and he sold the cloth himself.

This detailed procedure suffered from two drawbacks. First, not all parts of it were equally mechanized. The weaver's frame was an effective machine, but the spinning wheel was not. Anyone could spin who had the minimum sense of touch needed to draw an even thread and, as this needs little skill, spinning was therefore only a minor occupation of women—as the word "distaff" still reminds us. A weaver at work could keep many hands spinning: as Defoe described the Yorkshire villages of weavers, "Hardly any thing above four years old, but its hands are sufficient to it self." Second, there was little money in spinning, with its low productivity, and those who could gave it up whenever possible, in order to do the necessary work of house and farm. It was a seasonal occupation, which was dropped at seed and harvest time, and was therefore a bottleneck at the mercy of its occasional workers.

Another handicap in the organization of the woolen industry was economic. In principle, the weaver was his own master: he bought the thread, he owned the frame, and he sold the finished cloth himself. But he had little to fall back on if times were bad, and he got into debt. He had to borrow, that is to say he had to ask for credit, from the man from whom he bought either the raw wool or the spun thread. The only security he could offer for the loan was his weaving frame. In practice, therefore, even in the seventeenth century many weavers were in effect merely workmen for the wool merchant to whom their frames were mortgaged.[3]

The wool merchant commonly had his headquarters in a small town around which the weavers' villages were clustered. The weavers would come into town on a given day, often a Friday, and sell their pieces of woolen cloth: there still stand some of the market halls in which this was done, for example the Piece Hall in Halifax. With the money, the weaver would buy fresh wool; but if times were bad and there was a surplus of cloth, he would have to ask the merchant to keep the cloth and would have to get wool on credit against it. In

[3] Cf. *ibid.*, p. 207.

this way, the ownership of the wool, the weaving frame, and the finished cloth tended all to fall to the one merchant. Thus, in a practical sense, the weaver became a workman for wages—the uncertain wages made up of the difference between what he got for his cloth and what he paid for his wool.

This relationship became common in many industries: the woolen industry of Yorkshire, the cotton industry of Lancashire, nail and needle making around Birmingham, the making of gloves and stockings and hats, and many others. It became convenient for the merchant to send his agents into the working homes, to take the raw material there and to bring back the finished goods. He now had an investment in these homes, and needed to keep a sharp eye both on the tools and on the materials there.

He had to keep a sharp eye also on the return on his investment. The tools were in effect his, but he could not command the number of hours that the workman spent at them. In good times, the weaver wanted to work as little as was enough to earn the family's keep, but the merchant wanted him to make as much cloth as he could—for 12, 14, 16 hours a day. In bad times, the weaver wanted to work long hours in order to earn the family's keep, but the merchant wanted him to stop.[4]

In short, the organization of village industry had been appropriate to an age in which the cottager was truly his own master. It was not appropriate to an age in which he was becoming, under whatever disguise, a wage earner working for a central merchant. When the cottager was his own master, his output was controlled by the market. Once he became a workman, his output must be controlled by the investing merchant. The factory system was formed by this complex of pressures.

The early factories were organized in a number of different ways. In some, for example, the worker still brought his own tools—he did this in the Sheffield steel industry down to the present century—just as a skilled fitter brings his own tools today. Whatever the detailed organization, however, the factories turned out to have several advantages. They gave the owner control of the materials and the working hours. They enabled him to rationalize operations which needed

[4] Cf. *ibid.*, pp. 205–206.

several steps or several men. They made it possible to use new machines which could be worked by unskilled women and even children under supervision. And they allowed these machines to be grouped around a central source of power.[5]

III

The bottleneck which the factories first broke was in spinning. Several spinning machines had been invented which simulated the way in which the human fingers apply a growing tension to wool or cotton staples in order to form them into a continuous thread. In essence, these machines used a sequence of spindles or rollers to increase the tension stepwise. One such machine was invented by James Hargreaves in the 1760's; another was patented by Richard Arkwright. Within 15 years, Samuel Crompton had combined the two inventions in a machine which is, in its essentials, the spinning machine still used today.

Hargreaves and Crompton were real inventors; both tried to become industrialists, with no great success. Arkwright was an adventurer with a flair for finance; it is almost certain that he did not invent the machine which he patented, and likely that he stole it from two earlier inventors. He had been a barber by trade, but in his patent he described himself as a watchmaker, that is, in the setting of his time, as a man skilled in making the only precision instrument then used. But Arkwright's real skill was in raising money, in organizing production and sales, and in seeing at each step what innovation was needed in technique or in organization to meet the demands of new markets. With these skills, he went on to make a fortune even after his patent was overthrown.[6]

The market to which Arkwright and his competitors turned was not for woolen but for cotton cloth, which the growing trade with India had made popular. It was particularly difficult to spin a strong cotton thread by hand, and therefore English weavers had had to make a cloth in which the weft was indeed cotton, but the waft was of linen thread; for while the threads of the weft are stationary in the loom, those of the waft are constantly strained by the shuttle which

[5] Cf. *ibid.*, pp. 115–117.
[6] Cf. *ibid.*, p. 214.

throws them to and fro. Arkwright concentrated on spinning a cotton thread strong enough to be used for waft as well as weft: the example is characteristic of his sense of the market.

The market that he and other industrialists had in mind was in large part overseas. The pathos of the Industrial Revolution was that it did not see a market among its own workers for the new output which they were producing. Machines were first brought in to break the bottlenecks in industry; these bottlenecks were in unskilled processes such as spinning; and therefore the processes and the workers first brought into the factories were unskilled. It was not new that women and children should spin for long hours; Defoe had found just this in the Yorkshire villages in the 1720's, and had reported frankly that if they did not, the family must "fare hard, and live poorly." What was new was the setting: child labor was moved from the village home into the stony and brutal discipline of the factory.

These factories first used water power, and therefore they were built in the wild river valleys of such lonely counties as Derbyshire; only here could the streams be counted on to run full even in the summer. (Farther south in Europe all the streams dry up in the summer, and this may be an additional reason why few, if any, industries developed there.) Therefore, many children were brought from foundling homes and workhouses to live and work the year round in the factory. The village children who also worked there might have to walk five or ten miles to and from the factory; and there they might walk ten miles or more in a day between the machines at their trivial and desolate task, which was to join up the threads whenever they broke. It was reported that, in the factories, parents were the most cruel supervisors of their own children, because they depended on the children's earnings to keep the family alive.

These things began to change only when the steam engine took the place of water power in the factory, at the beginning of the nineteenth century. Then the factory moved to the source of labor, to the town; for power now—that is, coal—was no dearer in the town than in the country. And steam power was more massive and more dependable than water power, so that it could run heavier and more elaborate machinery. With the coming of the rotary steam engine, weaving also became a factory industry, and in time the more important factory industry.

IV

The steam engine had begun as a machine to pump water from mines. The problem of keeping a mine from flooding is very old; in struggling with it, Evangelista Torricelli had already discovered in 1644 that a suction pump will not raise water much more than 30 feet. In fact, Torricelli had taken the essential step toward the discovery of the pressure of the atmosphere.

Deeper mines had to be kept dry simply by bailing them out with buckets, or with a series of pumps. What was needed, then, was a source of power to run the chain of buckets or pumps.

There was an early steam engine, invented by Thomas Savery in 1698, which was really a force pump. It used steam directly to blow the water out of the cylinder to the surface and, not unnaturally, soon blew itself to pieces in the process. A more practical engine was invented by Thomas Newcomen, a smith working in the copper mines of Cornwall, at the beginning of the eighteenth century. In Newcomen's engine, the steam which drove the piston was condensed in the cylinder, and the piston was brought back by the resulting suction. This system worked; there are Newcomen engines still shown, with veteran pride, at two or three old mines in England. But the system had the disadvantage that the whole cylinder was chilled at each stroke when water was brought in to condense the steam.

A skilled instrument maker to the University of Glasgow, James Watt, repaired a model of Newcomen's engine about 1763, and became interested enough to make some rudimentary calculations of its efficiency. These calculations shocked him, and he asked himself how the heat loss might be avoided. As in many inventions, the answer was not difficult; what had been difficult was the insight to ask the question. The answer was to condense the steam in a chamber connected to the cylinder (by a pump) but not part of it. In this way, the condenser was always cold, and the cylinder always remained hot. This work of James Watt's was in part set off by the discovery in Glasgow at the same time by Joseph Black of the latent heat of steam.

The later history of Watt's steam engine is characteristic of the Industrial Revolution. His engine did not work well in Scotland because, he found, the workmanship of smiths there was not good enough. He looked round England for more precise engineering, and found it in the workshop of Matthew Boulton, a maker of metal trinkets in Birmingham. Accordingly, Boulton and Watt entered into a partnership agreement. Boulton, a thrusting industrialist, went out to look for markets for the engine. The engine had a distinct economic advantage over Newcomen's: it saved coal. This did not recommend it to coal miners, but it made it attractive to the metal mines of Cornwall, which were remote from the coal fields. Accordingly, Boulton sold the engine there, not at a fixed price, but at a rental calculated as a fraction of the coal saved each year. The saving, however, turned out to be so large that the owners of metal mines could seldom be persuaded to pay the rental for long.

As the demand for the steam engine by the mines became filled, the problem of finding new markets had to be faced. On a tour of Cornwall in 1781, Boulton wrote home to Watt:

> There is no other Cornwall to be found, and the most likely line for increasing the consumption of our engines is the application of them to mills which is certainly an extensive field.
>
> The people in London, Manchester and Birmingham are *steam mill mad*. I don't mean to hurry you, but I think that in the course of a month or two we should determine to take out a patent for certain methods of producing rotative motion.

The steam engine had so far only pumped up and down. Urged by Boulton, Watt had no difficulty in inventing several ways to make it turn a machine. In this form it became the new source of power for the factories.

V

With the entry of steam, coal and iron became the backbone of industry. Great advances had already been made in the working of iron. and were now made in engineering it. Until some date after 1700, iron ore could be reduced in the blast furnace only with charcoal. Wood to make charcoal remained plentiful in France, but it was scare in England; and the scarcity prompted technical invention. The natural raw material to use in England was coal. It was, therefore, a constant ambition to reduce iron ore with coal, and several inventors had hit on the notion of first ridding the coal of its gases and turning it into coke.

It is not certain when coke was first used with consistent success to make iron from the ore: the date was probably about 1709. The ironmasters who made the process work were Abraham Darby and his family, and they kept their success secret for over thirty years. The Darbys were, in fact, characteristic inventors of the Industrial Revolution. They were Quakers, they did not bother with the Royal Society and with public honors, they were shrewd and silent, and they ran their business by their own technical skill and foresight. They entered into an arrangement with other ironmasters to form what we should now call a ring, but they kept their secret to themselves.

The raw iron, when it has been made, has still to be purified and worked. Here also important discoveries were made which, in the 1780's, at last made it possible to cast iron so that it was strong and free from flaws. Cast iron became a universal material—the plastic of its age. Thomas Paine designed an iron bridge and exhibited a model of it. The greatest of the Shropshire ironmasters, John Wilkinson, who bored the accurate cylinders for Watt's engine, built such a bridge which still stands. He sailed under it in an iron barge and boasted that he would be buried in an iron coffin; and he was.

VI

The great drive in the Industrial Revolution was for a new organization of production: the factory system. This drive directed, and in turn was directed by, a number of technical inventions. The interplay of method and machines, of economics and techniques was constant and mutual. From it, rather than from any single cause, the Industrial Revolution grew. Like every historical movement it grew organically, cell by cell.

A major condition for its growth was a change in the economic climate of England. We are accustomed today to the idea that industry is financed by shareholders, who can choose to put their money into any one of many competing enterprises. We forget that this is quite a modern idea: it was almost unknown before the eighteenth century. The Industrial Revolution could only gather momentum when investment in industry—general as well as personal investment—became a familiar way for people to use their savings.

In the earlier age of Queen Elizabeth, inventors (and speculators in invention) had tried to protect their forms of manufacture by getting exclusive rights from the crown to exploit them. These monopolies did little to stimulate those to whom they were granted, and they became a positive bar to progress, and a scandal, in the years between the death of Elizabeth and the rise of Cromwell. Indeed, there is some ground for believing that the grant of long-standing patents slowed down the speed of development of some new ideas (for example, that of the steam engine) even in the Industrial Revolution itself.[7]

The Elizabethan inventor who wanted a monopoly had been trying to protect the money that he and others were investing in the invention. The inventor's great difficulty was to find backers with money. There was at that time no establised habit of putting one's savings into investment. On the contrary, all forms of investment (except lending money to governments) were regarded by the church as usury. For this reason, banking and money-lending were often in the hands of non-Christians. They had often begun life as goldsmiths and traders in luxuries, and they usually lent money to those who had a taste for luxuries.

In England, a wider class of savings began to accumulate after 1688, when political conditions became stable and overseas trade could flourish. These savings were used by the new king to finance his Continental wars, and in this way a public debt was established, and it became usual for merchants and landowners to have a share in the national debt and to draw interest from it. The Bank of England was founded in 1694 as part of this movement to finance government spending from the savings of all.

The movement to invest the savings which now accumulated grew, and would have grown faster had there not been a setback to it after 1711. In that year, a large company was set up to trade with the South Seas, and it offered such wild returns (indeed, it offered to displace the Bank of England as the main holder of government bonds) that it started a wave of speculation. Owners of capital were in fact looking for somewhere to invest it, but there was as yet no real source of new wealth for them: there was no Industrial Revolution yet. The boom of 1720 was expanding into a vacuum, and when it collapsed, it left behind great bitterness among those who had lost money in it. As a result, laws were passed which made it illegal to found joint stock companies. These laws turned out to be a severe handicap later in the century, when the solid expansion of industry needed capital.

In one way and another, however, capital did now flow into the

[7] Cf. *ibid.*, p. 107.

market. One condition for this was the progressive fall in the rate of interest. We can trace this fall most clearly in government loans, which had to pay as much as 8 percent before 1700, but which by 1727 were paying only 3 percent. As the rate of interest fell, it became profitable to undertake long-term work, such as the sinking of mines, the building of factories, and the construction of toll roads. When interest is high, the lender can get a quick return and therefore does not want to sink money into long-term projects which will only give a return over twenty or thirty years. But when interest is low, he cannot expect a quick return, and he is therefore encouraged to make solid investments which will give a return in the future.[8]

The expansion of credit and the general stimulus to investment was a major factor in the progress of the Industrial Revolution. The revolution would have been impossible without a new attitude by all owners of capital to the investment of money. Factory owners financed their own expansion, of course, by ploughing back their profits year by year. But they could not have expanded as they did, often they could not have begun at all, if they had not been able to find backers with money from outside their family circle. Financial skill was needed to start a new industry, and some of those who claimed to be inventors were in fact, like Richard Arkwright, skillful above all in finance and in industrial organization.

VII

We have also in the agriculture of the time an important movement —often called an agricultural revolution—characterized by the introduction of new methods in farming. Viscount Townsend introduced the rotation of root crops with grass and grain crops, and was called "Turnip" Townsend for his pains. Robert Bakewell brought scientific methods into cattle breeding. His work, coupled with other improvements, had such good effect that the average weight of oxen sold at Smithfield increased from 370 pounds in 1710 to 800 pounds in 1745.[9] There were other agricultural innovators such as Jethro Tull, Arthur Young, and Coke of Holkham. One result of their work was

[8] The relation between savings, the rate of interest and industrial expansion is well put by T. S. Ashton, *The Industrial Revolution* (Oxford, 1948). This admirable little book is recommended as a balanced and compact account of the main development of the Industrial Revolution. It also has a short but adequate bibliography, to which we refer the reader.

[9] J. H. Plumb, *England in the Eighteenth Century* (Harmondsworth, 1950), p. 82.

that farmers were enabled, probably for the first time in the history of the world, to keep their cattle alive through the winter. This, in turn, brought an increase in manure; and an increase in manure brought greater agricultural fertility and yields.

The whole period is a fascinating complex of crosscurrents, influencing and flowing into the tide of the Industrial Revolution. For example, there was a new enclosure movement at the end of the eighteenth century. This was the second great enclosure in English history. The point of the first enclosure movement in the sixteenth century had been to make the land carry more sheep for the wool trade; and Thomas More in *Utopia* had complained that "the sheep eat the men." Now the second enclosure movement was carried out by country landlords who wanted to grow food on the common land. Unhappy as it was, it helped to create efficient large holdings, and was in this respect like the movement to factories. At the same time, in addition to farm production it affected the supply of labor and the rise of population.

In the course of this movement (running from 1760 to 1850), over 6 million acres of common land were enclosed.[10] This made possible the growing of large, scientifically cultivated crops, and so formed almost an eighteenth-century TVA program—without the social benefits. The landlords in effect stole the common land, though they claimed (and in the long run rightly) that "a better income [for everyone] could be drawn by improved methods of exploitation."[11] One result of the enclosure movement was that it also helped to destroy village industry. For the home worker had only been able to make ends meet by grazing his few animals on the common land and getting his fuel there. Now the enclosure of the commons drove him from the village into the industrial towns, where he formed a source of cheap labor.

VIII

Food and population were linked in the thought of the eighteenth century. We know today that after 1760 the English population rose

10 This figure is especially impressive when one remembers that the total area of Great Britain is only about 95,000 square miles, of which only about 58 percent is arable or pasture land.

11 Paul Mantoux, *The Industrial Revolution in the Eighteenth Century*, rev. ed., tr. by Marjorie Vernon (New York, 1927), p. 190. This is the standard work on the whole subject of this chapter.

by leaps and bounds. At the time, however, no one could tell whether population was going up or down; until the first official census in 1801, everyone was merely guessing. An important authority, Richard Price, in his *Essay on the Population of England* in 1780, believed that the villages were being depopulated. Oliver Goldsmith wrote a famous poem, *The Deserted Village,* whose theme is enclosure and the resulting depopulation. They and others supposed that the population of England as a whole was falling; and Goldsmith in *The Deserted Village* intoned,

> Ill fares the land, to hastening ills a prey,
> Where wealth accumulates, and men decay.

Other observers thought that population was increasing, but drew equally pessimistic conclusions from this belief. Arthur Young, for example, in his *North of England,* thought that population would now outstrip food production. Thomas Malthus turned this into a general law, and when he wrote his *Essay on Population* in 1798, the rising population and rising distress in England seemed indeed to prove his point.[12]

Statistics have now shown that Malthus was right in believing that population was increasing. But his forecast of mass starvation was falsified by the unexpectedly large productive capacity of the new agriculture and, indirectly, of the factory system which yielded an industrial surplus large enough to open up the agricultural lands of North America. (To a small measure, it was falsified also by birth control; indeed, Malthus himself had advocated "moral restraint" as a solution.)

Even now, we do not really understand the factors responsible for the large increase in population, today often called the demographic revolution. We do know that the death rate fell, and that this did much to cause the surging upward of population. But was the death rate reduced directly by the Industrial Revolution? The economic historian Max Weber casts doubt on this explanation. "The growth of population in the west made most rapid progress from the beginning of the eighteenth century to the end of the nineteenth. In the same period China experienced a population growth of at least equal extent—from 60 or 70 to 400 millions. . . . This corresponds ap-

12 See *ibid.*, p. 354.

proximately with the increase in the West."[13] There was, we know, no Industrial Revolution in China at this time.

In short, we know little more than that population increased and that in England it shifted northwest.[14] Recent work in economic history has, in fact, tended to show that the industrial workers were newcomers, "an additional population which was enabled to grow up by the new opportunities for employment which capitalism provided."[15]

We also know that these people tended to be concentrated in towns, for a distinctive feature of the Industrial Revolution was the great increase in urban living. In turn, this crowding into new towns created new problems of sanitation. And we see the interconnections of all these factors when we remember that a famous medical report, written by Dr. Percival of Manchester in 1796, led to the first modern factory legislation.

Obviously, without going into further details, the Industrial Revolution was more than a change in the system of production; it was also a revolution in the conditions of life for entire sections of society. The main industrial growth was in wool and cotton textiles, and in the complex linkage of iron, coal, and steam.[16] The main social development was in the creation of a new locus for an increasing number of people—the factory and the city. More and more, men spent their working hours in the factory and their leisure hours in the crowded streets and tenements of the city.

These tremendous changes in the conditions of life had profound repercussions in the thought of the time. In addition to the newly created proletariat, an older class had become more important: the

13 Max Weber, *General Economic History*, tr. by Frank H. Knight (New York, 1927), p. 352.

14 The English people were free to move because serfdom no longer existed, by the eighteenth century, in England, as it did on the Continent. One result of this movement in England was the destruction of local ties and loyalties, with the worker often drifting rootlessly in his new urban surroundings.

15 Cf. *Capitalism and the Historians*, ed. by Friedrich A. Hayek (Chicago, 1954), and a review of it by H. Stuart Hughes in *Commentary* (April, 1954).

16 Some idea of the industrial increase in these fields can be given by the figures of raw cotton imports in Great Britain and of pig-iron output. In cotton, we have 1 million pounds in 1701, 3 million in 1750, 4,760,000 in 1771, 5,300,000 in 1781, 11½ million in 1784, and 32 million in 1789. In pig-iron output, we have 68,000 tons in 1788, 126,000 in 1796, and 250,000 in 1804. (All figures from Mantoux, *op. cit.*, pp. 259 and 313.)

middle class, their power visible in the Reform Bill of 1832, which, in effect, first gave the middle class a voice in public voting and affairs. A new man, the Captain of Industry, became a hero. Most important for our purposes—as we shall see in treating of the Lunar Society—these people had new ideas and values. A new sensibility entered into and began to change the arts, the literature, and the thought of the period.[17] The Industrial Revolution changed Western man from head to foot; and thence more deeply from head to heart.

[17] See J. Bronowski, *The Common Sense of Science* (London, 1951), Chap. 1, for a discussion of the new sensibility, which elaborates and extends some of what we say in the next chapter.

CHAPTER 18

The Lunar Society: Businessmen and Technicians

I

ISAAC NEWTON lived forty years after he published the *Principia*, and throughout that time his authority dominated science. His influence on scientific method was admirable, for he established what is essentially the modern method: the generalization from experiment to theory and then the testing of theory in further experiment, which make up the inductive approach. But Newton's influence on the aims of scientists in his day was catastrophic. His authority ran counter to all that the first Fellows of the Royal Society had set out to do; he was not interested in everyday inventions, and he was out of sympathy with those experimenters who loved to hunt for them. By the time that Newton died, in 1727, the inventor was no longer regarded as a reputable scientist; for now mathematics had become the model of the sciences. Since few fields of science were ready to be given mathematical form in the eighteenth century, the Royal Society came to a standstill as a group of men still making discoveries. Instead, it turned into a club of professors nursing their distinctions, particularly the distinction of hobnobbing with the amateurs of the nobility.

The new discoveries which created the Industrial Revolution in the eighteenth century were made by men outside the Royal Society and the universities. The many-sided, inventive men who had founded the Royal Society in 1660 still existed, as they exist in every generation—the Hookes and the Wrens, the Pettys and the Wallises. But they were no longer to be found in the respectable circles of professional scientists. Instead, they were found in the manufacturing towns, and in the academies of the dissenters, to which manufacturers sent their sons because they gave a more realistic education than the universities.

These dissenting academies were the first institutions which gave an education in the knowledge of their times: in medicine, in logic, in modern languages, and in science. They began as schools to train nonconformist ministers, who could not be trained in the universities, but they soon became broader; and the greatest of them, the Warrington Academy, from its foundation in 1757 set itself to give a modern education to laymen as much as to those who were going to preach. The spirit in these academies can be summarized in a single example. Joseph Priestley went to Warrington Academy, shortly after it was founded, as a teacher of English and languages. He instituted there some of the first academic courses in English literature and in modern history. But he himself was drawn, by the lectures of his fellow teachers, to take an interest in the new sciences of the day—in electricity and in chemistry. As a result, Priestley became the greatest chemist of his day, and, on his discoveries rests the modern system of chemistry.

Behind such examples stands a new generation of men who did not see the world as a dying classical tragedy: whose hopes were not bounded by the Gothic of Horace Walpole and the pathos of Thomas Gray's *Elegy*. The new men were manufacturers, and they drew their wealth from industries which they were creating and often inventing. They were agog for new ideas, they had the nonconformist conviction that they could make their own ideas, and they invented from a natural sense of practical curiosity which fired their outlook. When the Warrington Academy came to an end in the 1780's, men of this stamp who had been pupils there created the Manchester Philosophical Society: and the Quaker John Dalton became a great chemist in that society, though he had begun life as a weaver's son, just as Priestley had begun life as the son of a cloth shearer. Science was a natural and an essentially practical activity for such men, and they founded little scientific societies up and down the country.

II

A typical scientific society of the eighteenth century was the Lunar Society in Birmingham. It was an informal society; it had no exact membership or rules, and consisted rather of a set of friends with common interests, much like the Accademia dei Lincei and the Invisible College of Robert Boyle in the century before. The active mind in the Lunar Society, and most often its host, was Matthew Boulton. He was James Watt's partner in the making of steam engines, and was the most urbane and the most spirited in the group of men who made Birmingham great.

One of these men was John Roebuck, who left Birmingham before the Lunar Society was formed, but who played an odd part in bringing it into existence. Roebuck was a pioneer in the development of a chemical industry; he began the making in the Midlands of England of sulfuric acid and alkalis, which have remained the staple of commercial chemical processes ever since. From Birmingham, Roebuck went to Scotland in search of salt as a chemical raw material, and then became interested in coal and ironworks.[1] He ran into difficulties in the coal mines which fed his new ironworks, because they were flooded with more water than the primitive steam engines of the day could pump away. Hearing of James Watt's improvement in the steam engine, Roebuck got him to build an engine for him, and took a share in Watt's patent rights in his new engine. Watt's engine did not really help Roebuck, in part because the engine was not made well enough in Scotland; and in the end, Watt moved to Birmingham in order to get the better craftsmanship of Boulton's metal workers there. Roebuck sold his share in Watt's patents to Matthew Boulton, and so helped to create the partnership between Boulton and Watt. This partnership was also the formal beginning of the Lunar Society, since Watt was the most scientific mind in Birmingham when he came there.[2]

III

There is no exact day on which the Lunar Society began and no exact day on which it ended. There were probably informal and oc-

1 See Archibald Clow and Nan L. Clow, *The Chemical Revolution* (London, 1952), for an emphasis on the Industrial Revolution's orientation toward Scotland.

2 H. W. Dickinson, *James Watt, Craftsman and Engineer* (Cambridge, 1936), while not exciting, is a competent treatment of Watt.

casional meetings before 1770, and some perhaps went on until 1800. But the regular meetings of the Society probably began in 1775, and they probably petered out after the riot in Birmingham in 1791.[3] This was the time when Boulton and Watt were building up their business and were the focus of the rising industrial aristocracy of Birmingham.

The men who met in the Lunar Society came from many trades. John Roebuck had left the Midlands before the Society was formed, but another pioneer of the chemical industry remained. He was James Keir, and he seems to have been a founder member of the Society.

The best-known man who came from outside the metal industry of Birmingham was Josiah Wedgwood. He founded the Wedgwood potteries, and directed every part of them. He gave his pottery the air of the Greek vases newly discovered in the Etruscan country near Rome, and he even called his works Etruria. But his profound influence was, like that of other Birmingham men, in insisting on able organization in his works, on a high standard of workmanship, and on a scientific understanding of the process. Josiah Wedgwood had a wooden leg, and the story is told that when he found a piece of poor pottery on a workman's bench, he would smash it with his leg and would chalk on the bench, "This will not do for Jos. Wedgwood."

Wedgwood understood that good work depended on exact control, not only of his men and their working conditions, but of materials and of firing conditions. He studied the kilns in which his pottery was fired, he learned how to control their temperature and, what is as difficult, to measure it. Wedgwood invented the pyrometer which measures such high temperatures, and, in effect, he made the fundamental discovery that at high temperatures all materials glow in the same way—that color measures the temperature, whatever the material is.

[3] See Robert E. Schofield, "Membership of the Lunar Society of Birmingham," in *Annals of Science,* Vol. XII, pp. 118–136. Schofield's Ph.D. thesis, *The Founding of the Lunar Society of Birmingham (1760–1780): Organization of Industrial Research in Eighteenth-Century England* (Harvard Univ., 1955, unpub.), is the most exhaustive existing treatment of the Lunar Society; it also has an extensive bibliography. Samuel Smiles, *Lives of Boulton and Watt,* 2nd ed. (London, 1866), has a long but now outdated chapter on the Lunar Society.

In order to have his materials under his control, Wedgwood searched for sources of china clay, and found them far from the Midlands. He then became a pioneer in the improvement of canals and of roads to transport his clay and his pottery. He was, in fact, a business adventurer of the new kind, deeply interested in what he himself was making, but always willing to invest enthusiasm and money in ventures which surrounded it. Boulton turned to him for money for new enterprises, and persuaded him, for example, to become his partner in founding the new industry of mining copper in Anglesey, which finally killed the copper mines of Cornwall.

Since Josiah Wedgwood lived at a distance from Birmingham, he cannot have been very regular in coming to meetings. But the Lunar Society was arranged to make things easy for such distant members. Because roads were then bad, the Lunar Society met once a month on a night near the full moon, and it took its name from this pleasant and practical arrangement.

IV

In the eighteenth century the professional classes were made up largely of two sets of people: doctors and clergymen. Both were present in the Lunar Society. There was, for example, the remarkable Dr. William Withering, whose country estate can still be seen in Birmingham, and who made one outstanding discovery. He discovered the stimulating effect of digitalis on the action of the heart, together with the fact that digitalis is found in foxgloves.

The most lively and remarkable of the doctors in the Midlands was Erasmus Darwin. He was a poet as well as a scientist, and he put the new system of botany of Linnaeus into rhyme. Many young people learned their biology from his oddly elegant poems, and particularly from *The Loves of the Plants.* Because Erasmus Darwin, like others in his circle, was a man of radical opinions, he was ridiculed by conservatives after the French Revolution, and the poets of *The Anti-Jacobin* anthology made much play with a parody called *The Loves of the Triangles.*

Erasmus Darwin held advanced opinions in biology as well as outside it, and he certainly had some conception of the evolution of species. This was a prophetic thought, for the theory of evolution was formally put forward by Erasmus Darwin's grandson, Charles Darwin, nearly a hundred years later.[4]

The influence of the members of the Lunar Society, in fact, did not end with their own generation. Like other pioneers, they stimulated the education of their children, and some of the children married one another and continued the lively family life in which forward-looking minds grow up. The Darwins, the Galtons, the Wedgwoods, the Edgeworths were such families. Richard Edgeworth was a member of the Lunar Society, and he and his daughter Maria Edgeworth were influenced by the ideas of Rousseau to write important works on educational reform. So was another member of the Lunar Society, Thomas Day, who not only wrote about the young in Rousseau's shadow, but put his theories into practice by bringing up two foundlings in order to train a wife for himself.

V

It is not clear whether John Wilkinson came to meetings of the Lunar Society; he also lived at a distance. But it is worth coupling his name with the Society, because he was a man of the same stamp: a man of the stamp of Josiah Wedgwood. His father had been an ironmaster before him, and he and his brother William both were pioneers in the use of iron in the eighteenth century. His brother William had been a pupil at Warrington Academy, and in this way his sister Ann met Joseph Priestley when he was a tutor there. Priestley married Ann Wilkinson, and John Wilkinson stood by his brother-in-law throughout the difficult times that followed the French Revolution.

John Wilkinson, with Abraham Darby, designed and built the famous bridge made of iron at the small town called Ironbridge, where it was opened in 1779. As we said in the previous chapter, he built the first iron boat in 1787 and proudly sailed it under the bridge, and when he died in 1805, he was buried, as he boasted he would be, in an iron coffin.

Wilkinson's technical skill was equal to his robust and willful enthusiasm. He invented the horizontal boring machine which made it possible to work to much finer tolerances than ever before. In this way, it was possible to bore the cylinders of the new steam engine so

[4] Ernst Krause, *Erasmus Darwin,* with a preliminary notice by Charles Darwin (New York, 1880), is still the best book to be consulted.

that it gave an effective pressure. Boulton and Watt always recommended to their clients that they have their cylinders bored by John Wilkinson. It is characteristic of that competitive society that, meanwhile, John Wilkinson was quietly pirating James Watt's design of the steam engine when from time to time he had the chance.

One of the problems in organization which faced manufacturers at that time was a shortage of small change with which they could pay their workmen. Therefore, many ironmasters and others coined their own wage tokens, which were accepted as currency by local shops. Many of these, including John Wilkinson's, were minted by Matthew Boulton. All this became another ground for suspicion after the French Revolution; it is entertaining to read the secret report of an agent of the government at that time, who writes to the Prime Minister William Pitt about the wage tokens of John Wilkinson:

> The Presbyterian tradesmen receive them in payment for goods, by which intercourse they have frequent opportunities to corrupt the principles of that description of men by infusing into their minds the pernicious tenets of Payne's *Rights of Man*.

VI

The outstanding scientific mind in the Lunar Society was John Wilkinson's brother-in-law, Joseph Priestley. He was a later member of the Lunar Society, for he did not come to Birmingham until 1780, when he became what was often called a Presbyterian, that is, a Unitarian minister there.

The Unitarians held a view of the Godhead more impersonal than that of the orthodox religions, and they were sometimes accused of being deists and even atheists. They numbered the most advanced and the boldest thinkers of their time. For example, Isaac Newton's form of heresy in the established Church of England had been that he was a Unitarian; and at the very time of which we are speaking, the young poet Samuel Taylor Coleridge was training to become a Unitarian minister. Joseph Priestley had taken part in fundamental religious discussions at Warrington Academy, where these issues of the nature of the Godhead and of personal belief were much debated. He was a vigorous thinker on all such philosophic problems, and on their social meaning. In social matters, he was a pioneer of the new democratic view which inspired the manufacturing classes with the same ideals which were then agitating the colonies in America. In

1768, Joseph Priestley published an *Essay on the First Principles of Government,* in which he put forward for perhaps the first time the view that the happiness of the largest number of people is the standard by which social action must be judged. This is the theory which was taken up and elaborated by Jeremy Bentham to become the central doctrine of utilitarian philosophy.

Priestley's religious ideas were influenced by the fact that, like other Unitarians, he did not want to evade in his thinking what he believed to be the essentially material nature of the biological processes of life. Just as Erasmus Darwin saw the evolution of plants and animals as a single process, so Joseph Priestley saw all living processes as mechanisms. He did not think of life as something different from the physical sequences which govern the mechanics of atoms. In particular, his view of psychology was essentially materialist; he followed a simplified version of the psychology of association which had been propounded by the philosopher David Hartley. The poet Coleridge also was so impressed with this system of psychology that he called his eldest son David Hartley Coleridge.

This mechanistic view of all processes guided Joseph Priestley in his researches as a chemist. He had a strong practical knack: he was the first man to collect gases from chemical reactions systematically over water and over mercury. By this device, he was able to kill the confusion between different "airs" (that is, gases) which had hamstrung chemistry until then. Priestley's major discovery was oxygen, which he showed to be necessary to the burning of all substances. (Therefore when the poet William Blake included Priestley in his satire *An Island in the Moon,* he called him delightfully "Dr. Inflammable Gas.") Until Priestley's time, the process of burning was still thought of as the removal of a substance different in kind from the physical substances, and called phlogiston. Priestley's discovery implied, in effect, that what was thought to be the removal of phlogiston was the addition of oxygen. This discovery was taken by Lavoisier in France directly from Priestley, and Lavoisier made it the foundation for the modern system of the chemical compounds.

In 1791, there was an organized attempt in a number of English towns to stir up riots against the supporters of the French Revolution. In Birmingham it was said that Joseph Priestley was attending a dinner to celebrate the fall of the Bastille—though in fact he was not.

A crowd shouting for church and king went to the meeting house where Priestley preached and to his home and ransacked both; they destroyed most of Priestley's books and apparatus, and set what they could on fire. Priestley himself would have been in danger had he not been kept away by his friends in the Lunar Society. They opened a subscription for him, but Priestley left Birmingham; and in 1794 he went to America, in what turned out to be the unrealistic belief that his radical and materialist views would be welcome there. He died in America in 1804.

VII

The connection of the Lunar Society and other English groups with America was not accidental. There were in the Lunar Society three distinct links with American opinion. One was a general link: when such men as Benjamin Franklin were in England, they were welcomed by the scientific societies wherever they found themselves—and this not merely on scientific grounds, but because there was sympathy between the manufacturing interests and the American Revolution. But there were also two more specific links.

One link was in the ready exchange and movement of distinguished men between the two countries. The English universities were backward and reactionary, and men of cultural and scientific distinction from the nonconformist colleges of England often went to America to teach there. An outstanding example is Dr. William Small, who had a profound influence on Thomas Jefferson.[5] Small returned to England from his American teaching and settled in Birmingham. He died before the Lunar Sociey began to meet regularly, but there is evidence that in the deepest sense William Small was the spirit that set the meetings in motion. Coming from America, he brought to Birmingham the sense of personal exchange which had meant so much to Jefferson: the sense that science and culture are matters which are learned in conversation rather than in formal teaching, and that the meetings of neighbors on moonlit nights are the ideal occasions for this. And William Small, although a teacher of mathematics, also carried with him the practical sense which was upper-

[5] See further the interesting article by Herbert L. Ganter, "William Small, Jefferson's Beloved Teacher," *William and Mary Quarterly* (October, 1947), pp. 505–511.

most in American minds. He was instrumental in bringing the workmanship of Boulton to the attention of James Watt, and in drawing up the business arrangement between them.

Another link of practical sympathy between English manufacturers and American colonists was made by their common disfranchisement from government. Rising manufacturing towns, such as Birmingham and Manchester, had been mere groups of villages until the Industrial Revolution, and their parliamentary representation was still that of scattered villages. Therefore the new manufacturers had no voice in government, and no representation in Parliament.[6] Indeed, when the state of Massachusetts protested in the 1760's that the British government had no right to tax it because no representative of Massachusetts sat in Parliament, the government replied that Manchester had no representative in Parliament either. Thus, English manufacturers felt that Parliament stood for the past, and felt themselves at one with the young colonies who were in revolt against a backward government.

VIII

The Lunar Society in Birmingham was only one of a number of similar societies in England and Scotland. Among these were such formal bodies as the new Society of Arts in London, which was founded to encourage practical and applied science.[7] There were also bodies of working men who met in order to educate themselves and sometimes, in the Sunday schools which they began, to educate their children. All these societies had in common a number of aims and of sympathies, which we now summarize.

The new societies were run by practical men. A hundred years before, the Royal Society had been founded by men who had the aim of Francis Bacon in mind, to derive practical benefits from science. For odd and not wholly bad reasons, the Royal Society had long lost that aim. Under the powerful example of Isaac Newton, scientists had come to think of themselves as remote people. The new societies

[6] For a detailed study of representation in the British political system of the time, see L. B. Namier, *The Structure of Politics at the Accession of George III*, 2 vols. (London, 1929), Vol. I.

[7] Dereck Hudson and Kenneth W. Luckhurst, *The Royal Society of Arts, 1754–1954* (London, 1954), is the official history of the Society of Arts and Manufactures.

broke with this: their view was that which counsel for Richard Arkwright stated so roundly when he was appealing for him in a patent action.

It is well known that the most useful discoveries that have been made in every branch of art and manufactures have not been made by speculative philosophers in their closets, but by ingenious mechanics, conversant in the practices in use in their time, and practically acquainted with the subject-matter of their discoveries.

Not all the members of the new societies were manufacturers. Some of them were many-sided scientists like Priestley; and it is characteristic of Priestley's temperament that from the discovery of oxygen he went at once to the invention of what is, in fact, the oxygen tent in medicine. Some of them were themselves working men. The most famous of these was William Murdoch, who came from Scotland to Boulton's works in Birmingham in order to find work that was worthy of his skill, and who brought with him to show his skill only the hat that he wore—a wooden hat of elliptical shape which he had turned on a lathe. Boulton hired him, and in time made him an important person in his works, though Murdoch was probably not yet respectable enough during the time that the Lunar Society met to be a regular member. Among Murdoch's discoveries was that of gaslight, which he first used in Boulton's Soho Works in Birmingham.

Most of the manufacturers and others in the new societies were nonconformists. Here again, there is echoed the development of the Royal Society: it is the men out of sympathy with the existing university education and government control who make the new science. And the religious dissent of these men was coupled with some political dissent: certainly they were almost all in sympathy with the American Revolution. Later, the French Revolution caused a split of opinion everywhere in England, and this in the end killed many of the societies, as it killed the Lunar Society. But many of the rising manufacturers, and more of their workmen, held to the French Revolution too. It was a friend of Priestley's, another dissenting minister and famous Unitarian theologian and social thinker, Dr. Richard Price, who preached the sermon on the fall of the Bastille which prompted Edmund Burke's *Reflections on the French Revolution,* and thus gave a focus to the great British reaction from 1791 on. And Josiah Wedgwood himself, who was not a hothead, stated the economic views of the new manufacturing classes at the time of the French Revolution.

Politicians here say that we shall have no cause to rejoice of this Revolution, for if the French become a free people like ourselves, they will immediately apply themselves to the extension of manufactures and soon become more formidable rivals to us than it was possible for them to do under a despotic government. For my own part I should be glad to see so near neighbours partake of the same blessing with ourselves, and indeed should rejoice to see English liberty and security spread over the face of the earth without being over-anxious about the effects they might have upon our manufactures or commerce, for I should be very loth to believe that an event so happy for mankind in general could be so injurious to us in particular.

Until the French Revolution caused a bitter political division, and England went to war with France, the attitude to reform and to revolution in the new societies was prompted by such economic thinking. It was from such thinking that the philosophy of Jeremy Bentham began, first in his *Fragments on Government* in 1776. In the same year, there was published the most important economic work in modern history, Adam Smith's *Wealth of Nations;* and one reason for the impression which Adam Smith made on this generation was that the *Wealth of Nations* in effect foretold the American Revolution—and foretold that it would succeed. In the same vein, as Professor Paul Mantoux says, Wedgwood's attitude to the French Revolution "sets forth the principle that the real interests of nations are fundamentally identical, which lies at the basis of the whole of Adam Smith's political economy and of Bentham's utilitarian philosophy."[8]

By contrast, the economic methods of the government were long out of date. William Pitt, who was regarded as a financial genius because he had read the *Wealth of Nations,* and who became the youngest prime minister as well as the youngest chancellor of the exchequer, still taxed the country on the principles of the past. When he wanted to raise money, he taxed windows and hair powder; but the rising moneyed classes of the North did not build many-windowed houses and did not powder their hair. Only in 1798 did Pitt belatedly bring

[8] Paul Mantoux, *The Industrial Revolution in the Eighteenth Century,* p. 393; see further the entire chapter on "Industrial Capitalism," pp. 374–480.

in an income tax. He moved to new ideas in finance after reading the economic works of Dr. Richard Price, whose politics he hated so profoundly.

The gap between government and nation was a gap between minds still bound up with land and with windowed castles and hair powder on the one hand, and the rising industries of the North on the other. This was most deeply an intellectual gap. The trading past of the eighteenth century was done, and its place was taken not simply by different ways of making money, but by a different outlook on all nature. The change from the poetry of Alexander Pope and Thomas Gray to the poetry of William Blake is an example of this change, as characteristic (and as telling and as far-reaching) as is the decay of the Royal Society and the rise of independent bodies such as the Lunar Society. The men in the new societies were not merely manufacturers and scientists: they were men anxious to make a whole culture, and a new culture. Thus, Wedgwood turned the taste of the nation to the simple lines of the Greek vases, and Erasmus Darwin put science into poetry, both under the same impulse. They wanted to create a modern culture which would stretch all the way from abstract thinking to practical living, and from the literature of the past to the science of the present. They did not succeed completely, for no intellectual revolution succeeds in that fashion. But in their own way, the men of the Lunar Society did as much for their age as the men of the Renaissance did for theirs. In the spirit of such men, the political, economic, and social ideas of a Priestley or a Smith, the cultural ideas of a Wedgwood, went hand in hand with, and in essence were part of, the movement which created the new techniques and science of the Industrial Revolution.

CHAPTER 19

Adam Smith

I

As late as the 1790's, the view of the economic life of the nation taken by government in England (and England was far advanced over the Continent in this respect) was essentially the mercantilist view. In the mercantilist system the wealth of the community was thought to be produced only by favorable trade.

We ought not to despise this view of economics; it is the view still held to some degree by the man in the street, who works on the assumption that a nation cannot rise unless another nation falls, and that a nation cannot grow rich unless another grows poor.

It is of course natural for the individual to think that money in his pocket must come out of somebody else's pocket. This view is especially tempting in a stationary society, in which the sum total of goods is for practical purposes constant. In such a society, the individual could maximize his power by making a corner in that part of the fixed market which represented the most widely salable good: for example, spices. And such attempts were made. This was the individual's approach to mercantile economics.

At the level of government, the approach was the same, but the commodity was different. The government, assuming a constant total sum of goods, believed it should concentrate on those goods which were always sure to find a market; and the commodity in demand was the means of exchange itself: gold. Therefore, what the country did was to hoard gold. If the ruler, or even the individual, lived as frugally and as miserly as possibly with gold, he would eventually have a great accumulation and thus a great power over everybody else.[1]

In reality, there was no unified economic system or philosophy of mercantilism but only a loose set of beliefs. The core of mercantilism, however, was the developing territorial, or national, state, and the main purpose of mercantilist theory was to make economic activity subservient to the requirements of the state. The primary requirement was external power in relation to other states. Thus it was not consumption on the part of the subjects but production serving the authority and power of the state which was the aim of mercantilist economic activity.[2]

As to the means of achieving this end, different writers differed. In general, however, mercantilism resolved itself into a "monetary system" and a "protectionist system." It was under both aspects that Adam Smith, for example, attacked it, devoting almost a quarter of the *Wealth of Nations* to the task.

As a monetary system, mercantilism was concerned with the balance of trade. It was considered desirable to have a favorable balance of trade (more is exported than is imported and, therefore, a net gain in bullion is made), not only in general, but with every particular nation. To achieve this, the state fostered production (by subsidies, bounties, etc.) and commerce, hoping thereby to exchange manufactured goods for the precious metals.[3] To prevent the outgo of

[1] It is for this reason that the miser is a prominent figure in early literature. In a merchant community, it was the owner of gold who loomed large—hence all the stories about the power of bankers and about Shylock. Nowadays, in an industrial community, it is the captain of industry who dominates. Our literature and our thought, however, are so filled with misers and hoarders that we tend to forget that they are primarily part of a nonindustrial and trading economy.

[2] Cf. Eli Heckscher, *Mercantilism*, tr. by Mendel Schapiro, 2 vols. (London, 1935), Vol. I, p. 24.

[3] Linked with the interest in commerce was attention to the navy and merchant marine, which, in turn, meant not only trade but power. However, because Adam Smith did not concern himself with mercantilism as a system of power, we are refraining from doing so. Nevertheless, we ought to mention that Heckscher, in his monumental work *Mercantilism*, devotes most of the book to mercantilism as a unifying power; to what the German historian, Gustave Schmoller, referred to as mercantilism in the service of *Staatsbildung*, or state-formation. (See Heckscher, *op. cit.*, Vol. I, p. 28 and *passim*). For a fuller consideration of mercantilism as a monetary and protectionist system, at least as used by the French, see C. W. Cole, *Colbert and a Century of French Mercantilism*, 2 vols. (New York, 1939).

bullion, the state passed sumptuary laws and prohibited foreign luxuries. The furthest step in this direction was the total prohibition of the export of bullion.

As a protectionist system, the mercantilist doctrine favored the use of tariffs and direct prohibitions to prevent the country from being flooded with foreign goods (for which gold would have to be paid). The desideratum of this aspect of mercantilism was self-sufficiency; the fortunate country, one that was blessed with everything it needed.

In sum, the aim of mercantilism was neither to raise the standards of living of its own people nor to contribute to the well-being of other countries; its only aim was to regulate commerce and industry and to manipulate financial policy so that the power of the state might be promoted relative to other nations.

II

The mercantilist view first began to be undermined about the middle of the eighteenth century. It was attacked primarily by a sect of men in France who called themselves Economists, or Physiocrats, and whose motto was "laissez faire."

Impetus to the movement had come during the reign of Louis XIV. Outwardly glorious, France under the Sun King was described by one of its critics as a "rotting white sepulchre." In fact, Louis' wars and financial policies left France in 1715 with a debt of 3,460 million francs, of which over 3,300 million had been contracted since the death in 1683 of Colbert, Louis' mercantilist finance minister—a vivid proof that wealth was used in forwarding the power of the state and not the prosperity of the people.[4]

Under the absolutism of Louis XIV, it was difficult to protest. Some voices, however, as we remarked in Chapter 14 on Voltaire, were raised. Before the end of the seventeenth century, the priest Fénelon, the great military engineer Vauban, and his cousin, Boisguillebert, wrote in favor of reform; Boisguillebert, in fact, was the first to talk of "laissez faire" and "laissez passer." But their desires for reform were

[4] The worst features of mercantile economics were present. Government monopolies were enforced, and new types of manufacture were restricted and prohibited. Heckscher claims that the 1686 prohibition of cotton manufacturing cost 16,000 lives and led to the migration of the best workers (Vol. I, p. 173). The revocation of the Edict of Nantes, in 1685, as we have seen, further swelled the emigration of French workers.

not backed by a systematic analysis of economics, a task they left for the Economists, or Physiocrats.

By the middle of the eighteenth century a flurry of interest in political and social subjects, and especially in agriculture, was sweeping France. It was in 1748, for example, that Montesquieu published his *Spirit of Laws*. Further, the appointment of the liberal Malesherbes as censor permitted the publication of a *Journal économique* (1751), and that of Gournay to the important post of intendant of commerce gave a chance to the practical reformers to institute a laissez faire system.[5]

This time, however, economic reform was backed by an economic theory. Under the stimulus of English writers like Culpeper and Cantillon and of Tory free traders like Child and Hume, the French produced a sect, as Adam Smith described them, "distinguished in the French republic of letters by the name of The Economists."[6]

The leader of this group was François Quesnay. He was the physician to Madame de Pompadour (mistress of Louis XV) and used her salon to spread his new ideas. After contributing a few articles on economic subjects to the *Encyclopedia,* Quesnay printed his *Tableau oeconomique* in 1758. A year earlier, Quesnay had met and converted the Marquis de Mirabeau to his views, and around these two figures arose the sect of Economists. Dupont de Nemours, who had been convinced in turn by Mirabeau, wrote a summary of the new ideas, called *Physiocratie* (1767–1768), from which the sect derived its later description, "physiocracy," or "government by nature."[7]

Quesnay also strongly influenced Adam Smith. Indeed, Smith would have dedicated the *Wealth of Nations* to Quesnay, had the latter not died before its publication. In his masterpiece, Smith said of the Physiocratic system: "This system, however, with all its imperfections is, perhaps, the nearest approximation to the truth that has yet been

[5] This, however, met with severe opposition from entrenched interests, and the to-and-fro movement of the fight can be seen in the relaxing of the grain laws in 1754, a further extension in 1763, a suspension of the free-trade law in 1770 and then an even more liberal extension, under Turgot, in 1774.

[6] Adam Smith, *The Wealth of Nations,* Everyman's Library ed., 2 vols. (London, 1954), Vol. II, p. 172. This is a useful edition. The definitive edition is by Edwin Cannan; it is reprinted in the Modern Library series.

[7] The same Dupont de Nemours, after working for Turgot as inspector-general of manufactures and serving in the Constituent Assembly during the French Revolution, voluntarily exiled himself to America in 1793. It is from this Dupont that the present firm of chemical manufacturers is descended.

published upon the subject of political economy, and is upon that account well worth the consideration of every man who wishes to examine with attention the principles of that very important science."

It was characteristic of the Physiocrats that they attacked the mercantilist position by stressing the primacy of land and with it the notion that rent was the center of wealth in the community. They singled out the soil (from which rent was derived) and, accordingly, contended that agriculture was productive and industry sterile.[8] Such a view was typical of French landed society: no important doctrine of this kind arose in England or elsewhere.

A further sign of the originality of the Physiocrats was that they started with a kind of flow sheet of society. Even though in their flow sheet the total of goods was still kept constant, or nearly constant, they did begin to take a view of society which was quite different from the static mercantilist view. Their attention to the distribution of wealth and its "circulation" meant that, in this sense, their economics was dynamic.

Under the influence of the Physiocrats, Adam Smith pushed this view further. He concentrated, too, on the origin and circulation of wealth. What he changed was the central good that was being pushed around—the central source of wealth—hitherto regarded by the Physiocrats as rent. Smith, in the growing manufacturing and trading community of England, realized that, although land was fundamental to the creation of wealth, so, too, was labor. And on this basis he went beyond the Physiocrats and formulated his famous "labor theory of value."[9]

[8] It can be said that this marks the first modern analysis of economic classes, an analysis which was accepted and modified by Adam Smith and Karl Marx. The physiocratic problem of the conflicting interests of the "productive" and the "sterile" classes could be solved either by an acceptance of class conflict or by the control of that conflict through a coercive power. The Physiocrats favored the latter: what they called "legal despotism." Contending that everything was "for the best," and that the clash of interests was resolved by God into a harmony, the Physiocrats emphasized that the function of government was to preserve security of property. For this there was no need of representative government or a reliance on a "general will." The "legal despot" needed merely to uphold the natural order of things, not to interfere with it. (This Physiocratic doctrine of "legal despotism" should be proof that there is no *necessary* link between economic and political liberalism.)

[9] For a general consideration of the Physiocrats, see Georges Weulersse, *Le mouvement physiocratique en France*, 2 vols. (Paris, 1910), as well as his *Les*

III

Adam Smith was a Scotsman, and this in itself was not atypical. People like James Watt, Murdoch, and Hume had all gone south to England. Opportunity was small in Scotland, and the pressure on the bright young men to go out and make their way in the world was very large. Such minorities exist in many nations; the Scots occupied the same sort of "heterodox" role as the Huguenots, the Jews, and the Welsh, to name only a few. The Scots were, of course, frugal and nonconformist, and were very much disliked at this time by the English.[10]

Smith was born in the little town of Kirkcaldy, a few months after his father's death, in 1723. After preliminary schooling in his home town, he went, at 14, to the University of Glasgow, where he studied mainly mathematics and Greek, and where he attended the lectures of Dr. Hutcheson, to whose chair in moral philosophy he was later to succeed. At 17 he went to Oxford University, and there, instead of taking religious orders as his mother had wished, he spent six years studying literature (Greek, Latin, Italian, French, and English). Upon his return to Scotland, after two years of virtual unemployment, he eventually became professor of logic (1750) and then, in 1752, professor of moral philosophy at Glasgow.

Physiocrats (Paris, 1931); and, in English, Henry Higgs, *Physiocrats* (London, 1897). The labor theory of value was not original with Smith. Sir William Petty and, before him, Vauban, had stated the principle. What was original in Smith was his emphasis on, and development of, the idea.

[10] Indeed, we can see the English attitude toward Scotsmen vividly if we glance at the works of Dr. Samuel Johnson. Johnson, who compiled the first English dictionary, filled it with derogatory remarks about Scotsmen. For instance, he defined oats as food for horses in England and for men in Scotland. For a description of the prejudiced treatment of the Scots at Oxford when Smith was there, cf. John Rae, *Life of Adam Smith* (London, 1885), p. 26.

Unfortunately, there is no really first-rate, full-scale, modern treatment of Smith's life and works. Rae's is good, but it is limited and outdated. For special aspects of Smith, however, see C. R. Fay, *Adam Smith and the Scotland of His Day* (Cambridge, 1956), and William Robert Scott, *Adam Smith as Student and Professor,* with unpublished documents, including parts of the "Edinburgh Lectures," a draft of the *Wealth of Nations,* extracts from the Muniments of the University of Glasgow, and correspondence (Glasgow, 1937). See, too, Scott's *Studies Relating to Adam Smith During the Last Fifty Years,* ed. by Alec L. Macfie (from the *Proceedings of the British Academy,* Vol. XXVI, London, 1940).

Thus, Smith's early academic post was as a philosopher; and, indeed, his first important book, *The Theory of Moral Sentiments,* brought him fame in 1759 as a moral philosopher. He also wrote a number of other books on abstract philosophical subjects, but most of these were destroyed, on his instructions, after his death.

Gradually, however, Smith's mind seems to have turned from philosophical subjects *per se* to economics treated philosophically. It is true that, from the very beginning, Smith followed Francis Hutcheson's division of moral philosophy into four parts: Natural Theology, Ethics, Jurisprudence, and Political Economy, and gave his lectures accordingly. And we know that in 1763 he was already lecturing on material that was later published (although not until 1896) under the title *Lectures on Justice, Police, Revenue and Arms.* We have further proof of Smith's interests: the draft of another treatise, *On Public Opulence,* was made by him just before his trip to Europe in 1764.[11]

There is evidence to show that Adam Smith had formed for himself liberal and large views of trade early in life. We know that by 1752 he had become friendly with the philosopher David Hume, also a Scotsman. Hume, although he never wrote a major work directly on the subject of economics, wrote some essays on the subject and included a good many remarks about commerce in his general works—all of them in support of free trade.[12] It seems clear that Smith was strongly influenced by Hume.

What stimulated Smith to go further and write his great work on economics was the offer made to him to go to the Continent as a private tutor for the Duke of Buccleuch. Smith resigned his professorship at Glasgow and traveled, about two and a half years, on the Continent, spending most of the time in France. During that period he began writing his book on economics, the *Wealth of Nations,* or, in full, *An Inquiry into the Nature and Causes of the Wealth of Nations.*

The impulse to begin the book came, at first, from boredom. As he wrote to Hume from Toulouse (where he and his pupil were staying for eighteen months), "The life which I led at Glasgow was a pleasur-

[11] Like the *Lectures, On Public Opulence* was not published by Smith; in fact, it was not published until 1937.

[12] Hume's economic views were mainly expressed in a collection of essays, called *Political Discourses,* which appeared in 1752.

able dissipated life in comparison of that which I lead here at present. I have begun to write a book in order to pass away the time."[13] Thus came into existence the modern science of economics.

Hume later relieved Smith's boredom in another way, by introducing him to the prominent literary figures of the Paris salons. In this manner, Smith met such men as D'Alembert, Holbach, Helvétius, and the Abbé Morellet as well as Turgot and the founder of the Physiocrats, Quesnay. The influence upon Smith of the last two men was important; that of Quesnay, as we have indicated, was fundamental.

Upon his return from France, Smith retired to his home in Kirkcaldy, near Edinburgh. Here, almost in seclusion, he spent the next ten years writing his work on economics. The book, from the moment of its appearance, was hailed as a monumental work, and Hume wrote to Smith in great relief:

> EUGE! BELLE! DEAR MR. SMITH,—I am much pleased with your performance, and the perusal of it has taken me from a state of great anxiety. It was a work of so much expectation, by yourself, by your friends, and by the public, that I trembled for its appearance, but am now much relieved.

Relieved of his anxiety, Hume then added: "If you were here at my fireside, I should dispute some of your principles."[14]

The *Wealth of Nations* was published in 1776, a date easily remembered because the American Revolution broke out in that year. One of the things that made the *Wealth of Nations* nicely topical was that it contained a sentence saying that he, Adam Smith, expected the American nation beyond the sea to become one of the foremost nations in the economy of the world. This prophecy had a great effect in England when the book was published; and later events have proved Smith's foresight.

After his success, Smith had some schemes for going on with his ideas about economics and placing them in a general moral and philosophical system. He had intended, he said, to establish a system "of what might properly be called natural jurisprudence, or a theory of the general principles which ought to run through and be the foundation of the laws of all nations." Instead, however, he involved himself from 1778 on in the sinecure of commissioner of the

[13] Quoted in R. B. Haldane, *Life of Adam Smith* (London, 1887), p. 31.

[14] Quoted in *ibid.*, pp. 35–36. Hume, for example, insisted that high prices were the cause, not the effect, of high rents.

customs in Kirkcaldy; and he resorted often to Edinburgh society, especially at the club whose members included such figures as Dr. Black, the discoverer of the principle of latent heat; Hutton, the geologist; Dr. Adam Ferguson, a famous Scotch philosopher; John Clerk, the naval tactician; and Robert Adam, the architect.

IV

Smith, by writing the *Wealth of Nations,* had earned his relaxation. With that book, indeed, he had become the father of modern economics. As we have seen, his discovery of economics emerged from his general work as a philosopher, for he had originally treated the subject of political economy as part of his lectures on moral philosophy. We are, therefore, fully justified in saying that he founded the science of economics, as we know it, in the same way as the Greeks had founded the science of physics—out of philosophy.

Practical businessmen before Smith had elaborated a "political arithmetic," dealing with problems, as they arose, of credit, banking, or balance of trade; in the course of these treatments a mass of statistical information had been collected. On the other hand, enlightened thinkers in France had built up a speculative system, based on the notion that natural laws were in as much control of society's economic movements as of the movements of physical nature.

What Adam Smith did was to base his theory on the work of men who shared the notions of the enlightened climate of opinion—Quesnay was the counterpart, although not the intellectual equal, of Montesquieu, Voltaire, and Rousseau—and on the empirical data amassed to back up practical demands for reform. Like Newton before him, who combined the rational and empirical into one scientific method, Smith joined French "physiocracy" with English "political arithmetic" to form a superior science of economics.[15]

V

What were Adam Smith's own and specific contributions to economics and to thought in general? He made essentially two contributions. First of all, he introduced the historical method into

[15] Although Smith had reached many of his economic ideas before his trip to France, as evinced by his treatise *On Public Opulence,* we feel justified in believing that the *systematic* form of his science stemmed from the Physiocratic influence.

economic discussion, and secondly, he gave a central place in his economics to the value and to the division of labor. These are, however, two quite different achievements, and separate attention must be given to each.

First, the historical method. In the hundred years that preceded Smith, the deductive method of trying to set up physics or the laws of society or anything else was dominant. A glance backward to Hobbes or Locke shows us their method. They started with certain propositions about what human beings are like in a simple state of nature. These propositions, with almost no factual background at all, are in effect a set of axioms. They are not wild inventions, as we have tried to show, but they are essentially the axiomatic background from which everything is deduced.

Adam Smith did not believe in an axiomatic background; and his book is not presented so that it looks like a book of propositions. Instead, the *Wealth of Nations* was written in an amiable, chatty way, and for this reason, it is difficult to find one's way about in it. True, the book was ordered by a strong intelligence with a very clear idea of what he was talking about; but willing to be repetitious and long-winded, if necessary, to be understood. Thus, the style reflects the fact that Smith had little belief that an economic system consists of a set of fundamental axioms from which further propositions are deduced. As early as the *Theory of Moral Sentiments,* he had poked gentle fun at "A propensity which is natural to all men, but which philosophers in particular are apt to cultivate with a peculiar fondness, as the great means of displaying their ingenuity—the propensity to account for all appearances from as few principles as possible.[16]

[16] *The Theory of Moral Sentiments,* in Adam Smith's *Moral and Political Philosophy,* ed. with an introduction by Herbert W. Schneider (New York, 1948) (hereafter referred to as *Theory*), p. 33. Albion W. Small, *Adam Smith and Modern Sociology* (Chicago, 1907), is an unimaginatively executed attempt to investigate the imaginative idea that "the apostolic succession in social philosophy from Adam Smith is through the sociologists rather than the economists." Small's perception, however, is that "nineteenth-century economic theory was at bottom an attempt to discover the principles of honorable prudence, not to codify a policy of predatory greed" and that economic theory was, therefore, only one part of moral-social philosophy. *Adam Smith, 1776–1926* (Chicago, 1928), includes commemorative lectures by J. M. Clark, Paul H. Douglas, Jacob H. Hollander, Glenn R. Morrow, Melchior Palyi, and Jacob Viner. Douglas' lecture on "Smith's Theory of Value and Distribution" is especially valuable, as is Morrow's "Adam Smith: Moralist and Philosopher."

There were many reasons why Adam Smith did not think that economics was a pure, axiomatic science; we would like to point to one which is of especial interest. The great discovery that Hume made, and he made this as a very young man, was that, ever since the time of Hobbes, people had incorrectly supposed that the laws of cause and effect had the same kind of finality and rational certainty as the laws of logic. For attacking this notion he was at once called an atheist, and generally blackballed by all decent members of society, who had spent a hundred years struggling with the idea that the universe really worked by cause and effect and were not going to have some Scottish whippersnapper upset the notion that had been so difficult to learn.

Smith did not go as far as Hume in freeing himself from the bondage of cause and effect, but he achieved a healthy skepticism. Smith employed cause-and-effect relations when he said, "The constancy and steadiness of the effect supposes a proportionable constancy and steadiness in the cause" or "The suddenness of the effect can be accounted for only by a cause which can operate suddenly." But he also handled these relations in a more sophisticated way. He was aware that the same cause can have different effects: "The rise and fall in the profits of stock depend upon the same causes with the rise and fall in the wages of labour, the increasing or declining state of the wealth of the society; but those causes affect the one and the other very differently"; that effects may be mistaken for causes: "The carrying trade is the natural effect and symptom of great national wealth; but it does not seem to be the natural cause of it. Those statesmen who have been disposed to favour it with particular encouragements seem to have mistaken the effect and symptom for the cause"; that the mere fact of being posterior in time does not make a thing an effect: "Though the period of the greatest prosperity and improvement of Great Britain has been posterior to that system of laws which is connected with the bounty, we must not upon that account impute it to those laws."

In general, Adam Smith was wary of simple causal explanations and preferred to treat economics as an historical rather than as a logical or even as an empirical science. Hume had written a fine *History of England* and thereby given a strong impetus to the historical method. Smith followed his friend and filled the *Wealth of*

Nations with historical interludes. It will be well to give some idea of the subjects Smith covered. He traced the historical origin of freedom; the history of Rome; the origin of city independence; the rise of the third estate; and the changes in war and militias. He investigated the origin of money and gave an extensive disquisition on the history of price movements. He also devoted a large chapter to the history of colonies, in the process of which he formulated a refined theory of imperialism.

In part, Smith anticipated Marx in the theory of economic determination of history. Marx's formulation is more dogmatic, but was influenced by this idea of Smith (as well as by the idea of the labor theory of value, which we shall discuss in a moment). Smith, who had little respect for feudal society, contended that the great lords lost their power when they chose to sell their surplus produce to the towns in exchange for luxuries. In Smith's words: "But what all the violence of the feudal institutions could never have effected, the silent and insensible operation of foreign commerce and manufactures gradually brought about. These gradually furnished the great proprietors with something for which they could exchange the whole surplus produce of their lands, and which they could consume themselves without sharing it either with tenants or retainers." Thus the merchants unintentionally supplied the nobles with the instrument of the latter's own destruction: luxuries. Instead of using their surplus produce to maintain tenants and military followers, the real source of their power, the nobles sold the surplus to the towns for "trinkets and baubles." It was in this manner that, to borrow Hegelian terms, the "dialectic of history" played itself out. As Smith phrased it:

> A revolution of the greatest importance to the public happiness was in this manner brought about by two different orders of people who had not the least intention to serve the public. To gratify the most childish vanity was the sole motive of the great proprietors. The merchants and artificers, much less ridiculous, acted merely from a view to their own interest, and in pursuit of their own pedlar principles of turning a penny wherever a penny was to be got. Neither of them had either knowledge or foresight of that great revolution which the folly of the one, and the industry of the other, was gradually bringing about.[17]

[17] *Wealth of Nations*, Vol. I, pp. 369–370.

VI

Adam Smith's second major contribution to economics was his theory that the source of all wealth is the labor of the people who produce it. The mercantilists, as we have seen, had no real theory about the origin of wealth because they believed that wealth was more or less constant. The Physiocrats held, generally, that wealth resided in land; they believed that new wealth, i.e., agricultural products, was created as a gift of nature, and that man's industry merely transformed but did not add to this wealth. Adam Smith, however, held that wealth is created by man and he believed that man creates it by work.[18]

Thus, in the *Wealth of Nations,* Smith stated that the wealth of a community increases by the amount of work which it puts out. But he did not leave this as a superficial statement. He traced everything back to the original labor. He did not say merely that wealth was added to a bale of wool when you made it into cloth; he traced the labor back to the shepherd, and then to the man who sowed the grass on which the sheep fed, and so on in an exhaustive and complex fashion.

Adam Smith was aware that improved productivity comes most readily from the division of labor. In his analysis he pointed out that division of labor fosters increased dexterity on the part of the worker, saves valuable time, previously wasted in passing from one phase of work to another, and, through the use of machinery, facilitates and abridges labor. The famous example Smith used to illustrate these advantages was the pin-making trade.

Unfortunately, Adam Smith often treated labor as a commodity rather than an activity; very much like the things one buys and sells. For example, he writes: "It is in this manner that the demand for men, *like that for any other commodity* [our italics], necessarily regu-

[18] What we are propounding here is Smith's theory of the origin of wealth, i.e., labor. It must be distinguished from his theory of exchange value. When, nearly 100 years later, Karl Marx elaborated his labor theory of value, he really did little more in this respect than reproduce Adam Smith; he had an immense and open respect for Smith. As for Smith's theory of exchange value *per se,* which we are not discussing here, it wavered between the view that the value of a commodity is determined by the amount of labor required to produce it and the view that it is determined by the amount of labor it can purchase. On this whole question, see Douglas' article on "Smith's Theory of Value and Distribution," in *Adam Smith, 1776–1926.*

heart of his theory: a continuously advancing and expanding economy could only exist in an atmosphere of laissez faire. The worker, as well as the manufacturer, had to be free to pursue his own advantage, without let or hindrance.

VII

This leads us directly to what is called Smith's theory of "self-interest." One difficulty in following his account of self-interest is that he had discussed the matter thoroughly in the *Theory of Moral Sentiments;* and he assumed that the reader of the *Wealth of Nations* would not think that he, Smith, considered self-interest the only or even the main motive, or virtue, of humanity. His teacher, Hutcheson, indeed, had taught that the only virtue was benevolence; but Smith, while agreeing that this was the major virtue and the one which aimed "at the greatest possible good," felt strongly that the system of benevolent ethics was too simple and left no room for the "inferior virtues." Therefore he devoted himself to a more naturalistic theory of morals, in which man's nature was accepted as it was. He felt that it was not beneath the dignity of man to admit that "Regard to our own private happiness and interest, too, appear upon many occasions very laudable principles of action. The habits of economy, industry, discretion, attention, and application of thought, are generally supposed to be cultivated from self-interested motives, and at the same time are apprehended to be very praiseworthy qualities which deserve the esteem and approbation of everybody."[22]

Benevolence was undoubtedly the highest morality, but the humdrum, workaday world ran on the motives of self-interest. It was the latter motive which drove men to the division of labor, the basis of society's affluence. In a famous passage, Smith declared that: "It is not from the benevolence of the butcher, the brewer, or the baker that we expect our dinner, but from their regard of their own interest. We address ourselves not to their humanity, but to their self-love, and never talk to them of our own necessities, but of their advantage."

Smith's task in the *Wealth of Nations* was to analyze the butcher's, brewer's, and baker's world. From his work on morality, however, he

[22] *Theory,* p. 39.

brought one other very important notion to this task—a belief in Providence. According to Smith, "Human society, when we contemplate it in a certain abstract and philosophical light, appears like a great, an immense machine whose regular and harmonious movements produce a thousand agreeable effects." Even his belief in cosmopolitanism is grounded in his belief in Providence. Thus, the wise man realizes that the subordination of himself and of his country to the interests of mankind is for the "good of the whole"; that he must trust himself in this matter to the "great Conductor of the universe." On the subject of Providence, Smith could become lyrical. "The idea of that divine Being whose benevolence and wisdom have, from all eternity, contrived and conducted the immense machine of the universe so as at all times to produce the greatest possible quantity of happiness, is certainly of all the objects of human contemplation by far the most sublime."

In the *Wealth of Nations,* Smith combined the two doctrines: God's providential benevolence and man's earthly self-interest. The result is his famous "invisible hand" theory in which the individual, intending only his own gain, is led "to promote an end which was no part of his intention," the well-being of society.

The view that personal self-interest is the best regulator of public affairs had been put forward before; it is expressed in Bernard de Mandeville's phrase, *Private Vices Made Public Benefits.* When Smith wrote, this view was already familiar to eighteenth-century thinkers. What Smith did was to give it a reasoned economic exposition which made it acceptable and, so to speak, respectable. From then on, the inevitable benefits of self-interest became a doctrine to which rising manufacturers and owners of newly enclosed land constantly appealed.

A close inspection of Smith, however, shows how ridiculous it is to say that he favored the uncontrolled pursuit of self-interest.[23] Instead, he was constantly inveighing against the farmers, the workers, the manufacturers, and the banks on exactly this point and complaining that they did not understand their own particular interests. For example, on the question of paper money, after an exhaustive analysis of the real factors involved, Smith said: "Had every particular bank-

[23] Cf. Lionel Robbins, *The Theory of Economic Policy in English Classical Political Economy* (London, 1952).

ing company always understood and attended to its own particular interest, the circulation never could have been overstocked with paper money. But every particular banking company has not always understood or attended to its own particular interest, and the circulation has frequently been overstocked with paper money." He chided the mercantilists that their very cupidity, by imposing a heavy duty on certain goods, called into being a smuggling of the goods which ruined their business.[24] Country gentlemen were told that in their demand for a bounty on corn "they did not act with that complete comprehension of their own interest" which should have directed their efforts.[25]

We could extend this list almost indefinitely. One of the very practical purposes of the *Wealth of Nations* was to enlighten self-interest by pointing out its *real* interest. Smith was sophisticated enough to realize that it is the long-range effect of an action and not the immediate and obvious one which is to be attended to; and that this requires an understanding of principles. It was for this reason that Smith entitled his book an *Inquiry into the Nature and Causes of the Wealth of Nations.*

Smith was quite adamant that experience without explanatory principles was insufficient. For example, as he said: "That foreign trade enriched the country, experience demonstrated to the nobles and country gentlemen as well as to the merchants; but how, or in what manner, none of them well knew." None of "them" might know, but Smith did. His method was to form out of experience an abstract principle, to state this as a general rule and to give evidence and examples to support it. Thus, he and his science of economics could show "how" and "in what manner."

In order to discover such a science of economics, however, Smith had to posit a faith in the orderly structure of nature, underlying appearances and accessible to man's reason. It was essentially the same faith which, long before, had inspired the researches of Copernicus, Kepler, and Galileo—the faith that nature moves in a regular and harmonious fashion in spite of the bewildering variety of appearances. This, in our judgment, is what Smith really meant by the "invisible hand"; that, so to speak, an "order of nature" or a "structure of

24 *Wealth of Nations,* Vol. II, p. 152.
25 *Ibid.,* p. 17.

things" existed which permitted self-interest, if enlightened, to work for mankind's good.

Man's task, therefore, was to understand the nature or structure of things and to adjust himself harmoniously to the necessary results of this structure. On one level, this might mean the acceptance of a "natural" price of things (reached when the supply, whether of goods or of labor, exactly equaled the demand).[26]

On another level, Smith applied his faith in a structure of things when he said: "A nation of hunters can never be formidable to the civilized nations in their neighbourhood. A nation of shepherds may." This is true, he thought, because the nature of hunting is such that large numbers cannot indulge in it; the game would be exterminated. On the other hand, shepherds can grow in number as their flocks grow; and can carry war into the hearts of civilized nations because they carry with them their food supply.

On the highest level, there existed a "natural course of things" or, as Smith more frequently phrased it, "the natural progress of things toward improvement." The principle behind this "natural course" was "the uniform, constant, and uninterrupted effort of every man to better his condition." Thus, self-interest drives civilized society; the duty of the scientist and the legislator is not to castigate it but to accept and enlighten it. In so doing, they will act in harmony with the order or structure of things.

Smith was fully aware that a totally free-trade or "natural" society was a Utopia. The main obstacles to the realization of such a "natural" society were prejudices (i.e., unenlightened ideas) and partial interests (i.e., the desire for gain at the expense of the general good; ultimately, therefore, at the expense of one's own eventual good). He had no illusions that either of these barriers could ever be fully overcome. But he did believe that *gradual* change in the right direction was possible; and that his book, by truly exposing the nature and causes of the wealth of nations, would indicate to the legislator, scientifically, what that right direction was. Whether men would choose to proceed in that right direction was another matter.

[26] This, of course, was an equilibrium scheme, and Smith realized that, in practice, there would always be an over- or an under-supply in relation to demand; but, if capital and labor were left free to move about, they would inevitably tend toward equilibrium, i.e., harmony.

VIII

What effect did Smith's work actually have? First, it gave the rising manufacturers and merchants a rationale for their desire to change existing government policy. (Existing policy, as we have pointed out, favored the older trades, methods, and classes against the new "Lunar Society" type of individual and enterprise.) Thus, for example, it helped Pitt to pass a free-trade agreement, the Eden Treaty of 1786 with France, through Parliament.[27] And Pitt openly acknowledged his debt to Smith. The story is told that, at a dinner party, Smith was late and apologized. The whole company (men like Pitt, Grenville, Addington, and other British leaders) rose, and Pitt exclaimed: "We will stand till you are seated, for we are all your scholars."[28]

The second effect of Smith's work was in the shaping of thought. His influence in introducing historical method into political economy was far-reaching. He made the foundation of all subsequent economics the notion that wealth was created by labor. But, more than any of these things, he introduced science into the study of economics. Although he talked much about the "invisible hand" and the "natural course of things," Smith really freed man from the tyranny of chance by forming for him the analytical tools with which he might learn to control his economic activities. This he did in spite of the fact that selfish men later twisted his ideas so as to make it seem that Smith was subjecting man to cold, harsh economic forces beyond his power to control.

Smith provided the stalking horse behind which the new manufacturers of England could peacefully accomplish the "revolution" of freeing Britain from outworn and outmoded regulations and hindrances to economic progress. He was able to do this because his free-trade doctrines were fortunate in fitting the necessities of the time —a time when British industries were ahead of their neighbors', and were therefore sure to benefit by free rivalry.

This, in itself, was a momentous reason for Smith's importance. It

[27] The Eden Treaty worked to the benefit of England, because its industries were more advanced than the French; the latter, however, after a short trial of the "advantages" of free trade, abandoned the treaty and returned to a policy of protectionism. On free trade and protectionism in France, see especially Shepard B. Clough, *France: A History of National Economics, 1789–1939* (New York, 1939), Chap. 1.

[28] See Haldane, *op. cit.*, p. 49.

pales, however, in comparison with his second service to mankind. He provided a lasting contribution to man's ordering of his material life. Thus, Smith ranks high in the history of ideas as the philosopher who, exemplifying his own belief in the division of labor, founded the specialized science of economics and gave to the mind of man another means of controlling the world of matter.

CHAPTER 20

Benjamin Franklin

THERE IS no single type of eighteenth-century man; there is not even a type of eighteenth-century Englishman or a type of eighteenth-century American. Yet, there is a type of man associated with the rise of industry in the eighteenth century: direct, self-reliant, energetic, and humane, with a personal sense of dignity and of mission—a natural dissenter happily entering a new world which kept its promises. By a pleasant irony, the greatest of all such men came not from industrial England but from the New World: he was Benjamin Franklin.

I

One common feature of the new men of the industrial age is that they were given a start in their careers by a technical skill. The Industrial Revolution was an age of machinery, and in our minds we couple the rise of machinery with the displacement of craftsmen, and with the employment of unskilled women and children. This picture is just, but it is too narrowly focused: it concentrates on the disastrous foreground of the Industrial Revolution. Beyond the immediate displacement of the old crafts, however, there opened a development of new technical skills which were essential to the invention and the servicing of the machines. In the same way, we tend today to think of automation as merely displacing workers, and we forget that the new machines require in the long run a higher level of technical skill in the rising generation of men who must engineer and service them.

The leaders of industrial thought in the eighteenth century were men who had risen by their own skill, and it was a technical skill. It is characteristic that Benjamin Franklin, who was born in Boston in 1706 as the fourteenth child of a candlemaker, refused his father's trade, and at the age of 12 preferred to be apprenticed to his older brother as a printer.[1] Printing was a skilled and up-to-date occupation, with something of the same technical distinction that watchmaking and foundry work had. It gave a man the right to think of himself as a leader among working men, with a pride in his own status. Although Franklin gave up his printing business in 1748, when he sat down to write his will in 1788 he began it firmly with the words "I, Benjamin Franklin, of Philadelphia, printer." It was a statement that he still belonged to the community of skilled artisans and mechanics who had in his lifetime remade the outlook and the mode of living of the world.

Franklin describes in his delightful *Autobiography*[2] how he ran away from his brother, who bullied him. It is characteristic of Franklin that he adds the note, "I fancy his harsh and tyrannical treatment of me might be a means of impressing me with that aversion to arbitrary power that has stuck to me thro' my whole life"; and equally characteristic that he goes on to say of his brother that "he was otherwise not an ill-natured man: perhaps I was too saucy and provoking."

The young Benjamin became a printer's assistant in Philadelphia, attracted the patronage of the governor and, at the age of 18, was sent to England to buy type in order to set up his own printing house. This mission turned out to be a mere fancy of the governor's, who had no credit in London, so that Franklin found himself stranded there. But as a printer, he had no difficulty in getting work, and he stayed in London for a year and a half, until the summer of 1726.

[1] The facts of Franklin's life are admirably presented, and his achievement is assessed, in Carl Van Doren's biography *Benjamin Franklin* (New York, 1938). A readable account of Franklin seen through English eyes is provided in the biography by Evarts S. Scudder, *Benjamin Franklin* (London, 1939). There is also a stimulating and unusual assessment by Bernard Fay, *Franklin the Apostle of Modern Times* (Boston, 1929).

[2] Franklin's *Autobiography* is reprinted in Vol. I of *The Writings of Benjamin Franklin,* ed. by Albert Henry Smith (New York, 1905–1907). This ten-volume collection of Franklin's writings will be superseded by *The Papers of Benjamin Franklin,* which are now being edited for the American Philosophical Society and Yale University by Leonard W. Labaree and Whitfield J. Bell, Jr. We therefore give references to quotations from Franklin as they appear in books which will continue to be current, and chiefly in the Selection by I. Bernard Cohen, *Benjamin Franklin: His Contribution to the American Tradition* (New York, 1953).

II

Franklin's early stay in London gave him a sense of the unity of the intellectual world which sustained him all his life. He began to impress European thinkers and to make friends with them. For example, he met Bernard de Mandeville, whose *Fable of the Bees* is one of the most trenchant formulations of the eighteenth-century principle which is stated in its subtitle: *Private Vices Made Public Benefits.* This is the moral that every member of society, by the act of seeking his own advancement, contributes to the balanced functioning of society—in short, that competition is a better regulator of society than any form of planning. Since the rising manufacturing class was handicapped by many old government regulations and traditions, it came to make this doctrine of economic laissez faire central to its outlook later in the century. Franklin was at home with the new economics early, and later became a personal friend of many of its advocates—of Adam Smith, Joseph Priestley, Richard Price, and Jeremy Bentham in England, and of Quesnay and the Physiocrats in France. It is fairly certain that Adam Smith's remarks about America in the *Wealth of Nations* reflect the personal influence of Franklin on Smith in later life.

Franklin was much concerned about economic doctrines, as were all advanced thinkers at that time, when a new economic order was being created. In the debating club or junto which he founded when he returned to Philadelphia, such topics as "Is self-interest the rudder that steers mankind?" were prominent. There was also a lively argument in Philadelphia whether the amount of paper money in circulation should be increased. The division of opinion ran as might be expected. "The wealthy inhabitants opposed any addition, being against all paper currency." Those whose livelihood depended on the expansion of trade thought differently. "We had discussed this point in our Junto, where I was on the side of an addition, being persuaded that the first small sum struck in 1723 had done much good, by increasing the trade, employment, and number of inhabitants in the province, since I now saw all the old houses inhabited, and many new ones building." Franklin wrote and printed an anonymous pamphlet called *A Modest Enquiry into the Nature and Necessity of Paper Currency,* in which he supported the issue of more paper money by claiming that the true source of its value, and of all value, is labor. Thus, Franklin pronounced a labor theory of value before Adam Smith did; and his pamphlet was singled out by Karl Marx as one of the first to state this theory. In fact, however, Franklin's economic arguments in his pamphlet seem to lean heavily on the earlier writings of Sir William Petty.

It is fair to report that the advocates of more paper money won the day, and that they gave some credit for their success to Franklin. As he himself writes, those who had been on his side "thought fit to reward me, by employing me in printing the money, a very profitable jobb, and a great help to me."

III

It is not easy to define Franklin's religious and moral beliefs; yet it is important to do so, because they are representative of a large body of men of his time, whose worldly success certainly derived from their beliefs. D. H. Lawrence, who was angered by all success, treats Franklin simply as a hypocrite or, worse, as an amoral man who found the rules which lead to success and turned them into a religion.[3] This analysis is certainly false; but even if it were true, it would not take us far enough. For it would not tell us what made Franklin respected by men so different as his American friends, his English enemies, and his French admirers. There was something in Franklin's beliefs which had a symbolic quality for them all.

The charge that Franklin was a hypocrite can be presented simply. He advocated many virtues at a time when he undoubtedly lapsed into some vices. He began his marriage in 1730 by bringing an illegitimate son into the house. Indeed, he may never have been very vigorous in resisting the temptations of the flesh. As he charmingly wrote in one of his flirtations with the ladies of France, "I will mention the Opinion of a certain Father of the Church, which I find myself willing to adopt, tho' I am not sure it is orthodox. It is this, That the most effectual Way to get rid of a certain Temptation, is, as often as it returns, to comply with and satisfy it. Pray instruct me how far I may venture to practise upon this Principle?"[4] At the time that Frank-

[3] D. H. Lawrence, *Studies in Classical American Literature* (New York, 1930).

[4] *Mr. Franklin: A Selection from his Personal Letters*, ed. by Leonard W. Labaree and Whitfield J. Bell, Jr. (New Haven: 1956), p. 44.

lin wrote this he was 72, but it is not certain that his question was merely rhetorical.

These lapses from the conventions of family life today would not have outraged D. H. Lawrence if they had not been coupled with a certain priggishness in many of the household maxims to which Franklin gave currency. In 1732, Franklin began publishing *Poor Richard's Almanack,* which was by far the most successful work that he wrote, and in some ways the most influential. Like other almanacs, this is stuffed with those plums of wisdom which most people like to taste and few to digest—"hunger never saw bad bread," "well done is better than well said," "opportunity is the great bawd." It is these crystallized plums, so eminently homely and homemade, which have made Franklin's beliefs seem commonplace.

But this criticism confuses the manner in which Franklin expressed himself—and expressed himself at all times—with the content of his thought. Franklin had a special gift for putting a thought into a simple and earthy sentence. This is a gift of expression: a rare gift, and Franklin had it to perfection. At the end of his life, when he was asked what was the good of a new invention, he could still mint the phrase, "What good is a new-born baby?" The gift has a drawback: the thoughts that can be put into such sentences cannot be complex. In this form, Franklin's isolated thoughts do indeed wear a simple and sometimes a commonplace air. But it is a crude error to suppose therefore that the totality of Franklin's thoughts, the system into which the isolated thoughts lock and combine, is commonplace. In this respect, the simplicity of Franklin's sentences is as deceptive as the simplicity of Bertrand Russell's, and the outlook which they make up all together is equally complex.

The informality with which Franklin wrote and spoke is, however, just to his thought in one respect: he was opposed to formality and rigidity of belief. It is not merely that he did not care for the fine points of dogma; he thought it wrong in principle to wish to formulate religion in fine points. When he arranged to be kept up-to-date with what was published in England, he asked to have sent to him "such new Pamphlets as are worth reading on any Subject (Religious Controversy excepted)." He did not acknowledge any sectarian monopoly of truth; and when, at the age of 83, he stated his belief in God, he coupled it with another belief, "that the most acceptable service we render Him is doing good to His other children."

At bottom, it is this tolerance in Franklin's make-up which we must understand. It is possible to be tolerant of the beliefs of others because one is indifferent to them; or, at the other extreme, it is possible to be tolerant of them because one respects them. Franklin was never indifferent, but he was not much given to respect. Rather he was tolerant of others because he recognized in them the same humanity as he knew in himself. He never hid his own motives from himself, but neither did he belittle the motives of others. We feel him to be honest because we see him judging others exactly as he judges himself, with a realistic and generous sense of what can be expected of human beings. Sustained by this humanity, he could make friends with men as religiously diverse as the anticlerical Tom Paine and the evangelist George Whitefield. And when Whitefield thanked him for a kindness "for Christ's sake," Franklin firmly replied that the kindness was "not for Christ's sake but for your sake."

IV

Because he came from America, Franklin seemed to his European friends to have another grace: he was to them a child of Nature. The early eighteenth century had admired a classical formality in all the arts. But in the second half of the century there was a romantic reaction against formality. As we have seen, Jean-Jacques Rousseau attacked the society of his day as being overcivilized and remote from the fundamental impulses of nature. The poets of the Romantic Revival broke with city life and praised the feelings and the language of men brought up among lakes and mountains. These romantic currents were already strong when the American Revolution began, and they gave to the Revolution, and to Franklin as its European emissary, a pleasant aura of natural rightness. For example, when Louis XVI for the first time formally received the American delegates at Versailles in 1778, Franklin seems to have intended to wear court dress, and to have abandoned it only at the last moment because the wig did not fit. Yet his appearance in his own hair and in simple dress was treated as a sensational display of natural dignity, which commended the American Revolution to all France.

Franklin had no romantic illusions about nature, but he did have a strong sense of what was truly natural. By the standards of his time, he was remarkably free from self-deceit, and from the inhibitions with which self-deceit guards itself. For example, it is one of the pleasanter and neglected sides of his character that he liked practical jokes and not altogether proper stories. There is plenty of salty humor in his newspaper writings. And one of the most endearing features of *Poor Richard's Almanack* was Franklin's announcement (in the style of Jonathan Swift) of the impending death of a rival almanac maker, with all the hilarious disputes over his survival for the next eight years.

The feeling that the Americans had the simple virtues of nature was important in forming a public opinion on their behalf in England. It made it possible for Franklin to find friends in all classes of society, or at least of Whig society. The unity which dissenters and the new manufacturing men felt with the Americans was felt also by men as diverse as Edmund Burke, the elder Pitt, and the Duke of Richmond. For example, it is strange to find Horace Walpole, whose head was full of Gothic fancies, writing with mischievous glee about British failures against America. But Walpole's Gothic mannerisms were a part of the romantic movement, and of the wish to return to more ancient and more folklike ways, which captured the later eighteenth century. In all these forms, the return to nature was a revolt against authority, and was coupled (in the minds of those who longed for it) with the liberation of America from the tyranny of George III.

The feeling that man in earlier times had been free from the oppression of court government and state religion expressed itself, among other things, in societies which claimed to go back to a secret antiquity. The most important of these was the body of Freemasons. We think today of Freemasons much as we think of Rotarians, as men who meet over passable food in order to help one another and to undertake good works. This is indeed the form which Freemasonry takes in countries in which freedom of conscience and of expression is accepted. But in the less democratic countries of Central Europe, Freemasonry is still a strongly anticlerical and antiauthoritarian organization, which has need of the secrecy with which it surrounds itself. And this was the state of affairs in America and in Europe in Franklin's day.

At that time, the organized movement of Freemasons was new; the Grand Lodge of England had been founded in 1717, the Grand

Lodge of Ireland in 1726, and the Grand Lodge of Pennsylvania in 1728. Pennsylvania was thus the earliest home of Freemasonry in America, and one of the earliest in the world. Franklin became a Freemason there, and he used his influence with other Freemasons first in England and later in France. For example, Franklin presided at the ceremony in which John Paul Jones, after his raids on British shipping in 1777, was received into the Nine Sisters Lodge in Paris.

This background of Freemasonry was important to Franklin's private friendships, and was equally important in marshaling public support for the American Revolution. The opposition to George III in England was led, in practice as well as in sentiment, by men of whom several were connected with Freemasonry. The most violent opponent of George III, John Wilkes, was a Freemason; and the powerful leaders of commerce in the city of London, who used John Wilkes as their mouthpiece, were connected with Freemasonry.[5] It was these city leaders who gave support to the Gordon riots[6] in 1780 which expressed the popular dislike, among other things, of the American War and which, until they got out of hand, almost forced the abdication of George III.

V

We have referred, several times, to Franklin's wide friendships. He knew everyone who mattered, from David Hume the philosopher to John Baskerville the printer, from the Midland manufacturers in England to Beaumarchais and Voltaire in France. He was made a Fellow of the Royal Society in 1756 and, what was a rarer distinction, one of the eight foreign members of the Académie Royale in 1772.

In part, these friendships were personal; and in part, as we shall see, they grew from Franklin's scientific work. But behind these stood a larger and more robust reason, which made Franklin the symbol of the American Revolution in Europe. Franklin was the type of the new man; in him, his manufacturing and scientific friends saw exactly what they themselves wanted to be. Dissenters and tradesmen everywhere in England felt themselves to be closer to Franklin than

[5] See W. P. Treloar, *Wilkes & the City* (London, 1917).
[6] See J. P. De Castro, *The Gordon Riots* (London, 1926).

to George III's ministers. They identified themselves with Franklin and, through him, with the American cause, and remained open supporters of it throughout the War of Independence.

Thus Franklin was the ideal spokesman for the American cause in England, even though it was in part an accident that he was chosen for such a mission. When he had returned to Philadelphia in 1726, at the age of 20, after his first visit to London, he had simply set about his business. True, he had soon founded the debating junto, from which later grew both the first American subscription library and the American Philosophical Society. But in the main, he had worked as a printer, had opened his own shop, had married and published *Poor Richard's Almanack* and prospered. He had gone on thus for twenty years until, in his early forties, he was rich enough to sell his printing business and interested enough to turn to something else—to scientific work. It was largely an accident that almost at the same time, in the winter of 1747, he was drawn into public life. The war between France and Spain had already threatened the northern colonies, and was now threatening Pennsylvania. Franklin thought it necessary to organize a voluntary militia and, in order to counter the traditional opposition of the Pennsylvania Quakers to this, he wrote a persuasive pamphlet with the characteristic title *Plain Truth.* It recruited a militia of, ultimately, about 10,000 men, in which Franklin was offered the post of colonel.

As an outcome of the Pennsylvania crisis, Franklin began to speculate on a union between the colonies. He drew up a first plan for a union on his way to the Colonial Congress which the English government called at Albany in 1754, under the threat of the coming Seven Years' War. The early defeats of that war on the frontier made it essential to raise new taxes, but in Pennsylvania the proprietors of the colony, the Penns, refused to have their lands taxed. By the irony of history, the tension between England and America later became concrete in the refusal of America to accept taxes imposed by Parliament in England; but the real root of that quarrel reaches back to the bitterness of American smallholders when they were faced by the refusal of such great absentee landowners as the Penns to bear their share of the local taxes.

In 1757, Franklin was asked to go to England by the Pennsylvania Assembly to present their case against the Penns. It took him five years to score some success; and these five years gave him influence and prestige in England. In 1764, he was sent once more to England by several colonies to plead the American cause. This time he could not win a major success; George III was now on the throne, and was determined to impose an absolute rule on America as on England; and though Franklin steadfastly bore humiliation, and stayed until 1775, by the time that he returned to Philadelphia in that year the shots at Lexington and Concord had been fired. Franklin was elected a delegate to the Continental Congress, and was one of the five men appointed to draft the Declaration of Independence. He represented the newly formed United States in France, as no one else could have done, in the critical years from 1776 until 1785; and he died in 1790.

VI

Franklin gave up his business as a printer in 1748 in the hope of giving all his time to scientific research, and he expressed this hope again several times in later life. Moreover, his reputation in England and particularly in France rested in a large part on his scientific achievement. It is, therefore, important to decide how genuine his scientific talent was, and how original was the work which he was able to do.

Like other men of the new industrial age, Franklin was a handyman with an inventive mind. For example, since he was a great reader, he made himself bifocal spectacles and a gadget for getting at books on high shelves.[7] In 1744 he published a detailed description of a very practical new stove that he had invented—the junto had discussed, "How may smoky chimneys best be cured?"—and he refused to patent this and other inventions. And it was natural for him to turn his work on electricity to practical use by proposing the lightning conductor.

Like other men of his stamp also, Franklin was interested in any-

[7] These gadgets are described and illustrated in I. Bernard Cohen, *op. cit.*, pp. 206–210. Franklin writes, characteristically, that he finds his bifocal spectacles "more particularly convenient since my being in France, the glasses that serve me best at table to see what I eat, not being the best to see the faces of those on the other side of the table who speak to me; and when one's ears are not well accustomed to the sounds of a language, a sight of the movements in the features of him that speaks helps to explain; so that I understand French better by the help of my spectacles."

thing unusual in the workings of nature. On the first voyage to London he took with him a purse made of asbestos, which was a rarity in England and which can still be seen at the Royal Society. If at home he came across a small whirlwind, he followed it and then described it to his London correspondent, the Quaker merchant and naturalist Peter Collinson.[8] Even in his last years of negotiation in Paris, Franklin found time to watch the first balloon ascents and to write about them to other scientists.[9] And it is characteristic that for his journey to France in 1776, he provided for his expenses by taking on board a cargo of indigo, which was then a precious dye that the chemist had not yet been able to imitate.

But these practical interests are secondary to Franklin's true scientific achievement in the theoretical understanding of electricity.[10] Electricity was then a new subject, which came on the eighteenth century as unexpectedly as nuclear power has come on our century. For Franklin, an amateur far from the centers of research, to make discoveries in electricity then was as remarkable as if today an amateur in Tibet were to solve the outstanding problems of nuclear structure.

When Franklin began to be interested in electricity, the theory of the subject had recently taken some important steps. The distinction between electrical conductors and insulators had been made in 1732, and the principle of electrical induction had been discovered. In 1733, in order to account for electrical repulsion as well as attraction, it was suggested that there are two kinds of electricity—resinous electricity and vitreous electricity. In 1745–1746, the principle of the condenser was discovered, in the form of the Leyden jar.

These discoveries lend themselves to display, and public experiments became fashionable. A boy insulated by being hung from silk threads could be electrified so that his hair stood on end, and handsome sparks could be drawn from his nose. The Leyden jar, if heavily

8 See Leonard W. Labaree and Whitfield J. Bell, Jr., *op. cit.*, pp. 8–10.

9 See *ibid.*, pp. 55–58.

10 Benjamin Franklin's *Experiments and Observations on Electricity* has been edited by I. Bernard Cohen (Cambridge, 1941), with a full historical account and scientific appraisal, which establishes the nature and scope of Franklin's original work in electricity. The condensed account which we give here is based on Cohen's work; to avoid confusion, this work will be quoted as *Experiments*. See, too, Cohen's very important work, *Franklin and Newton* (Philadelphia, 1956).

charged, could give a shock to a row of 180 soldiers or even 700 monks, who leaped into the air as one man, to the great joy of Louis XV and the French court.

Franklin says in his *Autobiography* that he first saw electrical experiments at Boston in 1746, but it is possible that his memory was at fault, and that he saw the famous experiment of the suspended boy a little earlier, in 1744.[11] Soon after, in 1745 or 1746, the subscription library which Franklin had begun in Philadelphia received from Peter Collinson, who usually supplied it with books, a glass tube and a description of the electrical experiments that could be done with it. Franklin began his experiments with this piece of glass. "I eagerly seized the opportunity of repeating what I had seen at Boston; and, by much practice, acquir'd great readiness in performing those, also, which we had an account of from England, adding a number of new ones. I say much practice, for my house was continually full, for some time, with people who came to see these new wonders."

Franklin worked with several friends, and in 1747 he began to describe their findings in a series of letters to Collinson. One of their first discoveries was that pointed conductors were particularly effective "both in *drawing off* and *throwing off* the electrical fire."

In the letter in which he announced this discovery (which was only partly new, and only partly his own) in 1747, Franklin also criticized the current belief that there are truly two kinds of electricity. In his view, the electrified object can indeed be in two states, but these are merely positive and negative: there is only a single electrical fluid (what we now call "electrons"), and the two states are occasioned by a minus or a plus of this single electricity.

Some time later in 1747, Franklin received or heard of the new Leyden jar. In his analysis of the Leyden jar, the concepts of plus and minus of electricity came into their own; and this analysis is Franklin's most important scientific work. He showed that electricity is not created by friction, but is merely redistributed by it. The metal foil on the outside of the jar has a positive charge which is exactly equal to the negative charge of the water inside the jar: one side gains as much electricity as the other side loses. Moreover, the charge does not strictly belong either to the metal foil or to the water, but is fixed by the nonconductor between them—the glass of the bottle.

11 This possibility is discussed by I. Bernard Cohen in *Experiments*, pp. 47–56.

VII

The single fluid theory of electricity, and its demonstration in critical experiments, is for us Franklin's greatest work in electricity. It reached the Royal Society through Collinson early in 1748, and was quoted at length in its *Philosophical Transactions.*[12] But it was not the end of Franklin's work in electricity.

In 1749, Franklin sent to another of his English correspondents what he called "a new Hypothesis" connecting lightning with the electrification of clouds. This material reached Collinson, who wrote to Franklin early in 1750,

> Your very Curious pieces relating to Electricity and Thunder-Gusts have been read before the Society & have been Deservedly admired not only for the Clear Intelligent Stile, but also for the Novelty of the Subjects. I am collecting all these Tracts together, yr first account with the Drawings and your Two Letters ca 1747—and your Two last Accounts with Intention to putt them into some Printers Hand to be communicated to the Publick.[13]

Collinson did get Franklin's letters published as a pamphlet in 1751. The pamphlet came to the notice of the Comte de Buffon, the great French naturalist, who had it translated into French by another electrical experimenter, Jean François Dalibard.

Franklin thought that his identification of lightning with the discharge of electricity from clouds was new. This was not quite so: other scientists had speculated that this might be the nature of lightning. What was new was Franklin's proposal of a clear test of the hypothesis.

> To determine the question, whether the clouds that contain lightning are electrified or not, I would propose an experiment to be tried where it may be done conveniently. On the top of some high tower or steeple, place a kind of sentry-box big enough to contain a man and an electrical stand. From the middle of the stand let an iron rod rise and pass bending out of the door, and then upright 20 or 30 feet, pointed very sharp at the end. If the electrical stand be kept clean and dry, a man standing on it when such clouds are passing low, might be electrified and afford sparks, the rod drawing fire to him from a cloud.[14]

12 See *ibid.,* pp. 77–79.
13 Quoted *ibid.,* pp. 80–81.
14 Given *ibid.,* p. 222.

The French court was full of interest for electrical wonders, and at once eager to try what was called this "Philadelphia experiment." On May 10, 1752, Dalibard made the attempt, and announced his success triumphantly to the Académie Royale three days later. Other French scientists also succeeded, and Louis XV ordered that a note of thanks and compliments be sent to Franklin. His scientific reputation with the lay public had been made by a spectacular test, much as Einstein's public reputation was made by the confirmation in 1919 of his forecast that light bends toward a massive body. The Royal Society awarded Franklin the Copley Medal in 1753, even before it elected him a Fellow.

VIII

It would be pedantic to leave Franklin's scientific work without glancing at the two achievements which stand in every schoolbook. Some time in 1752, before Franklin knew that the "Philadelphia experiment" which he had proposed had already been carried out in France (and indeed, before he knew that his pamphlet had been translated into French), he thought of another test. This was to fly a kite during a thunderstorm, and so to draw a charge from an electrified cloud directly, without using a high building.[15] Franklin was able to make this test himself during 1752, and succeeded. The experiment was also repeated by others; and at least one scientist, Georg Wilhelm Richmann, killed himself in 1753 in confirming Franklin's work, either with a kite or a raised sentry box.

The death of Richmann in an electrical experiment with lightning helped to stimulate interest in the other proposal made by Franklin which is well known, that a metal rod which is earthed might be used to protect a building from lightning. Franklin had first thought that such a rod would draw the electric charge from the cloud without a lightning stroke, but he later found that the charge took the form of real lightning. He put an account of the lightning conductor into *Poor Richard's Almanack* of 1753.

Franklin had helped to discover that electric charges concentrate at and flow to points, and he therefore advocated that lightning conductors should be pointed. Some members of the Royal Society, however, opposed pointed conductors precisely because of "their great

[15] See *ibid.*, pp. 118–119, and Franklin's own account on pp. 265–266.

readiness to *collect the lightning in too powerful a manner*."[16] During the war with America, George III thought it a point of honor to deny whatever Franklin had proposed, and he insisted that royal buildings should carry blunt and not pointed lightning conductors. There is some evidence that the king tried to bully the Royal Society to cancel the recommendations of its recent committees (of which Franklin had been a member) in favor of pointed conductors; and that Sir John Pringle, although personally opposed to the American cause, resigned the presidency of the Royal Society rather than give way. This episode has a strangely modern ring, for it reminds us how other dictators—for example, Adolf Hitler and Joseph Stalin—have damaged science by intervening in it.[17]

IX

Thomas Jefferson began the first draft of the Declaration of Independence by describing the fundamental principles with which it opens as "sacred and undeniable." It was surely Franklin who deleted this phrase, and put in its place the word that now stands in the Declaration: the word "self-evident."[18]

The change tells us three things about Franklin's character. It tells us that he liked his language to be even simpler than Jefferson's—more homely and less elevated. It tells us also that Franklin did not want any principles, however ethical and religious they might be, to be received by sacred authority; he wanted ethics and religion to be accepted by the free acquiescence of the mind. And it tells us that

[16] This dissenting opinion is quoted *ibid.*, p. 135.

[17] For the sake of completeness, one other profound observation made by Franklin should be listed. He observed that when the outside of a hollow conductor is charged, there is no electric field inside the conductor. Franlin did not explain this observation, but he passed it to Dr. Joseph Priestley, who brilliantly drew from it the correct and difficult conclusion that electric charges obey the law of inverse squares: see *Experiments*, pp. 72–73. From Franklin's observation, in fact, most of the theory of electricity can be deduced—just as most of mechanics can be deduced from the assertion that perpetual motion is impossible: see Edmund Whittaker, *From Euclid to Eddington* (Cambridge, 1949), p. 59.

[18] This point is clearly discussed by I. Bernard Cohen in *Benjamin Franklin: His Contribution to the American Tradition* (New York, 1953), pp. 59–61. See also Carl L. Becker, *The Declaration of Independence* (New York, 1942), where the change is not necessarily ascribed to Franklin. More generally, see our discussion of what is "self-evident" in Chap. 11, "Hobbes and Locke," p. 202, and in Chap. 12, "The Method of Descartes," p. 221.

Franklin carried the habit of science into everything he did; the word "self-evident" in the eighteenth century was used as a scientific word, in exactly the sense in which we should now use the word "axiomatic."

These points are outstanding in the outlook which Franklin brought into the American Revolution. They show, as we have said, the kinship of Franklin with the rising manufacturers of England, with their inventive minds and their dissenting consciences. Franklin was most like the men of the Lunar Society, many of whom were his friends.

But there was a deep difference between England and America at that time: between the place which men of this stamp held in the two countries. In England, the rising manufacturers did not command the assent of those in power and of the leaders of the established culture. In America, Franklin's outlook was accepted and at bottom it was shared by all the leaders of thought. Jefferson did not object to Franklin's amendment, because it was as natural to him as his own words. This was the strength of the American Revolution: that aristocrats like Washington and Jefferson thought of man and nature in the same direct terms as did self-made men like Franklin. The unity of the best minds of America in the Revolution was not created by the common enemy in England: it was created by the inherent sense which they all shared that they were exploring together a new age and a new relationship among men.

CHAPTER 21

Thomas Jefferson and the American Revolution

I

"Did I not know that he was President of the United States," the Irish poet Thomas Moore remarked of Jefferson, "I would judge him to be a gentleman of landed property, with all the inclinations of a fox-hunting squire."[1] Such a remark was true of Jefferson in his early years as well as when he was President, and can be taken as typical for most of the American leaders in 1776.

Europeans, supported by the American myth, often think that the American Revolution was made by backwoodsmen who lived in log cabins. A visit to Washington's house at Mount Vernon or to Jefferson's at Monticello will go far to dispel this view. To walk around a house which was in its own way as grand as almost any house of that period in England or France—a house, in fact, of a distinguished landed proprietor who kept many horses in the stable, whose furniture was the best of the period, whose walls were covered with notable pictures, and whose silverware was as good as anything to be found in Europe at the time—is to obtain a different picture of the men and motives of the American Revolution.

In the case of Jefferson, his father was a self-made man, who married into the socially well-placed Randolph family, and left to his son more than 5,000 acres. To this initial inheritance of land and

[1] Quoted in Saul K. Padover, *Jefferson*, abridged by the author (New York, 1952), p. 146.

social position, Jefferson added over 11,000 acres when he married the daughter of a rich Williamsburg lawyer. In virtue of all his holdings, he, like his friend, Washington, was one of the largest landowners in Virginia.[2]

Jefferson was not only a landed gentleman; he was an educated one. Entering William and Mary College in 1760, he had the good fortune to be taken in hand by Dr. William Small, whom we have already discussed as the Scotsman who helped draw up the first agreement between Boulton and Watt. Small made Jefferson familiar with the general world of European science and culture.

Small introduced Jefferson to George Wythe, the professor of law at the college, from whom Jefferson learned the combination of lucid principle and the point-by-point application of it which marked so much of his later work. Wythe became Jefferson's lifelong guide in politics and law, and, indeed, it was in Wythe's office that the young scholar read for the law. Although repelled by legal aridity, when he set up by himself at the age of 24 Jefferson rapidly became a success. Within four years after he started practice, he had a great number of cases and many distinguished clients.

He did not use his legal talents solely for personal profit. As early as 1769, when only 26, Jefferson took his place, befitting to the station of landowner and lawyer, in the Virginia House of Burgesses. He was quickly forced to take a side in the growing internal divergence between "prerogative" men and "patriots." In this struggle the small group of "prerogative" men attached themselves to the court party of George III, while the majority of prominent Americans opposed them through the mechanism of state assemblies and self-formed "committees." Jefferson involved himself with the so-called patriots and, by 1773, had become a member of the Virginia Committee of Correspondence.

From this point on, events moved rapidly, and Jefferson with them. The Boston Tea Party led the British to pass a series of laws closing the port of Boston to commerce until the tea was paid for. The colonial response was a call for a general congress. To instruct the Virginia delegates to this Continental Congress, Jefferson wrote, in

[2] For details of Jefferson's landholdings, see Dumas Malone, *Jefferson the Virginian* (Boston, 1948), pp. 439–445. Malone's second volume, *Jefferson and the Rights of Man*, appeared in 1951. The two volumes represent an outstanding work on Jefferson.

1774, his first published work, *A Summary View of the Rights of British America*. Although not used by the Virginia delegates (it was considered too radical), it earned for Jefferson the reputation of a phrase maker and a polished writer.

The controversy going on in America mirrored, to a great extent, the contest in England between Whigs and Tories. The "new men" of England—the rising manufacturers and merchants—were sympathetic, as we pointed out in the previous chapter, to the aspirations of the American "patriots," for both the English "new men" and the American "patriots" were opposed to the oppressive, misguided, and backward economic policy of the existing British government.

Unlike their English counterparts, however, the American "patriots" were already in control of the provincial assemblies. The task for them, therefore, was not so much to capture the government as to defend their "rights" from the encroachments of that government. This explains why the American Revolution was begun by the upper class and never became a social revolution to the degree that the French Revolution did. Further, the American Revolution not only started at the impulse of the upper class but largely remained under their control.[3]

The source of upper-class strength was in extended families or kinship groups. As one student of the subject has put it: "Such families were needed in a society where reliance must be placed on one's own kinship group to support and advance the group interest. There were almost none of the social institutions we now rely upon to spread economic and social risks, such as insurance companies, banks, limited-liability corporations, and likewise none of the devices like holding-companies for pyramiding control."[4] The linking of the "great" families is shown by the example of Virginia. Thus, the Randolphs (Jefferson's direct link) predominated on the James River and were allied to the Byrd family around Richmond; the Lees, in a similar role on the Potomac, had an alliance with the Fairfax family. George Washington was a protégé of the Fairfaxes and the neighbor and intimate friend of the Lees; one of his closest friends was Peyton

[3] Cf., however, J. Franklin Jameson, *The American Revolution Considered as a Social Movement* (Princeton, 1926), for a stress on the social aspects of the Revolution.

[4] Robert K. Lamb, mimeographed copy of *The Business Man and His America*, Vol. I, "The Revolution Families: 1750–1789."

Randolph, whose resignation as a Continental Congress delegate permitted his cousin, Thomas Jefferson, to go to Philadephia as his alternate.

If we ask, in the broadest terms, what "caused" the American Revolution, this upper-class aspect of the 1776 period must not be overlooked.[5] It was clearly perceived by Adam Smith. In the *Wealth of Nations* he indicated that legal, economic, and constitutional arguments, while important, were often shields under cover of which men advanced to the real fight—the fight for position and power. This is a fundamental interpretation of the cause of the American Revolution, and it is worth quoting at length:

> Men desire to have some share in the management of public affairs chiefly on account of the importance which it gives them. Upon the power which the greater part of the leading men, the natural aristocracy of every country, have of preserving or defending their respective importance, depends the stability and duration of every system of free government. In the attacks which those leading men are continuously making upon the importance of one another, and in the defence of their own, consists the whole play of domestic faction and ambition. The leading men of America, like those of all other countries, desire to preserve their own importance. They feel, or imagine, that if their assemblies, which they are fond of calling parliaments, and of considering as equal in authority to the parliament of Great Britain, should be so far degraded as to become the humble ministers and executive officers of that parliament, the greater part of their own importance would be at an end. They have rejected, therefore, the proposal of being taxed by parliamentary requisition, and like other ambitious and high-spirited men, have rather chosen to draw the sword in defence of their own importance.[6]

II

In his *Summary View,* Jefferson employed much the same argument as Smith but in a broadened form; instead of attacking Parliament in the name of American parliaments, he attacked the British electorate in the name of the American electorate.

[5] There are, of course, many other "causes" of the American Revolution. For a handy summary of some of them, see "The Causes of the American Revolution," ed. by John C. Wahlke, in The Amherst Series, *Problems in American Civilization* (Boston, 1950).

[6] *Wealth of Nations,* Everyman's Library ed., Vol. II, p. 118.

Can any one reason be assigned, why one hundred and sixty thousand electors in the island of Great Britain, should give law to four millions in the States of America, every individual of whom is equal to every individual of them in virtue, in understanding, and in bodily strength? Were this to be admitted, instead of being a free people, as we have hitherto supposed, and mean to continue ourselves, we should suddenly be found the slaves, not of one, but of one hundred and sixty thousand tyrants.[7]

Jefferson was unconscious of what Smith would have called his "self-interest" in being a leading figure in an independent parliament. Instead, he introduced an appeal to the "importance" of *all* Americans. The democratic note was not hypocrisy with Jefferson; his belief in a "free people" was the core of all his thinking.

Jefferson's democratic quality was imbedded in his style of writing and is present in all the letters, documents, and books he ever wrote. One way to savor this is to read passages from him, not primarily for their sense, but for the way in which they are written.

If one bears in mind that Jefferson was born in 1743, and that he had formed his style and written the greatest of his works before Wordsworth and Coleridge published in *Lyrical Ballads* in 1798, or added the famous Preface, his prose takes on an added dimension. Wordsworth's point when he wrote the *Lyrical Ballads* was that the poetry of the eighteenth century, as well as prose such as that of Samuel Johnson, was decadent and outmoded. He attacked classical poetry and prose as being very formal and elaborate, and he demanded that the new poetry and the new prose be simple, forthright, and functional. The interesting thing is that, long before Wordsworth's strictures, Jefferson was writing almost as simply and directly as Franklin or the simplest manufacturer of the North of England.

It is hardly surprising that someone like Tom Paine wrote in that magnificent prose which still hums in our ears—all those phrases about the "summer soldier and the sunshine patriot." Tom Paine was a man who had taught himself everything that he knew—he had come up the hard way—and his blunt power derived from this: that, having taught himself, he was a superb instrument in the teaching of others. He could not write a difficult sentence.

[7] *The Complete Jefferson,* assembled and arranged by Saul K. Padover (New York, 1943), p. 11.

Jefferson, on the other hand, had been formally educated; yet his style was unaffected. He even remained untouched by the killing breath of legal style and mocked it as "lawyerish." We can see his model from his comment about Tom Paine. "No writer," Jefferson said, "has exceeded Paine in ease and familiarity of style, in perspicacity of expression, happiness of elucidation, and in simple and unassuming language."[8]

We make this point forcibly and anew. At the time in England, great works were being written, like Gibbon's *Decline and Fall of the Roman Empire* and Boswell's *Life of Johnson,* but these works were filled with rotund phrases, circumlocutions, and elaborate syntactical devices. To read these and similar books is to realize that the social cleavage in eighteenth-century England could be detected in literary style as well as in costume or manner of speech.

In America at this time, writings like Paine's *Common Sense,* Franklin's *Almanack,* and Jefferson's letters and documents were obviously being written from a "simple and unassuming" point of view. They reflect the way these men thought. Thus, at the Continental Congress of 1776, Jefferson, regarded as a sort of junior member of the company (he was only 33 at the time), was entrusted with the task of drafting the Declaration because he could express the direct thoughts of everybody simply and felicitously. The most striking proof of this statement is to look at Jefferson's own account of the Declaration of Independence.

> With respect to our rights, and the acts of the British government contravening those rights, there was but one opinion on this side of the water. All American whigs thought alike on these subjects. . . . Not to find out new principles, or new arguments, never before thought of, not merely to say things which had never been said before; but to place before mankind the common sense of the subject, in terms so plain and firm as to command their assent, and to justify ourselves in the independent stand we are compelled to take. Neither aiming at originality of principle or sentiment, nor yet copied from any particular and previous writing, it was intended to be an expression of the American mind, and to give to that expression the proper tone and spirit called for by the occasion. All its authority rests then on the harmonizing sentiments of the day, whether expressed in con-

[8] Quoted in Padover, *Jefferson,* p. 32.

versation, in letters, printed essays, or in the elementary books of public right, as Aristotle, Cicero, Locke, Sidney, &c.[9]

Perhaps nothing so much shows the spirit and genius of a society as the tenor of its style—whether it is in speech or art or writing. The Declaration of Independence is, therefore, almost as important for its style as for what it says.

III

If the American Revolution had failed, the Declaration probably would have been labeled in present history books as the "Seditious Document" and treated as an avowedly treasonable act. Indeed, the delegates to the Continental Congress were aware that they might hang. Thus, the debate at the Congress centered on the disadvantage or the advisability of forming either a foreign alliance or a domestic confederation before taking the risky step of a public declaration. The weight of the argument, however, favored the open and decisive step of declaring independence *as a means* of inducing France to favor the colonists against England. It was also expected to separate "the sheep from the goats" (the American Whigs from the Tories).[10]

Bearing these background circumstances in mind, let us now consider the Declaration itself. Its first declared purpose was to appeal to public opinion for judgment as to the righteousness of its cause. In the words of the Declaration, "a decent respect to the opinions of mankind" required that the colonists "should declare the causes which impel them to the separation."

[9] Letter of May 8, 1825, to Henry Lee, from *The Writings of Thomas Jefferson*, ed. by P. L. Ford, 10 vols. (New York, 1892–1899), Vol. X, p. 343. This work is being rapidly superseded by the fundamental *The Papers of Jefferson*, ed. by Julian P. Boyd (Princeton, 1950–), which carries through to November, 1789 in the recently published Vol. XV (1958). On details of the composition of the Declaration, see Carl L. Becker, *The Declaration of Independence* (New York, 1942).

[10] Ironically enough, the French had already pledged a million *livres* of French funds and arms for the rebel cause in May, two months *before* the Declaration of Independence; however, the poor communications of the time prevented the news from being known in America until October 1, or almost three months *after* the signing of the Declaration. (A similar breakdown in communications was to determine American actions leading to the War of 1812.) Actually, the formal alliance with France did not come until a year later; and how much part the Declaration of Independence played in the French decision is debatable. In any case, the French recognition gave the rebellious colonists the status of a new nation in the eyes of Europe, thereby finishing what the Declaration had begun.

These "causes" were, ostensibly, twofold: violations of general rights of mankind, and particular grievances of the colonists against the king. They reflect the combination of principle and application that Jefferson had learned so well from his former teacher, George Wythe. The violations of principle come first and are heralded by the majestic words: "We hold these truths to be self-evident. . . ." Then follow the specific grievances, so important at that time, but carrying no message to future generations or to other nations. These occupy the largest part of the Declaration and are introduced by the statement: "The history of the present King of Great Britain is a history of repeated injuries and usurpations. . . . To prove this, let *facts* [our italics] be submitted to a candid world."

Most commentary on the Declaration has been centered on the general principles and has somewhat slighted the particular grievances.[11] Thus, the view arose that the American colonists rebelled more out of a concern with natural rights than with historical rights; and it was this interpretation that had such profound repercussions in France. Yet, the opening words of the Declaration specifically declare that revolt is justified "When, *in the course of human events* [our italics], it becomes necessary for one people to dissolve the political bonds which have connected them with another. . . ." On this reading, the historical circumstances, not the eternal rights of man, were the prime cause of the American Revolt.

It is true that the Declaration goes on to say that the colonists have a right "to assume among the powers of the earth the separate and equal station to which the Laws of Nature and of Nature's God entitle them." But this, once again, only asserts that, when history places a country in a particular position, that country has a right to make that position a reality which is recognized by all other nations.[12]

Another important point about the Declaration is that it appears to set up revolution as an accepted method or mechanism of government. In solemn tones, it declares that "whenever any form of gov-

[11] Historians, however, have reaped a better harvest from the grievances. They have taken each grievance, investigated its basis, and embroidered an interpretation of the cause of the American Revolution from it.

[12] Indeed, the belief in America's "destiny" was part of the "climate of opinion" surrounding the second Continental Congress. Jefferson's other documents and speeches show that he moved very much in the same mental atmosphere regarding Providence as did Adam Smith.

ernment becomes destructive of these ends [life, liberty, and the pursuit of happiness], it is the right of the people to alter or to abolish it, and to institute new government."

That Jefferson, like Locke, was very serious about the right of revolution is supported by his comment, occasioned by Shays' Rebellion, that the tree of liberty needs to be watered by the blood of patriots from time to time. He also resorted to the threat of revolution against the American government when, in the Kentucky Resolutions of 1798 against the Alien and Sedition Acts, he said, "These and successive acts of the same character, unless arrested on the threshold, may tend to drive these States into revolution and blood."

Yet, during the first phases of the impending French Revolution, when Lafayette and others approached Jefferson for advice, he cautioned moderation rather than revolution. The clue to this scruple in Jefferson's mind resides, we believe, not only in his statement in the Declaration of Independence that "Prudence, indeed, will dictate that governments long established should not be changed for light and transient reasons," but in his belief in the "course of human events." In spite of his dislike of Hume, Jefferson shared with him the historical approach. In the case of France, Jefferson did not believe that either its history or its social structure had prepared it for revolution and democracy. "The Europeans are governments of kites over pigeons," he declared; it would take a long period of training for the "pigeons" to learn to govern themselves.[13]

IV

Americans, however, were prepared by history and training to govern themselves. What, however, should that form of government be? The Declaration had announced America's independence; the next great political document, the Constitution, was to give it its form of government. Was the Constitution a divergence from the line marked out by the Declaration? This has been the subject of much debate, and many modern commentators have followed Jefferson in

13 For an interesting discussion of Jefferson's attitude toward the French Revolution, see R. R. Palmer's article, "Jefferson: The Dubious Democrat," *Political Science Quarterly* (September, 1957). Palmer's point is that, at least initially, Jefferson's sympathies were with the aristocrats, and his great fear was that a strong executive (i.e., the king) would lead to tyranny.

believing that the Constitution was an "oligarchic" device to stifle democracy; a kind of conservative "counterrevolution."[14]

Actually, the Constitution represented, in many ways, the experimental proof of ideas enunciated by earlier theorists: men like Harrington, Montesquieu, and Hume.[15] It put into practice the political notions advocated by these thinkers, but adapted them to local needs. In doing so, it was pragmatic and avoided the possible excesses of theory. Thus, from their own political experience, Americans learned to develop the method of governing various separate political entities (states, in this case) through a federal union—which serves the rest of the world as a working "model" of a federal government operating on the principle of checks and balances.

American political philosophy has been singularly devoid of great theoretic writings. *The Federalist* is almost the only major work of American political philosophy; and it was the combined work of three men. Whatever genius in political thought America has possessed finds its expression in public documents, legal decisions, and political debates. Properly speaking, there is almost no pure political speculation in America. For better or for worse, American political thinkers are characterized by what Parrington said of Daniel Webster: "Immediate, domestic issues muddied his thought."[16]

The great American contribution to politics was not a piece of theoretic writing but the federal system of government. Federalism, however, has never been a "fixed" solution to the problem of linking the destinies of several states. The question of federal versus states' rights, of centralization versus decentralization, is still with us today. This lends contemporary relevance to the fact that, against the emphasis on centralized federal power advocated by men like Hamilton, Jefferson took his stand on decentralized states' rights.

Jefferson had had much experience with local government. During the revolutionary years of 1776–1779 he had worked with Wythe, George Mason, and James Madison to reorganize his native state of

[14] See "The Declaration of Independence and the Constitution," ed. by Earl Latham, in The Amherst Series, *Problems in American Civilization* (Boston, 1949), for a full discussion of the issue.

[15] Not of Rousseau, we believe; his influence was on the French Revolution but only slightly, if at all, on the American Revolution. Jefferson, of course, disliked Hume and Harrington, and favored unitary democratic control as against a balance of powers, but he had these ideas on his own and not from Rousseau.

[16] Vernon Louis Parrington, *Main Currents in American Thought*, 3 vols. (New York, 1930), Vol. II, p. 310.

Virginia. He and his friends reformed the state legal code, abolishing primogeniture in the process. Even more significantly, Jefferson drew up the bill for establishing religious freedom; this separated church from state in the fashion later followed by the First Amendment to the Constitution. From 1779 to 1781, Jefferson put his new laws into effect as governor of Virginia.

With this active state political experience, Jefferson could look with some misgivings on the extension of power by a distant and seemingly oligarchic federal government. These misgivings came to a head upon his return from France, where he had served as ambassador from 1784 to 1787 (so that he had been absent from the Constitutional debates), and discovered a Constitution giving powers to a federal government unchecked by a Bill of Rights. Temporarily assuaged by the addition of the first ten amendments in 1791, Jefferson joined Washington's cabinet as secretary of state.

He remained in this position for three years, becoming more and more uneasy as to the direction in which the federal government seemed to be heading under the inspiration of his cabinet enemy, Alexander Hamilton, secretary of the treasury. Finally, in 1793, Jefferson resigned from the cabinet in order to fight openly the Hamiltonian policies. Two years earlier, Jefferson and Madison had taken a trip through the Hudson Valley on a "botanical" expedition. In the course of this trip, they formed an alliance with George Clinton, the governor of New York State, and Aaron Burr, who controlled a society in New York City known as the Sons of St. Tammany. This alliance of Southern planters and New York City "bosses" became the basis of a new party—the Republican Party—and it was through this party that Jefferson waged his fight.[17]

Jefferson's first attempt to defeat the Federalists failed. He lost the Presidential election of 1796 to John Adams and, instead, because of the peculiar interpretation of the Constitution then prevailing, became Adams' Vice-President. Therefore, it was as Vice-President that Jefferson pursued his belief in states' rights to the point of drafting the Kentucky Resolutions of 1798. In these resolutions he announced the doctrine that the Constitution was a compact of states (we can see here the effects of "immediate, domestic issues muddying his thought"; in the Declaration of Independence, Jefferson had

17 This party is now known as the Democratic Party.

talked of "peoples," not "states"); that the federal powers were limited to those *expressly* stated in the Constitution, with all other powers reserved to the states; that the states had the right to revolt if these rights were infringed; and that the states had, in any case, the right to nullify any acts which they considered unauthorized by the Constitution.

Jefferson had his finger on a real difficulty. His accusation was "That the principle and construction contended for by sundry of the State legislatures, that the general government is the exclusive judge of the extent of the powers delegated to it, stop not short of *despotism*—since the discretion of those who administer the government, and not the *Constitution*, would be the measure of their powers." Who, then, was to judge when the general government exceeded its Constitutional powers? Jefferson's answer was that "as in all other cases of compact among parties having no common Judge, *each party* has an equal right to judge for itself, as well of infractions as of the mode and measure of redress."[18]

A showdown on the issue—which was to reappear in the Civil War—was avoided at this time. Adams' government never enforced the Alien Acts, and Jefferson's election to the Presidency in 1800 allowed the Sedition Act to expire harmlessly. Ironically, when he became President, Jefferson was forced to go along with the "course of events" and to strengthen the powers of the federal government. The political philosopher gave way to the statesman; or, as some would have it, to the politician. It was on his own authority, without consulting the states or even Congress, that Jefferson purchased the Louisiana Territory from France in 1803.[19]

[18] Jefferson's suggestions really meant a return to the spirit of the Articles of Confederation.

[19] In answer, however, to the question "Who was to judge the government's adherence to the Constitution?" another theory was put forth under Jefferson's Presidency. It was John Marshall, Chief Justice of the Supreme Court, who supplied the opposite viewpoint to the Kentucky Resolutions in his decision of Marbury v. Madison in 1803. In judging the case, Marshall asserted that the Supreme Court, not the individual states, should interpret the Constitution. This theory was repugnant to Jefferson, both because the Supreme Court at the time was Federalist-dominated, and because he believed that the democratic will of the people should not be thwarted by an unelected judiciary; and he opposed it to the point of trying to impeach one of the Justices of the Supreme Court. His failure in this attempt marked the emergence of the Supreme Court as a self-governing mechanism of the Constitution. It was a rebellion of the machinery not

V

Another motive to Jefferson's reluctance to strengthen the powers of the federal government was his fear that this meant the spread not only of the Leviathan state but of the commercial and manufacturing interests. And events have proved him right. What he did not at first foresee was the fact that the enlarged federal government could be captured and used by democratic as well as by oligarchic forces. When this possibility became apparent to him, through his own occupancy of the Presidency, he accepted the necessity of an enlarged central government. The responsibilities of office and the hard facts of political life—such as the Napoleonic Wars—further led him to abandon his previous distrust of the nonagricultural groups and to place manufactures on an equal footing with agriculture.

But Jefferson never, like Hamilton, placed manufacturing higher. Instead, he consistently followed the doctrines of Adam Smith. In a letter of 1790, he declared: "In political economy, I think Smith's Wealth of Nations the best book extant.[20]

It is of extreme importance for a correct interpretation of Jefferson and Hamilton (as well as of Smith) and of America's development after the Revolution to clarify this point. Most commentators assume that Smith, because he became the idol of the manufacturing class, must have been opposed by Jefferson and, therefore, supported by Hamilton. To take a typical example, Parrington says of Jefferson:

> He had read much in the works of the Physiocratic group, and was intimately acquainted with Dupont de Nemours; and the major principles of the school sank deep into his mind and creatively determined his thinking. . . . The sharp struggle between Jefferson and Hamilton must be reckoned, in part at least, a conflict between the rival principles of Quesnay and Adam Smith, between an agrarian and a capitalistic economy.[21]

The truth is that Adam Smith had little confidence in the manufacturing class and favored the landed gentlemen of his time. He is

only opposed by Jefferson but unintended by any of the constructors of the Constitution. That the Constitutional question was not fully settled by Marshall, however, was once again illustrated by the conflict in the 1930's between Franklin D. Roosevelt and the Supreme Court.

20 Letter of May 30, 1790, to Mr. Thomas Mann Randolph, in *The Life and Selected Writings of Thomas Jefferson,* ed. by Adrienne Koch and William Peden (New York, 1944), p. 496.

21 Parrington, *op. cit.,* Vol. I, p. 346.

constantly reminding his readers, in the *Wealth of Nations,* that the merchant-manufacturing class consistently placed their own particular interest above the general interest; that it was their high profits, and not high wages, which artificially raised prices; and that, if allowed to influence legislation, they would install the mercantilist system.[22]

Nevertheless, Smith became the prophet of the manufacturers. Why? We have tried to answer this question, in a previous chapter, by pointing out that Smith's doctrines did serve *some* of the manufacturers; that is, it served one party in a conflict between rival manufacturers and merchants. Thus, a first generation of "new" men—like Wedgwood and Wilkinson—fought against an older generation which had strong influence on government. The "new" men felt that existing government was hostile and had no understanding of their endeavors. They generalized this feeling into a belief that *no* government control of industry ought to exist. They were not, we believe, hypocrites, consciously rationalizing; they sincerely believed in the ideology of free trade. Yet the first spur to their desire to remove government control was that they did not manipulate that control.

The generations that followed them in England continued to mouth the same phrases, but their actions belied their words. For example, the "new" cotton manufacturers were very soon given a monopoly of the home market by prohibition of the import of printed cottons; nothing about free trade was mentioned in this case. In fact, the cotton manufacturers made frequent appeals for higher protection. When Pitt, however, in 1784 attempted to increase the excise duty on cotton, they defeated him under the cry: "Let Commerce flourish forever! Freedom restored! May Industry never be cramped."[23]

What the cotton manufacturers objected to was the attempt of government to regulate the production methods of the new trades. They were heartily in favor of such government measures as would increase the bounty on exports of cotton, or prohibit the combination of cotton workers to stabilize or raise wages. Thus they used Smith's

22 See *Wealth of Nations,* Vol. I, pp. 228–232.
23 See Mantoux, *op. cit.,* p. 265.

doctrines only as it suited their mood, and foisted an emasculated version of his ideas on the public.

In fairness to Smith's memory and his work, as well as for a better understanding of Jefferson's position, let us go "in quest of the historical Adam Smith." We discover, then, that the prime example Smith used of a nation pursuing the right path to wealth was America; and the right path was the *development of agriculture.* It is on the basis of this analysis that Smith made his famous prediction as to America's future greatness. In the "natural course of things," he said, improvement comes from agriculture, for an expanding agriculture accumulates a surplus which leads to a greater division of labor, and so forth. The policy of Europe, Smith claimed, had put the cart before the horse: it had favored towns and manufacturing over land and agriculture; it had followed the mercantile instead of the natural system of economics.[24]

In spite of these handicaps, not because of them, Smith asserted, Europe had been able to progress. Although weakened by poor economic policies, Europe had enjoyed political security, and this had permitted man's natural desire for improvement to function. Similarly, in America, the colonial prosperity had been achieved *in spite* of Europe's misguided policy. The real cause of American prosperity, said Smith, was what Burke called "salutary neglect," not the watchful (i.e., mercantilist) policy, of Europe.

Jefferson's views corresponded to those of Smith. This fact determined Jefferson's attitude toward economic questions, and influenced him in his desire for isolation (the new form of salutary neglect once the colonies had achieved independence) from Europe. Whether Jefferson at first derived his ideas from Smith or the Physiocrats (or even developed them by himself), is, at this point, unimportant; as we have seen, the principles of Quesnay were not "rival" but complementary to the principles of Smith. It was, however, in Smith that Jefferson's opposition to protective tariffs, bounties, subsidies, and funding operations could find reasoned support.

The man who favored these devices and opposed both Smith and Jefferson was Alexander Hamilton. He did not state his opposition openly, however. Instead, he skillfully used Smith's doctrines as a

24 See *Wealth of Nations,* Vol. I, especially Book III.

cover for his own. Where he could not do this, Hamilton, with consummate ability, picked the flaws in Smith's theory and tore apart the web of laissez-faire argument. The evidence of this is open to inspection in Hamilton's *First Report on Public Credit* (1790) and in his *Report on Manufactures* (1791).

Hamilton cleverly began by admitting the thesis of Smith: "It ought readily be conceded that the cultivation of the earth . . . has intrinsically a strong claim to preeminence over every other kind of industry." But—he went on—"that it has a title to any thing like an exclusive predilection, in any country, ought to be admitted with great caution." Then, Hamilton proceeded to turn the flank of the "agricultural" school by arguing strenuously that the prosperity of agriculture depends on that of manufacturing, which provides a domestic market for surplus crops.

If this argument is correct, it follows that anything which encourages manufacturing encourages agriculture and benefits the nation. In the earliest version of the slogan, "What's good for business is good for America," Hamilton declared: "It is a truth, as important as it is agreeable . . . that everything tending to establish substantial and permanent order in the affairs of a country, to increase the total mass of industry and opulence, is ultimately beneficial to every part of it." To increase the total mass of industry and opulence, Hamilton recommended tariffs, bounties, premiums, subsidies, a favorable balance of trade (including "pecuniary wealth, or money"), and a funded debt; in fact, the whole apparatus of mercantile policy. This policy was almost in complete opposition to Smith, who believed that the way to opulence was through the removal of these artificial and restricting measures.

Hamilton had Smith directly in mind, we believe, when he said, "A species of opposition is imagined to subsist between the manufacturing and agricultural interests. This idea of an opposition between those two interests is the common error of the early periods of every country." Then, delivering his body blow directly on Smith (and Jefferson), Hamilton declared accusingly: "Suggestions of an opposite complexion are ever to be deplored, as unfriendly to the steady pursuit of one great common cause, and to the perfect harmony of all the parts."

Hamilton, however, did more than play cat-and-mouse with

Smith's theories. He attempted to destroy them by attacking Smith's Achilles' heel. Smith, himself, had shown the way when he admitted that a system of perfect free trade could never exist. Hamilton pounced on this admission: "*If* [our italics] the system of perfect liberty to industry and commerce were the prevailing system of nations, the arguments which dissuade a country, in the predicament of the United States, from the zealous pursuit of manufactures, would doubtless have great force." But, he announced triumphantly, this was not the case. As a result, the only realistic policy was an "eye for an eye." "Considering a monopoly of the domestic market to its own manufacturers as the reigning policy of manufacturing nations, a similar policy, on the part of the United States, in every proper instance, is dictated, it might almost be said, by the principles of distributive justice; certainly, by the duty of endeavoring to secure to their own citizens a reciprocity of advantages."

Smith had made another exception to his theory of laissez faire. Where the necessities of military defense demanded it, he allowed government aid and interference. Following this lead, Hamilton prefaced his entire report on manufactures by saying that his suggestions were such as would "tend to render the United States independent of foreign nations for military and other essential supplies." Pursued far enough, Smith's exception in favor of national defense could easily, in a period of increasing "total wars," be made to cover the entire national economy.

Hamilton's attack on laissez-faire economics and on the claims of agriculture was superb. Where necessary, he had admitted the correctness of laissez-faire *principles,* but pointed out the exceptions present in them, as in "most general theories." The real point at issue, Hamilton claimed, shrewdly shifting the ground of argument, was one of political expediency. On this ground, he asserted, "The expediency of encouraging manufactures in the United States, which was not long since deemed very questionable, appears at this time to be pretty generally admitted."

Jefferson, relatively unversed in economic and financial matters, was unable to reply in the name of Smith to Hamilton's refined and sophisticated arguments (and perhaps there is no effective reply). As a result of this, he failed to impress his views as to the economic organization of society on America. Thus, his "agrarianism" lingered merely as an idyllic vision. Like a man of compromise, however, he accepted the "course of events" and went along with the "destiny" of his country.

VI

The truth is that Jefferson's real accomplishments were in other than the economic field. He himself realized this. When he died, on the fiftieth anniversary of the signing of the Declaration of Independence, he left his own epitaph, summing up what he considered to be his main achievements:

Here was buried THOMAS JEFFERSON Author of the Declaration of Independence, of the Statute of Virginia for Religious Freedom and Father of the University of Virginia.

The Declaration of Independence was in tune with Jefferson's motto that "Rebellion to tyrants is obedience to God." To prevent tyranny, he had turned to revolution, and on the cleared ground of revolution he had tried to build a government limited by state and individual rights. He defined his political desires in his First Inaugural Address of 1801: "A wise and frugal Government, which shall restrain men from injuring one another, shall leave them otherwise free to regulate their own pursuits of industry and improvement, and shall not take from the mouth of labor the bread it has earned. This is the sum of good government, and this is necessary to close the circle of our felicities."

The Statute for Religious Freedom emerged from his oath: "I have sworn upon the altar of God eternal hostility against every form of tyranny over the mind of man." Like the Declaration, the Statute has a long preamble as to why it was being enacted. In this, Jefferson pointed out that "Almighty God hath created the mind free, and that all attempts to influence it by tempt or punishments or burns or by civil incapacitations tend only to beget habits of hypocrisy and meanness, and are a departure from the plan of the Holy Author of our religion." Thus, it was *because* Jefferson believed in religion that he thought religious belief should be uncoerced. Religion itself, he contended, tells us to leave religious opinion and practice free. This theme he developed further in his *Notes on Virginia:* "The rights of conscience we never submitted, we could not submit. We are answerable for them to our God. . . . Subject opinion to coercion; whom will you make your inquisitors? Fallible men; men governed by bad passions, by private as well as public reasons. And why subject it to coercion? To produce uniformity. But is uniformity of opinion desirable? No more than of face and stature."

In sum, Jefferson's unquenchable greatness was that he devoted his life to resisting the attempt to enforce uniformity of belief upon men and fought with his pen unceasingly against "every tyranny over the mind of men." (It was to this end, too, that he founded the University of Virginia.) Because this struggle never ceases, the influence of Jefferson's thought will never die.

Above, and uniting, all these achievements, giving color to them, was Jefferson's own personality: an achievement in itself. In this, his life resembled that of the Renaissance geniuses: it was a "work of art." Like Leonardo and Michelangelo, Jefferson was a "universal man." His humanistic and scientific work included everything from writing *Notes on Virginia* (which contain important natural observations), inventing a plow, and writing down vocabularies of the Indian languages, to designing his own house at Monticello as well as the edifices of the University of Virginia.

Washington and Hamilton might be brilliant leaders and statesmen, but they were not "great men" in the sense that Jefferson was; they were not artists, or scientists, or philosophers, as was Jefferson. Nor, although they wished their country free, were they, fundamentally, in revolt against tyranny over the minds of men. Alone of the three, Jefferson was a "revolutionary character." His was a perpetual revolt, an eternal declaration of independence, against the forces which sought, and seek, to bind the spirits of men. It is in this sense that he is the special symbol of the American Revolution.

CHAPTER 22

The French Revolution and Its Napoleonic Sequel

I

THE INTELLECTUAL climate of the French Revolution can perhaps best be illustrated by a thumbnail sketch of one man. Concerning this man, Napoleon said that what he had written five years before the Revolution was already the Revolution in action. The document which Napoleon had in mind was a play, *The Marriage of Figaro;* and the man who wrote it was called Beaumarchais. His story is one of the interesting crosscurrents of which history is full.

Beaumarchais was born, as Pierre-Augustin Caron, in 1732, and lived beyond the Revolution's early stages to 1799. He was, like Rousseau, the son of a watchmaker. His grandfather had also been a watchmaker, and Pierre-Augustin followed in the family tradition.

At the age of 21 he invented one of the new escapements. Whether, by means of it, he made a wrist watch is uncertain, but he did make an exceptionally small timepiece. An incident in which his invention was almost stolen by another watchmaker resulted in a legal trial, which brought Pierre-Augustin to the attention of the Académie des Sciences and of the king, Louis XV. The young technician soon received many commissions from Versailles and, especially, one for a watch in a ring for Madame de Pompadour.

Pierre-Augustin's engaging face and manners led to his rapid rise at the court. He changed his name from Caron to Beaumarchais; ingratiated himself with the daughters of Louis XV by teaching them to play the harp (Beaumarchais had invented a new pedal mechanism); and made friends with the great financier, Pâris-Duverney, who assisted him in becoming a man of financial substance. Finally, in 1763, he purchased the position of *Lieutenant-Général des Chasses aux Baillages* and thereby acquired a title of nobility. No doubt, at this point he would have settled down to live a happy life, but his position was, in reality, too marginal and precarious, and unfortunately he became involved in a number of quarrels and legal suits.

One of these episodes, an exchange of blows with the Duc de Chaulnes, resulted in a short sojourn in jail; another, a money quarrel with the Comte de La Blache, involved Beaumarchais in a bribery case. Because of all this, Beaumarchais adopted a rather cynical and satirical frame of mind. His new attitude toward life was quickly given theoretical justification when, in an effort to redeem himself at court, he accepted a mission as secret agent to England to "buy off" a blackmailer, Theveneau de Morande. Through Morande, with whom he unexpectedly became friends, he was introduced to John Wilkes and the circle of extreme Whigs surrounding him. In this set Beaumarchais met the American Arthur Lee in 1775, and became infected with the "American principles of freedom."

To aid actively the American cause, Beaumarchais was instrumental in inducing the French government to supply arms and money to the rebellious colonists, and organized a company, Roderique Hortalez & Compagnie, to implement this aid.[1] He found another "revolutionary" outlet in writing two plays: *The Barber of Seville,* finally allowed to be presented in 1775, after a two years' delay, and *The Marriage of Figaro,* which was performed publicly in 1784, after a three years' prohibition.

Why were these plays called by Napoleon a preview of the revolution? To a modern audience, the two plays seem to contain no social criticism at all. Indeed, it is difficult for us to comprehend why the French king prohibited the performance of the plays for two or three years. The reason for our lack of comprehension is worth following: and it is the reason why we have brought in Beaumarchais. He and his plays stand as a corrective to the notion we may have of the French Revolution as a movement simply to depose the king.

The French Revolution was not directed originally against the king, but against the French nobility and the vestiges of French

[1] See Georges Lemaître, *Beaumarchais* (New York, 1949), Chap. 7.

feudalism.[2] And it did not reach its climax in the beheading of the king in 1793, but in 1789, with the victory of the commons over the nobility.

France under Louis XVI was a country in which the nobility dominated the court and demanded all the high posts; in which the nobility received enormous pensions and paid almost no taxes; and in which the nobility had preëmpted all the top clerical positions (*all* bishops in 1789 were of the noble class). The nobility might flirt with ideas of freedom, and, when the chips were down, it might accept taxation (as its cahiers of 1789 show); but it would not accept equality with the rest of the nation. This was the crux of the problem, and it explains why the French Revolution must be conceived of as primarily a social revolution.

The Marriage of Figaro is directly concerned with this problem, and it gives us a glimpse of the real content of the revolution, which one tends to forget when reading about Marie Antoinette or the flight of the king. Beaumarchais' play is filled with rebellion against the way the noble lord behaves with his servants. In a famous monologue the bold barber, Figaro, voices his challenge:

> Because you are a great lord, you think you are a great genius! . . . Nobility, fortune, rank, place; all that makes you so proud! What have you done to deserve all these blessings? You took the trouble to be born, and nothing more. Otherwise, a rather ordinary man! While as for me, good Lord! lost in the obscure crowd, I have had to employ more knowledge and more devices merely to exist than have been employed in the last hundred years to govern all of Spain. . . . I try to pursue an honorable career; and everywhere I am repulsed! I learn chemistry, pharmacy, surgery; and all the influence of a great lord hardly suffices to put in my hand a veterinarian's lancet.

Five years before the summoning of the Estates General, Beaumarchais had embodied in the character Figaro the voice of the third es-

[2] Cf. A. Aulard, *The French Revolution: A Political History, 1789–1804* (New York, 1910), Vol. I, pp. 9, 110, 121, and 225 for an exhaustive proof of this view. Aulard's is a standard Republican account of the Revolution. For a more left-wing version, consult the various works of Albert Mathiez. A synthesis of the two positions—Aulard's and Mathiez'—is often put forth by Georges Lefebvre, a brilliant and balanced scholar; see further his various works. For a long critical bibliographical essay on the French Revolution, see the excellent volume by Crane Brinton in "The Rise of Modern Europe" series, ed. by William L. Langer; Brinton's book is called *A Decade of Revolution* (New York, 1934).

tate. Figaro's successful marriage, against the intentions of his master, marks the servant's equality with the lord; soon the "career open to talent" will appear.

The "dull" Louis XVI understood the threat contained in *The Marriage of Figaro* better than did his clever courtiers. A plan to have the play read at Versailles was upset by the king's opposition and his reported exclamation: "Why, if this play were to be performed, the Bastille would have to be pulled down!"[3] Backed by Marie Antoinette and the courtiers of the Queen's Party, however, Beaumarchais went about openly boasting: "The King does not want *The Marriage of Figaro* to be played—therefore, it shall be played!"

We have, in the contention over this piece, a rehearsal of the events of the French Revolution. Just as the nobles in 1789 were to be instrumental in calling for the Estates General, so now they called for a performance of *The Marriage of Figaro;* just as Louis vacillated in prohibiting and accepting the acts of the Estates General, so he wavered about Beaumarchais' play. When the Comtesse de Polignac took it upon herself to present the *Marriage* in 1783, the king allowed the actors to learn their parts and the public to collect before calling off the play a half hour before performance. Finally, however, the king gave in and the play was publicly performed on April 27, 1784; a short time later, *The Barber of Seville,* Beaumarchais' earlier play, was given at the Royal Palace in Trianon, with Marie Antoinette acting the role of Rosine, and the Count of Artois as Almaviva. With the nobility itself playing at being "revolutionary characters," the Revolution stood nearby in the wings, waiting for its turn to occupy the stage.

II

What was the "setting" for the drama soon to be enacted? What was the position of France in the 1780's? Perhaps it can best be described by saying that France was a prosperous country with a bankrupt government. The fiscal system was in total confusion. No regular budget existed, and the king's income and the national income were confused. When the first budget of the old regime was compiled in 1788, expenses were estimated at 629 million livres and revenues at 503 mil-

3 Quoted in Lemaître, *op. cit.*, p. 274.

lion, leaving a deficit of 126 million livres; the dangerous factor in this situation was that 318 million livres, or more than 50 percent of all expenditures, were necessary for payment of the interest on the debt.

France was an agricultural country which had fallen behind in the industrial race but was not unprosperous. The government's difficulty was its inability to collect taxes, and thereby to tap the country's wealth. We have already described the rather odd tax system current in eighteenth-century England, and a brief glance at American history has shown how much taxation was resented by the colonists. The reason for this was that the English and American tax systems were constructed for a mercantile age; but they were being used in what was becoming an industrial age. The French system of taxation, on the other hand, went back to a feudal period; it was not only one century behind current economic developments, but at least two or three centuries. Whereas in England the rich nobles paid taxes on such things as hair powder, in France the nobles and clergy paid almost no taxes at all.

Because of the government's difficulty in raising taxes, they were "farmed" out, creating a class of wealthy farmers-general but little revenue for the government; it is estimated that over 60 percent of the gross revenue collected never reached the government.[4] Unfortunately, the device of making tax collecting a private business, run for profit, was a common one during the eighteenth century, where governments, having only a small staff of civil servants, farmed out any unpleasant job they had to perform by allowing the man who did it to make a profit on it.[5]

It was the financial situation that led to the calling of the Estates General in 1789 and, by thus providing a focus for the nation's discontents, to the subsequent Revolution. The nobility, taking advantage of the crown's need of money, rejected any solution to the problem by edict or by an Assembly of Notables, and asked for the calling of an Estates General which they thought they could dominate. Thus the nobility, by renewing the Fronde of 1648 against the king, opened the way to the Revolution.[6]

[4] J. M. Thompson, *The French Revolution* (Oxford, 1945), p. 190.

[5] To this inefficient and iniquitous system of tax farming was added an unfair tax assessment.

[6] For the Fronde, see Paul R. Doolin, *The Fronde* (Cambridge, 1935).

III

A striking feature of the French Revolution is its detailed recapitulation in the realm of action of the trends in thought which we have recently studied. Montesquieu and his ideal of an aristocratic government; Voltaire and the Physiocrats, with their notion of an enlightened despotism; Rousseau and the democratic and totalitarian implications of his rule by the general will: these political theories all worked themselves out in the course of the Revolution and its aftermath. Starting in 1789 and culminating in 1794, there was a movement of ideas to the "left." Concurrent with this was a movement of men; and we witness an emigration from France of nobles, after the failure of Montesquieu's views, and then of the upper bourgeoisie, after the failure of Physiocratic monarchical ideas.

Let us follow in some detail how the first of these three theories turned out in practice. The initial phase of the Revolution was, as the French historian Lefebvre points out, "an aristocratic revolution."[7] It involved an attempt by the nobles to assert, or regain, control of the state. For in spite of its social and tax privileges, the nobility had lost political power. Government officials, not the nobles, administered the rural areas, and it was said that France was governed by thirty intendants: although by 1789 the intendants were all noble, their ostensible loyalty was still to the state.[8]

The nobility itself was not homogeneous; it was divided into court and provincial nobles and into nobles of the sword and of the robe. The nobles of the robe were entrenched in the Parlements, a type of French law courts, which claimed the right to pass on the king's edicts and to register them. The parliamentarians tended at first to give a lead to the nation. In order to protect their own tax-free position, they advanced the idea that taxation should only be with the consent of the nation. Dissolved in 1771, the Parlements became martyrs to the cause of freedom; popular pressure on the king forced their recall. By the 1780's, the Parlements had linked themselves with the provincial estates in opposition to the king. Powerful in the pro-

[7] Cf. Georges Lefebvre, *The Coming of the French Revolution*, tr. by R. R. Palmer (Princeton, 1947), especially Part I.

[8] Cf. Alexis de Tocqueville, *The State of Society in France Before the Revolution of 1789*, tr. by Henry Reeve (London, 1888), for a full statement of the nobility's loss of real power in eighteenth-century France.

vincial estates, they assumed that they would control an Estates General and, through it, the king.

When, in 1788, the king demanded registration of a new tax to solve his financial difficulties, the Paris Parlement refused. To protect its members against royal reprisal, it condemned administrative arrest and declared that a natural right of all Frenchmen was liberty. Louis XVI is reported to have said that if the Parlements had their way, France would be "an aristocracy of magistrates."[9] Nevertheless, Louis was forced to yield, and, in the hope of obtaining money, he agreed to the calling of the Estates General.

Having conquered the king, the Parlements immediately lost the nation. Thowing aside the veil of disinterested love of freedom, the Parlements proposed that the Estates General should be constituted as in 1614, the last date of its meeting, with each of the three orders voting as an order; it was obvious that the clergy and the nobility would vote together against the third estate.

When the third estate asserted itself in opposition to the other two estates, it took over from the nobility or aristocracy valuable lessons in resistance. The aristocratic class, in its opposition to the king, had developed means of communication through salons, cafés, and agricultural societies, had organized political groups which exchanged correspondence and instructions, had enunciated declarations of fundamental laws, and had asserted the principle that taxation could only be levied by consent.[10] All of these lessons were taken to heart by the third estate, when, following Abbé Sieyès' advice, it became the "nation."

IV

The failure of the "aristocratic revolution" opened the way to an attempt at enlightened despotism. The nobility had been unable to lead the nation because it refused to give up social privileges in return for political power; we have seen from Beaumarchais' plays that Figaro no longer was willing to accept an inferior lot.

Worse, the nobility had lost faith in itself and in its principles; Montesquieu had been superseded, even among the nobility, by more "enlightened" thinkers. Thus, in 1789, at the meeting of the Estates

[9] Lefebvre, *op. cit.*, p. 30.
[10] *Ibid.*, pp. 33–34.

General, there was no forceful exponent of the aristocratic point of view, with the possible exception of Cazalès, the son of an adviser to the Parlements. Almost all the really talented nobles—Lafayette, Mirabeau, Talleyrand—had passed over into the camp of the third estate.[11] In a sense, therefore, the aristocratic revolution had gone by default.

Outstanding among the "other-class" leaders was Mirabeau. The son of the famous Physiocratic thinker, the Marquis of Mirabeau (who styled himself the "friend of man" but was the enemy of his child, shutting him up for a time in prison), the younger Mirabeau pursued a dissolute, passionate life. He was active but aimless until 1789. Rejected by his fellow nobles of Provence as a delegate to the Estates General, he then offered himself to the third estate and was successfully elected. It was Mirabeau, the noble, who dramatically replied to the king's command for the third estate to disperse: "We will not leave except by force of the bayonet."

Having thus identified himself with the third estate and asserted its rights, Mirabeau proceeded to embrace the cause of the king. The Spanish philosopher Ortega y Gasset has made the penetrating observation that "All revolution, inexorably—whether red or white—provokes a counterrevolution. The politician is he who anticipates this result, and makes at the same time, by himself, the revolution and the counterrevolution. The Revolution was the Assembly, which Mirabeau dominated. It was also necessary to dominate the Counterrevolution, to hold it in his hand. He needed the King."[12] A political realist, Mirabeau (like Edmund Burke, as we shall see) focused on practical issues, and opposed drawing a Declaration of Rights out of thin air. He declared: "We are not savages recently arrived from the banks of the Orinoco to form a society. We are an old nation, perhaps too old for our time. We have a preexisting government, a preexisting king, preexisting prejudices. It is necessary, as far as possible, to accommodate all these things to the Revolution and to attenuate the suddenness of the change."

[11] Lafayette and Talleyrand, unlike Mirabeau, had been elected by their respective orders, the nobility and the clergy; this did not prevent them, however, from supporting the position of the third estate.

[12] José Ortega y Gasset, *"Mirabeau, o el politico," Obras Completas*, 6 vols. (Madrid, 1950), Vol. III, p. 619. This brilliant essay should be consulted in its entirety, if possible; unfortunately there is no English translation.

It was Mirabeau who led the second phase of the Revolution—the bourgeois revolution—and attempted to consolidate the victory in a hybrid of constitutional monarchy and enlightened despotism. He advised the king to accept the National Assembly and a constitution; he advised the Assembly to allow the king an absolute veto over its future actions. Mirabeau's idea was that the king would use the royal veto and the royal power in the interests of the people.[13]

Like Voltaire and the Physiocrats, Mirabeau preferred reform to a freedom which might degenerate into license.[14] He did not trust the mob. He realized, in a remarkable flash of prophecy, that the Revolution was to a large extent a continuation of old-regime trends, but in an accelerated form. He wrote to the king: "The idea of forming a single class of all the citizens would have pleased Richelieu; this equality of the surface facilitates the exercise of power. Several successive reigns of an absolute monarchy would not have done as much for the royal authority as this one year of revolution." So it was imperative, in Mirabeau's view, that the heightened power provided by equality be in the hands of a royal authority, backed by the bourgeoisie.

Mirabeau's policy was the realistic acknowledgment of existing power in France as well as of the demands voiced in the cahiers, or instructions, drawn up for the delegates to the Estates General. French opinion of 1789 was not antimonarchical, and there were no voices raised at the time in favor of a republic. Why should there be, when the king, by calling the Estates General, seemed to be inaugurating the reign of reform? In accord with this frame of mind, the French Constitution of 1791 set up a limited monarchy, controlled by the middle class through a restricted suffrage. It was not until August 10, 1792, that a democracy was established with universal

13 His enemies accused him of supporting the royal veto because he was in the pay of the king. Mirabeau, from 1790 on, *was* in the pay of the king; he received a salary of 300 pounds per month, and his debts of 10,000 pounds were paid for him. But his ideas were not for sale, and Mirabeau favored the royal veto for genuine political and not for venal reasons.

14 Was it that, while rejecting his father, Mirabeau retained his ideas? Cf. Tocqueville, *op. cit.*, p. 136, on the idea that "the French aimed at reform before liberty."

suffrage, and not until September 22 that the monarchical form was abolished and a republic proclaimed.

Why did Mirabeau's ideas fail? It is easy to say why Mirabeau as a person came short of his goal. He was distrusted because of the ill repute of his past life and of his financial connection with the king. Nevertheless, he maintained a form of control over the Assembly by his matchless oratorical abilities; it was primarily Mirabeau who introduced the parliamentary rhetoric which was the source of power until the coming of Bonaparte. But Mirabeau's premature death, at the age of 42, on April 2, 1791, ended the one force which could hold together the bourgeoisie and the monarchy.

The main obstacle to the success of Mirabeau's ideas—both during his life and after—was Louis XVI. His personality was pivotal to the entire French Revolution. His vacillation between force and concession undermined any consistent policy. When he disowned the third estate, Louis ended the possibility of a peaceful revolution. By attempting to crush the Estates General, he provoked the decisive intervention of the Paris mob. Louis XVI caused Bastille to be the point of no return.

It was Louis XVI who made both enlightened despotism and constitutional monarchy impossible solutions for France. This is a point worth insisting upon; it was the same flaw of personality which faced Plato in Sicily when the prince whom he attempted to form into a philosopher-king turned out to be unworthy. The French Revolution was primarily a social revolution against the aristocratic hierarchy; the vacuum caused by Louis' refusal to play the part assigned him led to a republic. Condorcet's ingenious proposal, that an automaton of a king be substituted for the real puppet, was not enough.

As late as July 13, 1791, Robespierre had said: "As for the monarch, I have never been able to share the terror with which the title of king has inspired almost all free peoples. . . . I should not fear royalty; not even the hereditary nature of the royal functions in a single family."[15] But Louis had already demonstrated, by his attempt at flight on June 20, 1791, that he was not willing to stand in alliance

[15] Quoted in Aulard, *op. cit.,* pp. 309–310.

with the bourgeoisie. Alone against the mob, the moderate Monarchicals gave way step by step to the Feuillants, the Feuillants to the Girondists, and eventually the Girondists to the Jacobins, led by the same Robespierre who had earlier been willing to accept an hereditary monarch.[16]

V

It was now time for the theories of Rousseau to be put into practice. The other *philosophes* had unintentionally done their part in undermining monarchical principles by accustoming the people to question the origins of sovereignty and authority. Rousseau went further and directly placed the authority of government in the hands of the people. The satire of the *philosophes* had corroded the foundations of the monarchy and prepared for its easy toppling; Rousseau laid the cornerstone on which a government of the people, or a government of the general will, could be constructed. The great question was whether the democratic or totalitarian implications of Rousseau's political philosophy would come to the fore.

It was the military invasion of France by Austria and Prussia, to support the king, which settled the issue. The conflict, starting in 1792 as a war of defense (although it was France that declared war), rapidly turned into a war of conquest and propagation of revolutionary principles. At home, the war led to a republic and, eventually, to a military dictatorship. In the beginning there seemed some possibility that Lafayette, the hero of two revolutions, might convert his position as head of the militia and general of the French armies into a dictatorship. At least this was the fear of Robespierre, who cited Cromwell as an example of the danger of military dictatorship of the republic. "I would rather see a popular representative Assembly, with the citizens free and respected, under a King, than an

[16] The first breach in the moderate position occurred when the advocates of a bicameral legislature (with either an hereditary or elective upper chamber) and an absolute royal veto, the Monarchicals or Anglophiles as they were called, were voted down by the majority of their own so-called patriot party. Headed by Mounier, and with men like Lally-Tollendal and Clermont-Tonnerre in its ranks, this group of moderate monarchists gave up its parliamentary opposition and took to emigration; Mounier, for example, left France around 1789. This amputation from the patriot party in turn left the remaining moderates (who, under the leadership of Lafayette and Barnave, rejected an upper chamber because the nobility might regain its power therein) weakened in the face of the radicals who wished to push the Revolution further.

enslaved, degraded people under the rod of an aristocratic Senate and a dictator. I love Cromwell no more than Charles the First."

But Lafayette was not the man to play at Napoleon. He lacked political genius or realism, and contented himself with cutting a romantic and generous figure. Further, he sincerely believed in freedom and, with Barnave, headed the Feuillant party which sought a constitutional monarchy. His effort to "save" the revolution for the upper bourgeoisie failed when he could not persuade his troops, in August, 1792, to march against the more extreme revolutionaries, who were in power. Fleeing to Belgium, Lafayette took with him any chance of the Feuillants succeeding in their aim. His removal as commander of the French troops also marked the end of aristocratic leadership of the Revolution; henceforth, the French *levées en masse* were to be led by new "revolutionary" generals.

The totalitarian implications of Rousseau's doctrines now began to appear in full force. Robespierre, taking advantage of the dismay and confusion produced by the revolutionary war without, and the civil war within, instituted "government by terror." His strategy was constantly to accuse his opponents of being in the pay of Pitt and the English.[17] His tyrannical methods Robespierre justified as the reign of virtue advocated by Rousseau. The "incorruptible" leader of the French Republic in 1793 even went beyond his philosophical predecessor and expounded the totalitarian principles of "a revolutionary government":

> The theory of the revolutionary government is as new as the Revolution itself, from which this government was born. This theory may not be found in the books of the political writers who were unable to predict the Revolution. . . . The goal of a constitutional government is the protection of the Republic; that of a revolutionary government is the establishment of the Republic. The Revolution is the war waged by liberty against its foes—but the Constitution is the regime of victorious and peaceful freedom. . . . Under Constitutional rule, it is sufficient to protect individuals against the encroachments of the state power. Under a revolutionary regime, the state power itself must protect itself against all that attack it.[18]

17 See Albert Mathiez, *The French Revolution,* tr. by Catherine Alison Phillips (New York, 1929), p. 360.

18 Maximilien Robespierre, "Report on the Principles of a Revolutionary Government," *Introduction to Contemporary Civilization in the West,* 2 vols. (New York, 1946), Vol. I, p. 1089.

In theory, the revolutionary government was checked by the natural rights of individuals, which even the central government might not violate. These natural rights soon gave way in the face of *raison d'état* and of "cruel necessity" (to recur to Cromwell's phrase). Free speech and press and freedom of person—all were violated in the time of the Terror on the grounds of public safety. And Robespierre, who headed the Committee of Public Safety, merely formulated the theory to describe the reality.

When would the "revolutionary government" give way to "constitutional rule"? Only when the revolution was complete and all its foes defeated. To insure this outcome, Robespierre insisted, resort must be had to a reign of terror. One of the most competent present-day students of the Terror has stated that "In the circumstances no government could have maintained itself without very severe measures of repression. The Terror, after all, was inevitable."[19]

According to the best evidence, about 20,000 people were direct victims of the guillotine and another 20,000 died in the prisons or were executed without trial. To this total of 40,000 dead, we must add the imprisonment, at one time or another during the Terror, of about half a million Frenchmen. Yet apparently this new type of state repression, which was most active in the area of the Vendée and the war frontier, did assist the military success of the Republic.

Was the Terror also a class weapon, a tool for the achievement of social or economic ends? The answer appears to be in the negative; the Terror, on sober consideration, was a political weapon. The social revolution had already been accomplished by the bourgeois Assembly.[20]

The other major political innovation created by the French Revolution was government by "clubs and committees." Real power in Revolutionary France resided, not in the legislature, but in the

[19] Donald Greer, *The Incidence of the Terror During the French Revolution* (Cambridge, 1935), pp. 115 and 137. Even so judicious a historian as Georges Lefebvre has asserted that the Terror institutionalized justice and prevented massacres (such as the one of September, 1792, in which over 1,000 victims were claimed). In our view, however, a justification of the Terror on the grounds of *raison d'état* merely opens once again the question whether any *raison d'état* justifies state action.

[20] Cf. *ibid.*, pp. 1, 14, 81, 85, and 87.

machinery of the Jacobin clubs.[21] These had their origin in the literary societies, or *sociétés de la pensée,* and in the secret societies, such as the Freemasons, which existed before the Revolution. The source of the Jacobin Club was the Breton Club, formed during June, 1789, to serve as a center for representatives of the third estate. The Breton Club members quickly discovered that, in order to succeed in their fight against the king and the other two orders, it was necessary to mobilize public opinion from outside the Assembly. This they did through the formation of the Society of the Friends of the Constitution, which became better known as the Jacobin Club, named after the Convent at Paris in which its meetings took place.

An early Jacobin, Count Alexandre de Lameth, has left us a description:

> The aim of the Society of the Friends was to discuss questions which were already on or were about to be placed on the calendar of the National Assembly. . . . It was a great advantage for the popular party to determine, by preliminary ballots within the society itself, the nominees for president, secretaries, and the committees of the Assembly. For from that time the elections were almost always carried by the left wing, although up to that time they had been almost entirely controlled by the right.

After telling us the purpose of the Club, Lameth proceeds to indicate how its influence spread:

> Very soon the place of meeting became insufficient, and permission was obtained of the friars of the convent to meet in their library and later in their church. About the month of December, 1789, many leading inhabitants of the provinces, having come to Paris either on private business or to attend more closely the course of public affairs, were presented to the society and expressed the desire to establish similar organizations in the chief cities of France.[22]

Both in their origins and in their careers, the Jacobin Clubs were middle-class bodies, whose members had standing in their local com-

[21] This was conclusively demonstrated by the elections of 1791 to the Legislative Assembly. On the basis of a restricted suffrage, the Feuillants gained 264 seats to the Jacobins' 136; 345 representatives considered themselves as independents. The parliamentary victory of the Feuillants, however, was misleading; without support in the clubs (which had organized the political activity of the country), the Feuillants remained, as Crane Brinton says, a "mere parliamentary caucus."

[22] From Lameth, *History of the Constituent Assembly* (1790).

munities. According to one authority, there were probably about 500,000 enrolled Jacobins in France, or about 2.2 percent of the French population.[23] Almost all were from the more prosperous levels of society. With the fall of the old government hierarchy, the Jacobins necessarily had to take over the task of local administration and government.

Within the Jacobin societies, two groups formed. One, outside the Assembly, was headed by Robespierre; it was this group which appropriated to itself the name of Jacobin. The other group was led, initially, by a journalist named Brissot, who was a member of the 1791 Assembly; these Brissotins, as they were at first called, split away from the Jacobins after the formation of the republic, and then became known as Girondists. Eventually led by Danton, the Girondists fought for power against Robespierre and the Jacobins. The victory of Robespierre marked the final point of the trend to the "left"; the limits to which philosophic theory might be tried in practice had been reached.

VI

It was not that theory had been exhausted; far from it. There were socialistic ideas to be found in some of the *philosophes,* like Mably and Morelly, which were not put to the test. The middle-class Jacobins who supplied the power behind Robespierre had no intention of giving up the rule of private property.

Their view on property and its rights had been carefully enshrined in the Declaration of the Rights of Man and the Citizen. Article II declared that "The aim of all political association is to preserve the natural and imprescriptable rights of man. These rights are liberty, *property* [our italics], security and resistance to oppression." To make this quite clear, it was again stated in the last article, Article XVII: "Property being an inviolable and sacred right, no one may be deprived of it except for an obvious requirement of public necessity, certified by law, and then on condition of a just compensation in advance."

[23] See Crane Brinton's detailed study, *The Jacobins* (New York, 1930), especially pp. 40–43.

The exigencies of war, of counterrevolution, or of poor harvests and unchecked food speculation might lead to agitation for state limitation of the rights of property; but not to socialism. Thus, the *enragés,* led by men such as Jean Varlet and Jacques Roux, took advantage of the period of economic distress in 1793 to advocate governmental regulation of the grain trade and to demand price fixing. On September 29, 1793, the Convention passed a "law of the Maximum," which fixed the highest price for certain essential articles; this law also involved the fixing of wages. Some rationing was also attempted, but this was never very effective. Hoarding, however, was pronounced a capital crime by a law of July 26, 1793, and accounted for a number of executions during the Terror.

Robespierre's right-hand man, Saint-Just, did propose the distribution of confiscated land to the revolutionary proletariat in February of 1794. But he made it clear that the property of patriots was sacred, and that it was only the property of conspirators (like enemy property today) which was subject to confiscation during wartime. Even the successors to the *enragés,* the Hébertists, or ultrarevolutionaries, who clustered around the assistant procurator of the Commune, Hébert, only advocated stronger social measures to relieve the misery of the poor. They did not advocate the establishment of a different social order. And even the Hébertists were eliminated by the Jacobins in March, 1794, as going too far.[24]

Socialism appeared in the French Revolution only once, "as a last convulsive effort of the principles of the great French Revolution to work themselves out to their logical ends"—in the Conspiracy of Equals.[25] This movement arose after the fall of Robespierre, during the period of the Directory, and was led by a former registrar of seignorial rights, who took the name Gracchus Babeuf (to recall the role of Gaius Gracchus, who wished to distribute land to the people in Roman times). Babeuf, who had held only minor posts during the Revolution, did not become a "leader" until after Thermidor. At that time, he gathered around himself, in a society of equals, the

[24] Cf. Greer, *op. cit.,* pp. 98–99.

[25] Cf. Edmund Wilson's discussion of this in *To the Finland Station* (New York, 1940), pp. 69–79. Also David Thomson, *The Babeuf Plot* (London, 1947), which contains a critical bibliography on Babeuf.

radicals who were discontented at the conservative turn of the revolution. They plotted to achieve power and the realization of their plans by the method of "insurrection."[26]

Babeuf and his followers envisioned a society in which economic equality would take its place alongside the political equality achieved by the Jacobins. This, Babeuf claimed, had been implicit in the ideas of Mably, Helvétius, Diderot, and Rousseau; it was in their tradition that Babeuf claimed to place himself. In the *Manifesto of the Equals,* written to express the ideals of the society by Babeuf's friend, Sylvain Maréchal, we are told that "The French Revolution is but the precursor of another, and a greater and more solemn revolution, which will be the last!"

The Conspiracy of Equals was acute in drawing the real line between "agrarian law" and true socialism. They noted that the agrarian law aimed simply at the partition of lands; the Babouvists demanded "something more sublime and more equitable—the common good, or the community of goods. No more individual property in land; the land belongs to no one. We demand, we would have, the communal enjoyment of the fruits of the earth, fruits which are for everyone!" The conspiracy aimed at the elimination of the distinction between rich and poor, and between great and small. Taking their cue from Rousseau, they announced: "Perish, if it must be, all the arts, provided real equality be left us."

The Conspiracy of Equals, like the Fifth Monarchy Movement or the Diggers of the Puritan Revolution, did not amount to a great deal during the course of the French Revolution; and the execution of Babeuf and the transportation of some of his followers in May, 1797, ended the conspiracy as such. Thus, it cannot be said that socialism, either as an idea or as a movement, played any significant part in the French Revolution. Even the idea of "agrarian law" did not advance beyond the emergency proposal of Saint-Just to distribute

[26] Babeuf's ideas on the method of achieving power had a strong influence on Louis-Auguste Blanqui, who inherited them through personal acquaintance with Babeuf's friend, Buonarroti (a descendant of Michelangelo), as well as through a reading of the latter's *Conjuration des égaux* (1828); according to Blanqui, power could only be achieved through a secret elite, or minority, willing to use violence. Cf. Edward S. Mason, *The Paris Commune* (New York, 1930), pp. 19–27.

the property of traitors to the poor. The French Revolution was political and social; but it was not socialistic.[27]

VII

Robespierre's "revolutionary government" had successfully established the Republic at home and protected it from its foes abroad; his execution on the 10th of Thermidor (July 28, 1794) marked the transition to "constitutional rule." The constitutional government of 1795–1799 which followed, known as the Directory, strongly affirmed the individualistic, bourgeois nature of the Revolution. It abolished the Terror and attempted to institute government by law; it began work on the revision of the law codes; it repealed the law of the Maximum; it attempted to solve the financial problem and to bring economic stability to France.

Unable, however, to maintain itself securely against the unregenerate followers of Robespierre, on one side, and the returned royalists, on the other, the Directory turned increasingly to the Army. The Directors were forced to suppress a royalist uprising in Paris on October 5, 1795, by employing Napoleon Bonaparte and a "whiff of grapeshot"; it used General Hoche to chastise the royalist forces in the Vendée and Brittany. It was Bonaparte again who closed down Babeuf's club, the Society of Equals; and it was Bonaparte's nominee, General Augereau, who prevented a rightist coup d'état on September 4, 1797. These episodes showed that there could be no stable settlement in France unless the government could either conciliate or suppress the radicals and the reactionaries; and the Directory by itself could effectively do neither.

27 In fact, the economic consequences of the French Revolution, rather than favoring socialism, were on the side of individualism. As a result of the Great Fear and the stirring renunciation of feudal rights by the National Assembly on the night of August 4, 1789, France became a nation of small proprietors. Even the large-scale capitalist farming, characteristic of England after the enclosures, never came to dominate in France. (Cf. Henri Sée, *Economic and Social Conditions in France During the Eighteenth Century,* tr. by Edwin H. Zeydel, New York, 1927, pp. 189 and 229.)

So, too, the effect of the Revolution upon industry was, more or less, to free it from the medieval and mercantile restrictions which had previously been imposed. For example, a broader scope to individual endeavor was given by the abolition of the guilds. In fact, the Le Chapelier Law of 1791, prohibiting all labor coalitions, was, theoretically, a consequence of pure laissez-faire doctrine rather than of antilabor bias.

It was again the war which, having first brought the totalitarian implications of Rousseau to the fore, now brought forth the solution to the Directory's problems: Bonapartism. In the person of Napoleon, we have a new version of the condottiere. This comparison is lent color by Napoleon's Italian ancestry and his affectation of Roman attitudes and modes. Napoleon was like that other condottiere, Cesare Borgia, in his astuteness and Machiavellism. Unlike his prototype, however, he covered over his naked self-interest and desire for personal gain with a cloak of revolutionary ideology.

The secret of Napoleon's greatness was that he linked the doctrine of self-interest with the ideology of revolution, and turned the power which their fusion gave him to his own egoistic ends. To the French, Napoleon preached the bourgeois Directory's doctrine of self-interest. We detect the new note in Napoleon's first proclamation to the army in Italy of March 27, 1796:

> Soldiers, you are naked, ill-fed: the Government owes you much, it can give you nothing. Your patience, and the courage which you have shown in the midst of these rocky crags, are admirable; but they have won you no glory; no glamour clings about you. I wish to lead you into the most fertile plains of the world. Rich provinces, great cities, will be in your power; you will find honor, glory, and riches there. Soldiers of Italy! do you lack courage or constancy?

With Napoleon, the French revolutionary army took on some of the loutishness of the old mercenary armies. But the revolutionary note was not forgotten. To the countries he entered, Napoleon held out the gains of the Revolution. For example, he promised the peoples of Italy that they "should be counted among the free and powerful nations" and reminded them that before his coming they had been "divided, and bent down by tyranny." Then, without the assent of the Directory, Napoleon set up Cisalpine and Ligurian republics in Italy.

A conquering hero abroad, Napoleon returned to France in 1799 to give it the same "liberty" he had proclaimed elsewhere. In the "Proclamation to the French People," delivered immediately after the successful coup d'état of the 19th Brumaire against the Directory, Napoleon declared: "On my return to Paris, I found division among all authorities, and agreement on just one fact, that the constitution was half destroyed and could not save liberty. Every party came to me, confided to me their designs, imparted their secrets, and requested my support; but I refused to be the man of any party." Then, having established himself as being the representative of France, a man "above party," Napoleon gave a lurid, and fictitious, account of the coup d'état and concluded, in a masterly hash of contradictory verbiage: "Conservative, protective, and liberal ideas have resumed their sway."[28]

Nevertheless, in spite of the logically contradictory nature of Napoleon's conservative and liberal ideas, and of his appeal to self-interest and revolutionary idealism, the *historical* logic of his position was accurate. Napoleon had profoundly sensed the needs of his time; he seemed to embody its dialectic spirit. It was this political genius of Napoleon which led Hegel to proclaim Napoleon as a "world historical figure."

In the judgment by Hegel, we have a recurrence of the Renaissance admiration for the brutally successful man by the great thinker. The new element is Hegel's glorification of the hero, not only as a superior man, but as the embodiment of historical forces. Taking his cue perhaps from Napoleon's synthesis of self-interest and revolutionary idealism, Hegel stated:

> Such world figures have no consciousness of the general idea they are unfolding while prosecuting their own private aims. On the contrary, they are practical political men, but possessed of an insight into the requirements of the time, an understanding of what is ripe for development. It is theirs to realize this nascent principle; the next step forward which their world is to take. It is theirs to make this their aim and spend their energies promoting it. They are the heroes of an epoch; must be recognized as its clear-sighted ones. Their deeds, their words, are the best of their time.

We have encountered this sort of man before in Mirabeau. Napoleon followed him in the attempt to combine the revolution and the counterrevolution in one policy. He endeavored to achieve a balance of social forces while at the same time doing away with any governmental division of powers which might check or balance his own. Napoleon's method of achieving unchecked power was through direct appeal to the people in a plebiscite; thus, like Hitler later, he claimed to embody the general will.

His method of achieving the balance of social forces was continuously to play off against one another the bourgeoisie and the

28 From Napoleon, "Proclamation to the French People" (1799).

workers, the liberals and the royalists, in a masterly game of opportunistic politics. He held out to the bourgeoisie the promise of stability and order, to the lower classes of France the lure of national glory and power, to the liberals the slogans of equality and fraternity, and to the rightists a revival of Catholicism and of unity and authority in the government. On top of all this, he attempted to create "new men," dedicated to Bonapartism; he inserted a "marshal's baton in every soldier's knapsack," established the Legion of Honor, and set up a new, upstart nobility.

Napoleon recognized that the broad mass of Frenchmen were generally satisfied with the social reforms of the Revolution, and that the desire was for stability and order. Thrusting political liberty to one side, Napoleon used alternating force and conciliation to achieve his ends. He secured a measure of domestic peace by pacifying the Vendée; he initiated fiscal reform and set up the Bank of France; he ordered a legal code (embodying a conservative interpretation of the Revolution's social gains) drawn up, which influenced all Europe and has lasted until today; and he tried to heal the religious wounds of France by effecting a concordat with the pope.

Unlike the enlightened *philosophes* and bourgeois of the Revolution who had seen in religion the enemy of all reform, Napoleon regarded religion as a mainstay of the state. The only difference between his view and that of the old-regime supporters was in the nature of that state. For Napoleon, the state was the centralized, nationalistic state enhanced by the Revolution and based on the social dominance of the bourgeoisie. Religion was merely one more social force to be used in maintaining this Bonapartist state. As Napoleon commented: "For my part, I do not see in religion the mystery of transubstantiation but the mystery of social order."[29]

Nevertheless, like Mirabeau, Napoleon failed in his effort to achieve a stable synthesis of the revolution and the counterrevolution. In his case, of course, the reasons were somewhat different from those that defeated Mirabeau. One cause of Napoleon's failure was his political illegitimacy. Louis XVI could have held the French na-

[29] Quoted in Pieter Geyl, *Napoleon For and Against*, tr. from the Dutch by Olive Renier (New Haven, 1949), p. 360. This is a splendid book of historiography, treating the various treatments of Napoleon with a very critical eye. For a good general survey of the Napoleonic period, see Geoffrey Bruun's volume in "The Rise of Modern Europe" series, *Europe and the French Imperium* (New York, 1938); like Brinton's book, it has an exhaustive bibliography.

tion together, as the embodiment of legitimate authority. Napoleon was forced to do so by means of his charismatic personality. In fine, Napoleon was never accepted either by the liberals or the royalists in France. The first considered him as the betrayer of the Revolution, and the latter thought of him as the impediment to the rightful Restoration. Thus, Napoleon did not so much enjoy their support as dexterously thwart their opposition; this, however, was not a firm basis for extended political rule.

If Napoleon was like a Renaissance condottiere in his political illegitimacy, he was also like one in placing personal ambition above the needs of his people. There is some justice in saying that Napoleon was Machiavelli's Prince come to renewed life and to increased success; and the very success of Napoleon proves the fundamental weakness in the Prince. It is as if we hear the voice of Machiavelli again in Napoleon's letter to his brother Joseph, King of Naples, in 1808: "I wish the Naples mob would attempt a rising. As long as you have not made an example, you will not be their master. Every conquered country must have its uprising." The thin covering of Napoleon's politics, provided by Revolutionary slogans, is pierced by his candid conception of the people as a conquered beast to be ridden by a master.

The basic flaw in Napoleon's attempt to bring domestic stability to France was that it conflicted with his grandiose personal ambition for empire. Napoleon was not really interested in the welfare of the French people, but only in governing them so that he might use them and the nation as a tool to realize his plans. Napoleon was a condottiere who combined with this role the attributes of Marlowe's Tamburlaine, and was constantly "overreaching" himself. Or, to use the terms of nineteenth-century romanticism, Napoleon was filled with the spirit of endless striving toward an unreachable, infinite goal. Upon the altar of his lust for power, Napoleon sacrificed the synthesis of revolution and counterrevolution which he might have effected in France.

VIII

The French Revolution had destroyed the old regime in France: Napoleon, in turn, had used the momentum of the Revolution to conquer much of Europe. Between them, the Revolution and Napoleon changed both the map and the mind of the Western world.

By carrying the concept of the nation in arms, the slogans of liberty, equality, and fraternity, the legal code, and the whole paraphernalia of revolutionary reforms into all the conquered countries, the Revolution and Napoleon toppled not only the feudal structure of Europe, but undermined the foundations of the monarchical system as well. Then, by awakening the spirit of nationalism, in reaction to French aggression, the Revolution and Napoleon laid out the lines upon which the changed European nations might grow.[30]

The result was a "new world," with its roots deep in the fecund soil of the old Europe. The ideas of the *philosophes,* often distorted and twisted almost out of all recognition by the pressure of events, served to mold a new version of society, but in the process the French Revolution may be said to have partially exhausted these ideas. The result was that the minds of men were prepared to receive new thoughts and philosophies, some of which we shall discuss in further chapters. In sum, the French Revolution had not introduced the millennium, but instead had solved some of the old problems and raised new ones in their place.[31] The task of post-Revolutionary thinkers became one of searching for new solutions to the new problems.

[30] For example, by his famous *Reichshauptdeputationschluss* of 1803, Napoleon fused more than 300 German principalities and free cities into less than 100 and made Bismarck's "blood and iron" unification of 1870 possible.

[31] The argument whether the Revolution was a success or failure is an inexhaustible and unanswerable one. As to the question whether Napoleon was a fulfillment or a miscarriage of the Revolution, that argument has also occupied the attention of a great number of historians of the French Revolution. (Cf., for example, Peter Geyl, *op. cit.*, and Geoffrey Bruun, *op. cit.*) Napoleon himself helped create the legend which pictured him as the son of the Revolution who consolidated its gains. Writing his *Mémorial de Sainte-Hélène* to justify his policy in the eyes of posterity, Napoleon contended that he had maintained equality, restored order, and brought peace; whatever war he had indulged in had been forced upon him by English imperialism. Even then, Napoleon added, his conquests had merely brought the blessings of the Revolution to other countries.

Napoleon's critics have retorted that he destroyed freedom and set up tyranny in its place, corroded the glorious doctrine of equality where he could, ruled France for his own benefit, set up his family as little kings all over Europe, and, in general, perverted and corrupted the aims and ideals of the Revolution.

CHAPTER 23

Edmund Burke

I

HISTORY DOES not consist of isolated events, and it is not made by isolated people. The purpose of historical study is to show the connections between events, and if (as in this book) the study is presented in terms of people, the latter must illustrate the struggles and divisions of their times. We are not presenting a portrait gallery of heroes and villains, but an account of the historical changes in which sensitive and intelligent men have been involved, and through which they have tried to find their way.

Edmund Burke has been seen most often either as a hero or as a villain—and at different times in his life, he was a hero and a villain to different sides. But what makes him fascinating is that he lived the contradictions of his age, and was more outspoken about them than most men. His contemporaries thought his speech too violent, and most of them sooner or later thought him unbalanced. In this they were partly right, but it is exactly this violence, this unbalance, which makes him the tormented conscience of his age.

Burke and his contemporaries lived through three revolutions, and the three revolutions through which they lived were not isolated from one another. True, neither Burke nor any other politician understood how the Industrial Revolution was entering into the lives of men, and so into the affairs of states. But Burke understood very well that the American and the French Revolutions were somehow linked, though the knowledge threw him into a frenzy of denial. His later years were dominated by a compulsion to find differences between France and America which would justify, in retrospect, his past support for an antiroyalist policy which now frightened him.

But of course, the common ground of the revolutions did not lie in this or that detail of political constitution. The ground common to them—and common to all three revolutions—was at bottom the changing outlook of men, their discontent with the tradition of living which had satisfied earlier generations. There was a ferment in the way that men looked at accepted authority, and what made Burke unhappy in his old age was that he knew that he had helped to generate the ferment. Long before he became a political orator, Burke had written about literature and the arts in a way which made concrete the revolt of creative minds against the formalism of the first half of the eighteenth century. Burke's politics are, in many ways, very ordinary; but his critical writings are a pioneering statement of the romanticism which meant much in the political and intellectual landscape of the revolutions.

II

Edmund Burke was born in Ireland in 1729. His mother was a Roman Catholic, and he in turn later married a Catholic. It was therefore rumored by political opponents that Burke was a Catholic himself, and that he kept this a secret only because it would have barred him by law from public life. But the accusation was false, and Burke's strong and constant advocacy of greater freedom for Roman Catholics, and for Ireland as a whole, was a genuine act of political vision.

Burke's father, who was a lawyer, sent his son to London to study law, and cut off his allowance when the young man took to literature instead. For some years, Burke had to live by his wits, and he remained preoccupied with money all his life. All his life he had to find patrons, and had to try by some gambler's throw to turn their occasional gifts into a capital large enough to give him and his dependents a permanent income.

Meanwhile, the penniless young man had fulfilled the dream of all penniless young men: at the age of 27, in 1756, he had written a book which had made him the talk of the literary world. The book is called *A Philosophical Inquiry into the Origin of Our Ideas of the Sublime and Beautiful;* it is almost unreadable today, but it was a milestone in the transformation of taste during the eighteenth century. It marked dramatically for its generation the shift from the classical formalism of the first half of the century to the romantic violence of the second half.

The title of Burke's book is itself significant, for it reveals a new approach to literary appreciation—a psychological approach. Earlier in the century, no one would have thought of asking for "the origin of our ideas" of the beautiful; that is, no one would have thought of approaching the definition of beauty by way of the psychology of the appreciator. The Augustans at the beginning of the eighteenth century were sure that they knew what is beautiful, because they believed that what is beautiful is so in itself, by virtue of its inherent nature—and is bound to be beautiful to all beholders of real sensibility in all ages. Burke's book did not necessarily challenge the permanence of our ideas of beauty, but it did treat them as personal ideas; it implied that what is beautiful is so only by virtue of the human mind that perceives it. From Burke's time, beauty has come to be analyzed in psychological terms. This was a new approach, which only became general at the end of the century, in the literary criticism of August Wilhelm von Schlegel in Germany and Samuel Taylor Coleridge in England.

A psychological approach was necessary to Burke because he wanted to distinguish between the different ways in which literature may move us. It would not have occurred to the Augustan critics at the beginning of the century to play off one literary emotion against another; to them the work was either beautiful, or it failed. But Burke's theme is that a work of art can move us in different ways, and in particular that it can be sublime rather than beautiful. The romantic movement was not yet ready to challenge the formal definition of beauty of the Augustan age, but it was taking a first step in Burke's book by formulating a less calculated criterion than beauty: the criterion of the sublime. The *Inquiry into the Origin of Our Ideas of the Sublime and Beautiful* is a cautious but critical challenge to the absolute conception of beauty which the Augustans had set up, by proposing to rank with it the new concept of the sublime.

At the same time, Burke appropriated the most powerful emotions for the sublime, so that the beautiful was left by default only with the finicking and the genteel. We feel ourselves in the presence of the sublime, said Burke, whenever our emotions derive from pain or from danger. It is clear that Burke's definition of the sublime is related to Aristotle's definition of the tragic; for the tragic, in Aristotle, gives rise to the emotions of pity and terror. Thus, the sublime stands for the profound emotions which man feels in face of the vastness of what is outside himself; as in every romantic movement (in existentialism today, for example) the accent is on the emotion of *Angst* or anguish, the human sense of being dwarfed and lonely. With this great range given to the sublime, Burke could define the beautiful only by its smallness, its smoothness, it loving finish; the implication is clearly that beauty is a minor attribute.

Burke had not made a frontal attack on the Augustan canons of absolute beauty; he had outflanked them. The consequences of his ingenious rebellion reached far. For the Augustans, nothing had been beautiful which was not clear; now the sublime had to have vague and shadowy outlines. For the Augustans, nothing had been beautiful whose language was not well measured; now the sublime had to be rhetorical and wild. For the Augustans, beauty was civilized; now the sublime created a feeling for ancient and primitive poetry. One Scottish poet, James Macpherson, almost at once found splendid examples of Gaelic folk poetry, which fitted Burke's analysis of the sublime to perfection. Soon after, the young Thomas Chatterton discovered equally romantic poems of the Middle Ages. Alas, the perfection and the romance were simulated; the *Ossian* of Macpherson and the ballads of Chatterton, able as they are, are largely fakes.

III

In 1759, Burke became private secretary to a minor politician, and six years later he was made secretary to a leading Whig. Burke was a Whig who might be taken to lean to the radical wing of that party, except that (unlike most radicals) he was not a rationalist; and this difference turned out to have important consequences in the long future. A seat was found for Burke in Parliament, and next year, in 1766, he began to speak there. His first speech was on the American question.

Burke was a steadfast and voluble advocate of the cause of the American colonies. He attacked the dictatorial ambitions of George III, and he supported John Wilkes in his long feud against the king. For a time, he was member of Parliament for the city of Bristol, in the west of England, which lived by its trade with America. Burke's support of the American cause was by this time not disinterested, for it was known that he was being paid by the colony of New York; but such help to a needy politician was usual enough in the Parliament of that time.

Burke's views on the quarrel with America were straightforward. First, he charged the members of George III's government with bungling their own policy.

> They never had any kind of system, right or wrong; but only invented occasionally some miserable tale for the day, in order meanly to sneak out of difficulties into which they had proudly strutted. And they were put to all these shifts and devices, full of meanness and full of mischief, in order to pilfer piecemeal a repeal of an Act which they had not the generous courage, when they found and felt their error, honourably and fairly to disclaim.[1]

This is the accusation which Burke puts succinctly elsewhere, "that we know neither how to yield nor how to enforce."

And second, Burke insisted that the feelings of America had to be placated if government was to remain possible. This was consistently his major theme: that whatever the rights or the wrongs of taxation, England had now to face the reality of American refusal. "No man ever doubted that the commodity of tea could bear an imposition of

[1] Edmund Burke, *On American Taxation* (1774). The fundamental text for Burke is his collected works. See *The Writings and Speeches of Edmund Burke*, 12 vols. (Colonial Press, Boston, no date given). There are two other essential texts for Burke: *The Correspondence of the Rt. Hon. Edmund Burke: Between the Year 1744 and the Period of his Decease in 1797*, ed. by Charles William, Earl Fitzwilliam, and Lt.-Gen. Sir Richard Bourke, K.C.B. (London, 1844); and Arthur P. I. Samuels, *The Early Life, Correspondence and Writings of the Rt. Hon. Edmund Burke LL.D.*, with an introduction and supplementary chapters on Burke's contributions to the *Reformer* and his part in the *Lucas Controversy* (1748) by the Rt. Hon. Arthur Warren Samuels (Cambridge, 1923). The existing *Correspondence*, however, is being superseded by a projected eight-volume work, *The Correspondence of Edmund Burke*, ed. by Thomas W. Copeland, of which one volume (Cambridge and Chicago, 1958) has so far appeared; judging by the first volume, it is an excellent and definitive edition.

three pence. But no commodity will bear three pence, or will bear a penny, when the general feelings of men are irritated, and two millions of people are resolved not to pay."[2]

In a later and greater speech on the American question, Burke put this argument with penetrating force.

> The question with me is, not whether you have a right to render your people miserable, but whether it is not your interest to make them happy. It is not what a lawyer tells me I *may* do, but what humanity, reason, and justice tell me I ought to do. Is a politic act the worse for being a generous one? Is no concession proper, but that which is made from your want of right to keep what you grant? Or does it lessen the grace or dignity of relaxing in the exercise of an odious claim, because you have your evidence-room full of titles, and your magazines stuffed with arms to enforce them?[3]

If these sensible and expedient arguments had been listened to—if George III had been capable of listening to argument—there would have been, in all probability, no breach with America.

IV

Burke opposed George III's policy on grounds of expediency; it is difficult to find in his speeches any assertion of principle on behalf of America. At bottom, Burke always distrusted those who claimed to be inspired by defined principles. He was in some sense a forerunner of that modern school of historians who believe that we read principles into history after the event, but that at the time we act by expediency. All acts of state, say these historians, are particular acts; they do not conform to principles but rather, one by one, combine to form the principles which we then discover in them.

Yet Burke at this time certainly had, behind his arguments, one constant and fundamental thought: the thought that authority derives from popular consent. Indeed, he justified his belief, that policies should be formed by good sense rather than by principle, on the very ground that the body of people can be trusted to be sensible but not to understand principles. "It is very rare indeed for men to be wrong in their feelings concerning public misconduct; as rare to be right in their speculation upon the cause of it." If the body of people

[2] *On American Taxation*. Burke may have based his estimate of the population of the American colonies on the forecast Benjamin Franklin made in 1751, of "this million, doubling suppose once in twenty-five years."

[3] Edmund Burke, *On Conciliation with the Colonies* (1775).

felt that their government was at fault, then it was at least likely that their feelings were justified. "I am not one of those who think that the people are never in the wrong. They have been so, frequently and outrageously, both in other countries and in this. But I do say, that in all disputes between them and their rulers, the presumption is at least upon a par in favour of the people." This is the only ground which Burke is willing to trust—that in the end all authority must derive from the people who are governed. There can be no appeal to any absolute right, either on the part of the rulers or of the ruled; there can be only the certainty that the rulers derive whatever right they exercise from those whom they rule. "The King is the representative of the people; so are the Lords; so are the Judges. They are all trustees for the people, as well as the Commons; because no power is given for the sole sake of the holder; and although government certainly is an institution of divine authority, yet its forms, and the persons who administer it, all originate from the people."[4] This is not a comfortable doctrine, even on grounds of expediency, because it can be interpreted in too many different ways. And Burke certainly interpreted it differently at different times in his life. But it was the one ground on which he founded his steady support of dissident America.

V

The English Parliament in the eighteenth century governed the American colonies directly, but it administered the subcontinent of India only indirectly. India was still in theory ruled by her princes, and it was only her commerce which was in the hands of a huge English monopoly, the East India Company. In practice, however, the East India Company was so powerful that it dominated Indian politics, and had a strong influence on politics in England too. For example, it was in order to help the East India Company, which had got into difficulties in the slump of 1772, that Parliament allowed the company to send tea to America without taxing it in England first, and thereby set off the Boston Tea Party.

Burke had attacked the power of the East India Company, and he now led a strong prosecution of its senior official, Warren Hastings. There was some substance in Burke's main charges: Warren Hastings, like other offiicals of the company, had in some ways exploited his

[4] Edmund Burke, *Thoughts on the Cause of the Present Discontents* (1770).

position in India to play the despot and make himself rich. Burke had the support of the Whig Party in attacking these abuses, and Warren Hastings was impeached by Parliament. But as the preparations for the trial mounted, and as the trial went on (it lasted for seven years, from 1788 to 1795), Burke turned it more and more into a personal feud against Hastings. His speeches became more intemperate, his manner grew more vindictive and his charges wilder. In 1789 Burke was formally censured by Parliament for one of his more extravagant accusations. In the long run, Burke's violence may have helped Hastings, who was acquitted.

There were two grounds for the distrust which Burke's colleagues felt for him. One was the violence with which he spoke and acted. He flew into rages, he seemed to have no sense of proportion, and even his greatest speeches were (as his modern biographer has tactfully put it) "marred by extravagant abuse and lapses of taste."[5]

This wildness in Burke's oratory made his colleagues uncomfortable. To Burke, it was naturally a mark of the sublime, a Gothic exuberance on a par with, say, Thomas Gray's poem *The Bard* and the prophetic books of William Blake. But to his colleagues, Burke's language smacked of the evangelical rages with which, in their opinion, John Wesley and George Whitefield disfigured their sermons. This is what the eighteenth century called "enthusiasm," a word of disapproval which stood, roughly, for revivalism and hysteria, and which was considered to be in vulgar taste. There was, indeed, always something vulgar about the imagery of Burke. Here is a characteristic example from one of his earlier speeches.

> At the beginning of the century some of these colonies imported corn from the mother country. For some time past the Old World has been fed from the New. The scarcity which you have felt would have been a desolating famine, if this child of your old age, with a true filial piety, with a Roman charity, had not put the full breast of its youthful exuberance to the mouth of its exhausted parent.[6]

And when Burke came to attack Warren Hastings, his flights into the sublime lost all reality.

[5] Philip Magnus, *Edmund Burke: Selected Prose* (London, 1948), p. 12.
[6] *On Conciliation with the Colonies.*

His cruelty is beyond his corruption: but there is something in his hypocrisy which is more terrible than his cruelty; for, at the very time when with double and unsparing hands he executes a proscription, and sweeps off the food of hundreds of the nobility and gentry of a great country, his eyes overflow with tears, and he turns the precious balm that bleeds from wounded humanity, and is its best medicine, into fatal, rancorous, mortal poison to the human race.[7]

As a result, members of Parliament no longer listened to Burke's literary oratory. When he spoke on India, they simply got up and left, and he became so effective in emptying Parliament that he was nicknamed "the dinner bell."

The second ground on which Parliament distrusted Burke was more solid and earthy. There was, again in the words of his modern biographer, "a bad financial smell about the Burkes."[8] The facts have now been carefully uncovered, and they give a remarkable picture of the politics of a high-minded man in the eighteenth century.[9]

Edmund Burke was devoted to his family: to his wife and son, and also to his brother Richard and to a distant cousin, William Burke, who all lived with him and shared his money. Richard and William Burke were speculators who lived, or tried to live, for the most part by fraud, and Edmund Burke's fortunes were bound up with their manipulations. For example, at the time that Burke attacked the government for its policy in America, he was privately trying to induce it to grant Richard a large and preposterous claim in the West Indies. At the time that Burke attacked the government's policy in India, he declared that he had no interest in the stock of the East

7 Edmund Burke, *Speech on the Sixth Article of Charge in the Impeachment of Warren Hastings* (1789).

8 Magnus, *op. cit.*, p. 14.

9 See the biography by Philip Magnus, *Edmund Burke* (London, 1939), for one account. Until now, this work, which employs the so-called Wentworth papers, has been the best, although by no means the definitive, life of Burke. Carl B. Cone's *Burke and the Nature of Politics: The Age of the American Revolution* (Lexington, 1957), uses the same papers and is intended to supplant Magnus; a second volume is promised. Thomas W. Copeland, *Our Eminent Friend Edmund Burke: Six Essays* (New Haven, 1949), gives a detailed picture of Burke's financial affairs, especially in Chap. 2; in general, this work breaks new ground in Burke studies. The definitive account of Burke's financial dealings, however, is Dixon Wecter, *Edmund Burke and His Kinsmen: A Study of the Statesman's Financial Integrity and Private Relationships* (Boulder, Colo., 1939).

India Company; but, in fact, his family was speculating in the stock. And most remarkable, at the time that he impeached Warren Hastings, Burke was trying hard to help his cousin William (whom he had found a job in India) in two schemes to rob the public purse there, which were both clearly dishonest.

Burke was not a dishonest man himself; he did not deliberately suit his politics to his desire to save his finances. But he hid from himself the strain which financial need imposed on his own motives, and in the end he blinded himself to his motives. He did not see himself as others saw him, and he was indignant that they should think that his political vision could be distorted by private advantage. In short, he lacked judgment and perspective, and thus no one would put him into a cabinet. Particularly he lacked the judgment to see it was not tactful to roar in public about the evils of money in politics.

> We dread the operation of money. Do we not know that there are many men who wait, and who indeed hardly wait, the event of this prosecution, to let loose all the corrupt wealth of India, acquired by the oppression of that country, for the corruption of all the liberties of this, and to fill Parliament with men who are now the object of its indignation? To-day, the Commons of Great Britain prosecute the delinquents of India: to-morrow the delinquents of India may be the Commons of Great Britain.[10]

VI

In 1789 the French Revolution began; the Paris crowd stormed the Bastille on July 14 of that year. Many Whigs were delighted that absolute government in France had fallen, and that representative assemblies were being called; and they hailed the Revolution as the beginning of democracy on the continent of Europe. Charles James Fox, the leader of the Whig party, wrote enthusiastically, "How much the greatest event it is that has ever happened in the world! and how much the best!"[11]

There was in England a Whig club called the Revolution Society, which met in November of each year to glorify the bloodless Revolution of 1688, when England had brought William and Mary from Holland to displace her autocratic king. This year, the Revolution Society was addressed by one of the intellectual leaders of rationalist

[10] Edmund Burke, *Speech on the Impeachment of Warren Hastings, Fourth Day*, (1789).

[11] Quoted in Edward Lascelles, *The Life of Charles James Fox*, 2nd ed. (London, 1939), p. 213.

dissent, Dr. Richard Price. He promptly drew a likeness between the English Revolution of 1688 and the French Revolution of 1789, of which he said, "I have lived to see the rights of man better understood than ever; and nations panting for liberty which seemed to have lost the idea of it. I have lived to see thirty millions of people, indignant and resolute, spurning at slavery, and demanding liberty with an irresistible voice. Their King led in triumph, and an arbitrary monarch surrendering himself to his subjects."[12] At the end of the meeting, the Revolution Society sent an address of congratulation to the National Assembly in France.

Edmund Burke had been uneasy at the French Revolution from the outset, and Dr. Price's address, coming at a time when there were disturbing reports of mob violence in France, made him explode. At the first Parliamentary session in 1790 he quarreled publicly with his leader, Fox. The impeachment of Warren Hastings was going on at this time, and Burke was under great emotional strain. He now turned all his anger against the French Revolution, and at the end of 1790 he published his *Reflections on the Revolution in France*. This is the book which outraged Tom Paine, and caused him to write *The Rights of Man* in 1791.

VII

Burke's *Reflections on the Revolution in France* probably did not do much to form public opinion in England, which was already turning against France. But like his earlier book on *The Sublime and Beautiful,* it had a profound effect in fixing the change in public opinion. It had the same approach of romantic enthusiasm to its subject. There is a famous passage in the *Reflections on the Revolution in France* which shows this at its most eloquent, and which is worth quoting again.

> It is now sixteen or seventeen years since I saw the Queen of France, then the Dauphiness, at Versailles; and surely never lighted on this orb, which she hardly seemed to touch, a more delightful vision. I saw her just above the horizon, decorating and cheering the elevated sphere she just began to move in—glittering like the morning-star full of life and splendour and joy. Oh! What a revolution! and what an heart must I have, to contemplate without emotion that elevation and that fall! Little did I dream, when she

[12] Quoted in *ibid.*, pp. 214–215. Richard Price was regarded as an authority on population, but here he overestimates the population of France in 1789.

added titles of veneration to those of enthusiastic, distant, respectful love, that she should ever be obliged to carry the sharp antidote against disgrace concealed in that bosom! Little did I dream that I should have lived to see such disasters fallen upon her in a nation of gallant men, in a nation of men of honour, and of cavaliers! I thought ten thousand swords must have leaped from their scabbards to avenge even a look that threatened her with insult. But the age of chivalry is gone. That of sophisters, economists, and calculators has succeeded; and the glory of Europe is extinguished forever. Never, never more, shall we behold that generous loyalty to rank and sex, that proud submission, that dignified obedience, that subordination of the heart, which kept alive, even in servitude itself, the spirit of an exalted freedom![13]

Burke's closest friend, who was then associated with him in the impeachment of Hastings, made him angry by calling this writing "pure foppery." But what is the meaning, for example—the meaning to the French—of the assertion that some sentiments "kept alive, even in servitude itself, the spirit of an exalted freedom"? Well might Burke's friend say, with rough scorn, "I wish you would let me teach you to write English."[14] However, the public liked the way that Burke wrote English. The public liked to weep, as Burke said that he had wept, over his sentences in praise of Marie Antoinette. And the public was now alarmed by the French Revolution.

The French Revolution created a new alignment in English politics, of which Burke's change of sides was characteristic. For a hundred years, since the bloodless Revolution of 1688, there had been two political parties, the Whigs and the Tories. This division had held through all great issues, including, for example, the American Revolution: most Whigs had sided with the Americans. But now it was clear that there were two groups of Whigs: the romantics like Burke on the one hand, and, on the other hand, men of rationalist and nonconformist temper like Price and Paine, who were the last heirs of the age of reason. There was, of course, much overlap and

[13] Edmund Burke, *Reflections on the Revolution in France* (1789).

[14] This remark by Philip Francis, who shared the main burden of the impeachment of Hastings with Burke, is quoted by Magnus in *Edmund Burke: Selected Prose,* p. 13. See also Burke's passionate defense of what he had written in his letter to Philip Francis, 1790, which is quoted on pp. 81–83. W. Somerset Maugham has contributed a specialized study of Burke, in an essay, "The Style of Burke," to be found in his book *The Vagrant Mood* (Garden City, 1953). More along this line may be found in Gilbert Highet's *The Classical Tradition* (Oxford, 1949), which, while not devoting a separate chapter to Burke, treats in passing of him and his style.

many individual exceptions, but, on the whole, the rationalists most often came now from the new manufacturing class, or were in sympathy with it; the romantics belonged to the older Whig tradition of the great aristocratic houses. After the French Revolution, the two sides of the Whig party broke apart, and the romantics joined George III and the Tory government. The others, led by Fox, suffered 12 years of political humiliation and persecution, during which for a time they refused to attend Parliament altogether.

The Whigs who went over to the government did as those who change sides always do—as ex-Communists have done in recent years—they protested their antirevolutionary sentiments more loudly than the Tories. The Duke of Richmond, when he presided at the trial of the poet William Blake for sedition in 1804, belabored the court with royalist sentiments.[15] And it was Burke himself who, early in 1790, for the first time in his life refused to support a Whig motion for religious toleration, because (he said) every measure of reform would now strengthen the influence of the French Revolution, and would help to undermine the traditional state of things in England. Thereafter, no reform could succeed. Even William Wilberforce could no longer get support for his campaign against slavery.

VIII

Edmund Burke was a complex and fascinating character who straddled a watershed in history. On one side was the American Revolution, which had wide support by public opinion in England. On the other side was the French Revolution, which alarmed men of property in England, and in time, after the rise of Napoleon, had the whole weight of English opinion against it.

Burke did not try to reconcile his support of one revolution and his opposition to the other on grounds of principle. He felt, as many other men felt (in America as well as in England), that the American Revolution was within the historical tradition by which democratic constitutions had evolved, and that the French Revolution was outside it.[16] In part, he and others felt this because the American Revolu-

[15] For a discussion of this episode in the context of its time, see J. Bronowski, *A Man Without a Mask* (London, 1944).

[16] For Burke's tergiversations on the subject of revolution, as well as for further discussion of the effect of romanticism on his political thought, see Bruce Mazlish, "The Conservative Revolution of Edmund Burke," *The Review of Politics* (January, 1958). No first-rate, full-length modern critical treatment of Burke exists. Lord Morley's *Burke* (New York, 1879), points in the right direction; it is, how-

tion was farther away, and was therefore less likely to spread to England. In part, they felt that the American Revolution was run by men like themselves, and not by a barefoot and starving crowd. And in part—a most important part—they felt that the American Revolution was concerned only with the rearrangement of political power, but that the French Revolution was a design to take property away from men of substance.

Burke did not advocate a political principle. In the letter in which he justified his break with the Whigs, he said clearly what had been implied in his outlook from the beginning. "Political problems do not primarily concern truth or falsehood. They relate to good or evil. What in the result is likely to produce evil is politically false; that which is productive of good, politically true." But the implications of this hedonistic view of government are awkward. It is all very well to point out, as Burke did,

> The moment you abate anything from the full rights of men each to govern himself, and suffer any artificial, positive limitation upon those rights, from that moment the whole organisation of government becomes a consideration of convenience. This it is which makes the constitution of a state, and the due distribution of its powers, a matter of the most delicate and complicated skill. It requires a deep knowledge of human nature and human necessities, and of the things which facilitate or obstruct the various ends which are to be pursued by the mechanism of civil institutions."[17]

It is all very well to use these fine and negative phrases, but the crux of them remains that "the whole organisation of government becomes a consideration of convenience." Whose convenience?

Burke's antirationalist outlook did not allow him to set up the convenience of the people as a principle; he could not, like Jeremy Bentham, propose the greatest happiness of the greatest number.

---

ever, now outdated both as to facts and judgments. Woodrow Wilson's essay, "The Interpreter of English Liberty," in *Mere Literature and Other Essays* (Boston, 1896), sheds light on both the author's and on Burke's political philosophy. The relationship of Burke to Rousseau is explored in A. M. Osborn, *Rousseau and Burke* (London, 1940). Alfred Cobban analyzes Burke's relationship to Locke, and Burke's influence on the Lake poets, in *Edmund Burke and the Revolt Against the Eighteenth Century* (London, 1929). Ernest Barker has an essay on "Burke and his Bristol Constituency" in his *Essays on Government* (Oxford, 1945), which is very well done and sheds further light on Burke's concept of "virtual representation."

[17] *Reflections on the Revolution in France.*

Instead, he had at one time held that it was a matter of expediency to consult the convenience of the people—that "in all disputes between them and their rulers, the presumption is at least upon a par in favour of the people." But now, when Burke attacked the French Revolution, he was denying the one claim to which he had held in the American Revolution: the claim of people to have a voice in settling their discontent with their rulers. People were no longer simple and noble to him; and carried away by his own rhetoric, he wrote.

> The occupation of a hair-dresser, or of a working tallow-chandler, cannot be a matter of honour to any person—to say nothing of a number of other more servile employments. Such descriptions of men ought not to suffer oppression from the state; but the state suffers oppression if such as they, either individually or collectively, are permitted to rule.

This was a tactless remark in the age of Richard Arkwright, who had been a hairdresser, and of Benjamin Franklin, whose father was a tallow-chandler.

Burke thus gave up his early notion, that it is expedient for a government to have the support of its people. Instead, he held at last that constitutions are evolved by a long process of adjustment between government and people which should not be violently disturbed. Like other men of his time, he moved toward a romantic reverence for history as the great healer—the goddess of change, but of imperceptible change. He saw the past as a providential source of good sense, and the constitution of England as its finest creation. He believed that the changes brought by the Bloodless Revolution of 1688 had made the constitution so perfect that it could only be ascribed to the direct intervention of God. The old Tories of the Augustan age would have thought this mere sentiment, and would have insisted that either one must have good reasons against change, or good reasons for it. But the age of reason had collapsed, and Burke was the representative of new passions. The Conservatives of the nineteenth century, and particularly their leader Benjamin Disraeli, took over Burke's mystic view of the past. At the end of a troubled life which had never fulfilled its promise, Burke became in fact what in character he had always been: the founder of romantic Conservatism.

CHAPTER 24

Jeremy Bentham

I

DURING THE period of reaction in England after the French wars, a new generation of reformers came into being who abjured the belief in natural rights, discredited by the Jacobin excesses. They turned, instead, to a different doctrine: Utilitarianism, and found their leader and guide in a rather colorless and seemingly unpolitical man: Jeremy Bentham. Largely under his name and doctrine, the English middle class moved forward to the capture of political power without a revolution. Adam Smith had supplied the ideology for a peaceful economic revolution; it was Jeremy Bentham who furnished the philosophic banners under which the "legislative revolution" was carried on.

In Bentham's time, the English legal system had been allowed to ossify for many years. Under the scare of the French Revolution and the threat of the Napoleonic Wars, the slightest breath of criticism was construed as a step to sedition.[1] There were judges in England, like Lord Eldon, who held that to touch anything of the ancient fabric of laws was an invitation to revolutionary sentiment: things were to be left exactly as they were. Perhaps we can best convey the state of opinion by two examples.

The first concerns William Wilberforce and his attempt, during the course of the Napoleonic Wars, to fight the slave trade. Wilberforce invited one of the most liberal English lords of the time to support his

[1] Cf. Philip Anthony Brown, *The French Revolution in England* (London, 1923). The story that follows refers to Lord Ellenborough's letter, printed in *Private Papers of William Wilberforce,* ed. by A. M. W. Wilberforce (London, 1897).

cause. The latter responded with a long, backing-and-filling letter. He declared that he completely agreed with Wilberforce, that no Christian could believe otherwise; *but* that in the middle of such a dreadful period the slightest tinkering with the law would open the door to revolution in England. Therefore, in conclusion, the great lord begged to state that he could not attend a meeting to oppose the use of English ships in the slave trade because he feared that the whole system of established law and order might be placed in jeopardy.

Another, even more striking, instance relates to the use of hair powder. Around 1795, bread became very scarce in England. George III, who was a farmer, wanted to set an example in the economy of wheat. He decided to give up powdering his hair because hair powder is made from starch and starch is made from wheat. This was a reasonably warm-hearted suggestion on the part of the monarch. But it was rather angrily opposed by some of Pitt's men on the grounds that the distinction of dress between the upper and lower classes was fundamental; if people stopped powdering their hair, the distinction between classes would disappear, and Jacobinism would be let in at the door.

It was in this atmosphere of semi-hysteria that Bentham elaborated his Utilitarian doctrines and attempted to effect a legal reform. In place of vague "natural rights" or mystical "historical rights," Bentham offered a "new science," the science of law, to justify his proposed changes. He taught men to govern by the simple rule of "the greatest happiness of the greatest number," which, in practice, could be discovered by a "felicific calculus." Thus, he sought to establish an external standard, mathematically calculable, whereby to measure the legislator's accomplishment. His contention was that he had made legislative reform a matter, not of "caprice" or of unenlightened benevolence, but of logic.

II

What sort of man accomplished this peaceful "revolution"? Born in 1748, Bentham was something of a child prodigy. He devoted himself to Latin at the age of 4, and the violin and French shortly thereafter. At the age of 12 he entered Oxford University, where he stayed for three years. At the end of that time, bowing to his father's wishes,

he took up the study of law. From that point on, he was always concerned with legal matters, but his father's hope that he would become a judge was never realized; instead, the precocious son became an outstanding critic of law.

Bentham, in possession of an independent income, was able to devote himself wholeheartedly to his chosen work. He lived a simple, somewhat severe life, never marrying and never seeing anyone except for "some specific purpose." He took no wine and had no taste for poetry: "Prose," he said, "is when all the lines except the last go on to the margin. Poetry is when some of them fall short of it." His only relaxations were music (he played Handel and Corelli on the organ) and gardening. He knowingly labeled himself "The Hermit of Queen's Square Place."[2]

By the quality of his life, as the pure and disinterested "Hermit of Queen's Square Place," Bentham attracted disciples and followers; a modern counterpart might be Mahatma Gandhi. Ironically, however, he scorned asceticism as a theory. From the utilitarian point of view, "saints," were idle.[3] In fact, of the two principles which were directly opposed to the principle of the greatest happiness of the greatest number, one was asceticism, because in effect it glorified unhappiness.[4] Yet, it was as the altruistic, almost ascetic "Hermit of Queen's Square Place" that Bentham charmed men like James Mill and Samuel Romilly.

Let us return to Bentham's earlier life. A turning point in his development came in 1763 when he heard Blackstone's lectures on the law, later published in 1765–1769 as the famous *Commentaries*. According to Bentham, he "immediately detected Blackstone's fallacy concerning natural rights" and was so absorbed by his discovery that

[2] *The Works of Jeremy Bentham,* published under the superintendence of his executor, John Bowring, 11 vols. (Edinburgh, 1838–1843) (hereafter referred to as *Works*). Vol. X, p. 509. Vol. X covers the *Memoirs* of Bentham, including *Autobiographical Conversation,* and correspondence. On Bentham's life, see, too, Charles Milner Atkinson, *Jeremy Bentham: His Life and Work* (London, 1905), and Charles Warren Everett, *The Education of Jeremy Bentham* (New York, 1931).

[3] Jeremy Bentham, *An Introduction to the Principles of Morals and Legislation* (New York, 1948) (hereafter referred to as *Introduction*), p. 12.

[4] The other principle opposed to utility was the principle of sympathy and antipathy, i.e., the principle which made simple caprice the standard of right and wrong.

he was unable to take notes.[5] He wrote, however, a *Comment on the Commentaries* and, in 1776, published a part of the *Comment* anonymously, as *A Fragment on Government.* In the *Fragment,* Bentham mercilessly exposed what he took to be Blackstone's empty but sonorous mixture of common law and natural rights and tore into the legal rigidity for which the *Commentaries* stood. Typical of his view of Blackstone was the comment: "His hand was formed to embellish and to corrupt everything it touches. He makes men think they see, in order to prevent their seeing."

Jeremy Bentham attacked the concept of natural rights throughout his writings. In the *Introduction to the Principles of Morals and Legislation* (1789) and in his *Anarchic Fallacies* (collected by Bentham in 1785, but not published until much later), he pressed home the attack. He declared that both the American Declaration of Independence and the French Declaration of the Rights of Man and the Citizen were meaningless hodgepodges, similar to Blackstone's glittering but specious generalities. Of the American Revolution, which he favored, Bentham commented: "Who can help lamenting, that so rational a cause should be rested upon reasons, so much fitter to beget objections, than to remove them."[6] Every coercive law, he pointed out, would necessarily violate a declaration which claimed that men had a "natural right" to life, liberty, and property. If only, he sighed as late as 1827, the American colonists had appealed to utility. "The American colonies really said nothing to justify their revolution. They thought not of *utility,* and *use* was against them. Now, utility was the sole ground of defence. What a state the human mind was in, in those days!" To reform the "state the human mind was in," Bentham expounded his doctrine of utility.

Essentially, there was nothing new in this doctrine. Bentham had borrowed the doctrine of utility from David Hume, and was merely continuing the latter's work in undermining the belief in "natural laws" and "natural rights." It was after reading Hume's *Treatise of Human Nature* that Bentham told how "scales had fallen from his

[5] See further Leslie Stephen, *The English Utilitarians,* 3 vols. (New York and London, 1900), Vol. I, p. 174. Of this splendid book, the whole of Vol. I is devoted to Bentham. In John Plamenatz, *The English Utilitarians* (London, 1949), Chap. 4 is on Bentham.

[6] *Introduction,* p. 336. See Stephen, *op. cit.,* Vol. I. p. 180, for an account of Bentham's initial opposition to the American Revolution.

eyes" and how he had "learned to see that *utility* was the test and measure of all virtue." It was not only Hume, however, who influenced him. Locke and, as Bentham admitted, "Montesquieu, Barrington, Beccaria, and Helvétius, but most of all Helvétius, set me on the principle of utility."

Gradually, Bentham grew dissatisfied with the use of the term "utility." Under the inspiration of Priestley and Beccaria, he began "to pronounce this sacred truth:—that the greatest happiness of the greatest number is the foundation of morals and legislation." Although supplanted in Bentham's usage, the term "Utilitarian" remained to distinguish the members of his school.

Clearly, there was nothing original in Bentham's doctrine of utility (by any name). As John Stuart Mill, the son of his disciple James Mill, admitted: "In all ages of philosophy, one of its schools has been utilitarian, not only from the time of Epicurus, but long before."[7] Why, then, its power in Bentham's hands?

We believe the answer to lie in two factors; and these two factors taken together make for Bentham's greatness. The first involves that vague but essential statement—the times were ripe. Before the French Revolution, the *philosophes* were able to proceed under the banner of natural rights, a long and satisfying tradition going back to the Romans; there was no need to appeal to utility. The events of 1789, however, showed that natural rights might be a "blank check" on the future, waiting for the masses to write in the overturn of any and all political order. A defense of reform based on other than "Jacobin" principles was needed, and Bentham's utility doctrine was right at hand.

The second reason why "utility" suddenly came alive under Bentham was his use of it. Where other thinkers tended to apply utility only to the individual, he took the axiom of the greatest happiness of the greatest number as the measure of *social* morality. Out of the ancient philosophical doctrine of hedonism, a political philosophy of Utilitarianism was created. Applying the standard of utility methodically and relentlessly to the penal and constitutional laws of his time, he found them wanting; then, on the basis of utility, he drew up, in their place, complete and logical codes of law.

[7] John Stuart Mill, *Dissertations and Discussions,* 3 vols. (Boston, 1865), Vol. 1, pp. 370–371.

Thus, Bentham supplied a doctrine, meticulously and fully worked out in the minutiae of practice, to a group, the English middle class, who were at that moment looking for an ideology other than "natural rights" as a basis for gaining control of political life. The coalescing of a movement of social forces and a movement of ideas turned the "Hermit of Queen's Square Place" into a powerful leader of political and economic reform.

III

Let us examine Bentham's application of the "greatest happiness" doctrine carefully. Bentham, so to speak, asked the question: "What is a government and its legal system for?" He answered that the object of the law was not, as Locke had said, to protect the life, liberty, and property of the individual; indeed, the object of government was not really to deal with the individual, *per se*. Nor was its object, as Rousseau said, to express the general will. Instead, in Bentham's view, the object of government and its laws was to supply the population at large with its greatest possible happiness. And inasmuch as government has to choose between what pleases one person and what pleases another person, it must estimate and choose, scientifically, "the greatest happiness of the greatest number."

Philosophers have subjected this idea of Bentham's to two major criticisms. They ask, first, what ground is there for accepting utility as a fundamental principle, and happiness as a fundamental value? And, second, if the principle is accepted, how would one actually measure happiness?

Bentham, in typical philosophical fashion, tried hard to answer the first criticism by showing that all other principles than utility were wrong. But, as he himself ultimately recognized, the only real answer was that the principle of utility was an axiom not itself subject to proof. As Bentham put it: "Is it susceptible of any direct proof? It should seem not: for that which is used to prove every thing else, cannot itself be proved: a chain of proofs must have their commencement somewhere. To give such proof is as impossible as it is needless." This may not have been a satisfactory defense for unsympathetic critics of Bentham, but for him and his followers the principle of utility was like a vision which guided them through the labyrinths of government. As a vision, it was not to be questioned in the ordinary way.

What about the mathematics of pleasure and pain? How was a "felicific calculus" to work? Bentham spent much of his *Introduction to the Principles of Morals and Legislation* explaining how the value, as he put it, of "a Lot [bundle] of Pleasure or Pain" was to be measured. Intensity, duration, certainty, propinquity—such factors as these were to be considered in estimating the value of a pleasure or a pain.[8] When the value of each pleasure-pain was calculated, a "balance" might then be struck. "Sum up all the values of all the *pleasures* on the one side, and those of all the pains on the other. The balance, if it be on the side of pleasure, will give the *good* tendency of the act upon the whole, with respect to the interests of that *individual* person; if on the side of pain, the *bad* tendency of it upon the whole." Finishing off his bookkeeping approach, Bentham explained how, by summing up the good and bad tendencies of an act for each individual concerned and adding all these sums together, we may obtain the tendency of any act to affect the happiness of the community. Much of his *Introduction* is devoted to a catalogue of pains and pleasures and their values.

To express the aim of such a "felicific calculus," Bentham coined the words "maximize" and "minimize"; we seek to "maximize" our pleasures and to "minimize" our pains.[9] This theory of weighing

[8] Bentham borrowed this analysis of the elements which make up the gravity of a pain directly from Beccaria. See Elie Halévy, *The Growth of Philosophic Radicalism,* tr. by Mary Morris (Boston, 1955), p. 21. This is a deep, penetrating treatment of Bentham and his followers.

[9] The temptation for Bentham's followers was to give a monetary estimate of happiness. And Bentham himself, in his "Codification Proposal," took money as representing pleasure and showed how it could be calculated like a sum put out to interest. (Cf. *Works,* Vol. IV, p. 540, and Stephen, *op. cit.,* Vol. I, p. 250.)

At first sight, Bentham's appears to be pretty crude "science." But, consciously or unconsciously, we all make some sort of similar calculus when we are wavering about an act; Benjamin Franklin, for example, used to draw up a balance sheet of an action's pros and cons. And it is a commonplace today for legislators to hear various opinions on how much "happiness" or "unhappiness" a bill will cause before passing it; this, in fact, is the purpose of legislative hearings. Thus, however we may laugh at Bentham's "calculations," his idea accords with our own daily actions and has a strong influence upon our view of how a government should legislate.

A more elaborate "minimax" principle, as it has been named, is the basis for the important book, *The Theory of Games and Economic Behavior* (Princeton, 1947),

pleasure and pain gave Bentham his key to the method as well as to the purpose of government. If the primary function of government was to maximize happiness and to minimize pain, it could do this by the use of punishments. In accord with this idea, Bentham worked out a theory of "sanctions." There were four "sanctions," or pains and pleasures, annexed to actions: the "physical or natural," "political," "moral or popular," and "religious." Of these, only the political sanction was imposed by the legislator.[10] He, by attaching artificial consequences, i.e., punishments or "political sanctions," to an action which was inimical to the greatest happiness of the greatest number, might prevent that action. This was the sole duty of government; otherwise, government was to "be quiet."[11]

Obviously, government, for Bentham, was not a positive agency; it was merely a coercive force. He believed that punishment is in itself an "evil," but necessary "in so far as it promises to exclude some greater evil." As punishment is government's major purpose, it follows that "All government is in itself one vast evil."[12] Only a set of laws drawn up on the principles of utility could "justify" such an "evil" government. A good penal code, Bentham declared, "is so contrived that the very place which any offense is made to occupy, suggests the reason of its being put there. It serves to indicate not only that such and such acts *are* made offenses, but *why* they *ought* to be.

---

by John von Neumann and Oskar Morgenstern, in which the authors seek to explain economics by using, as a model, games in which each participant, in a situation where he does not control all variables, seeks to maximize his gains. Although employed in somewhat the same effort as Bentham, the two authors, von Neumann and Morgenstern, have nothing but scorn for the crude mathematics of Bentham. But they misunderstand him, and their short criticism is not to the point. "A particularly striking expression of the popular misunderstanding about this pseudo-maximum problem is the famous statement according to which the purpose of social effort is the 'greatest possible good for the greatest possible number.' A guiding principle cannot be formulated by the requirement of maximizing two (or more) functions at once" (p. 11).

[10] Cf. *Introduction,* Chap. 3.

[11] We can see from Bentham's use of *economic* images that he thought punishment to be the government's major "business." For example, he declared: "The value of the punishment must not be less in any case than what is sufficient to outweigh that of the profit of the offense" (*Introduction,* p. 179). And again: "The greater the mischief of the offense the greater is the *expense* [our italics] which it may be worth while to be at, in the way of punishment" (*ibid.,* p. 181).

[12] *Introduction,* p. 170. Cf. *Works,* Vol. IX, p. 48.

. . . To the subject then it is a kind of perpetual apology: showing the necessity of every defalcation, which, for the security and prosperity of each individual, it is requisite to make from the liberty of every other."

IV

Now, perhaps, we can see why Bentham was considered by some to be a "dangerous" character and why the booksellers were afraid to print his works during the period of repression from about 1811. And we can understand why the Spanish minister, Jovellanos, to whom Bentham applied for a passport to Mexico, decided, as Lord Holland told Bentham, "that the character of *Jurisconsultus,* and writer on criminal law, might possibly be considered as a bad recommendation" and, therefore, ventured to ground Bentham's petition on a "love of botany, and of antiquities."

Bentham, a lover of antiquities! This was exactly what he was not and why he was dangerous. Indeed, he was a challenge to governments based on antiquities or, as Burke put it, on the "wisdom of our ancestors." Bentham's principle of utility meant that governments were to be judged solely on performance and not to be accepted merely because they had existed since time immemorial. This was the hidden dynamite in Bentham's legal "theodicy." Admitting that governments were coercive agencies, he insisted that their coercion was to be accepted only if they could explain logically why offenses were offenses. Obedience to government was required only if government was useful.

The political consequences of Bentham's utilitarian "polidicy" (if we may coin a word in typical Benthamite fashion) were significant. To attack the traditional laws of England was also to attack the class which "traditionally" benefited from those laws: the landed interest. Bentham's proposal that the laws be codified in a rational fashion and measured against a standard of usefulness meant, really, the triumph of the commercial middle class.

This was the "radical" aspect of Bentham.[13] By going to the roots of the law, he dug away the foundations upholding the traditional

[13] See Halévy, *op. cit.*, p. 261, for the historical derivation of the word "radical." For a general treatment of Bentham and the law, see *Jeremy Bentham and the Law: A Symposium,* ed. by George W. Keeton and Georg Schwarzenberger (London, 1948).

social structure. By shifting attention from the origin of law to its operation, from precedence to consequence, he effected a far-reaching change in political and legislative action as well as theory.

Bentham did not immediately perceive the radical nature of his projected codification of the law. He naïvely assumed that the people in power—the king and the Parliament—would at once embrace his notions and call upon him to reform the English law. In fact, for some time the illusion persisted with him that he was about to be asked to be the Solon of England. Gradually, he came to realize that the "governors" did not identify their happiness with the happiness of the greatest number.

Bentham, whose family were extreme royalists and who was himself a Tory the first part of his life, began to turn against what he called the "sinister interests."[14] The sole means of identifying the interests of the governed and the governors, he concluded, was to make the latter responsible to the former; this could only be done by a reform of Parliament. Bentham's mental evolution, however, was slow. As late as 1793 he wrote to Pitt's cabinet, admitting that some of his tracts "might lead you to take me for a Republican—if I were, I would not dissemble it: the fact is that I am writing against even *Parliamentary Reform,* and that without any change of sentiment."

What Bentham really desired was an administrative reform and not a parliamentary reform. His science was for statesmen and philosopher-kings, whom he would teach to govern for the greatest good of the greatest number. Therefore, if government would reform itself, there would be no need to change the mechanics of political representation. In 1809, he wrote to his friend, Mr. Mulford:

> I am hard at work, trying whether I cannot get the public . . . to turn its attentions to the law department. . . . It is perfectly astonishing to see how, by comparatively trifling instances of misgovernment, the current of public opinion has been turned against the Ministry, or rather against all Ministries, and in favour of Parliamentary Reform as the only remedy.

But Bentham, too, finally became convinced that the Ministry would not abolish corruptions of its own accord. He turned toward

[14] For Bentham's early Tory sentiments, see Stephen, *op. cit.,* Vol. I, p. 180. Bentham's shift away from the Tory position after 1807 was to a large extent influenced by James Mill.

Parliamentary reform. As a logical man, he went the whole way. When asked by Sir Francis Burdett to draw up a bill for Parliamentary reform in 1818, Bentham replied: "I never can bring myself to put my name to any plan of Parliamentary Reform, under which suffrages would not be free; nor do I see it possible how they ever can be free, otherwise than by being placed under the safeguard of secrecy." Bentham added to his request for a free and secret ballot a demand for annual Parliaments, an inclination to give the vote to women, and a disinclination toward a king, house of peers, and an established church. Having opposed the theory of natural rights, Bentham nevertheless ended by asking for the extreme Jacobin reforms.

As we can judge from the limited Reform Bill of 1832, he had gone too far for his middle-class supporters. They wished the suffrage for themselves and not for the masses. It was not until 1885 that universal manhood suffrage prevailed in England and not until 1918 that women obtained the vote. Even his immediate disciples, James Mill and his son John Stuart Mill, doubted whether it was good for men to be governed by an absolute, unchecked majority.[15] As J. S. Mill put it: "Is it, at all times and places, good for mankind to be under an absolute authority of the majority of themselves? . . . Is it, we say, the proper condition of man, in all ages and nations, to be under the despotism of Public Opinion?"[16]

Bentham's "democratic consequences," however, followed from his view of government as simply a "police agent." He had no concept of the state as a moral and cultural agency, educating and stimulating its members to a "way of life." For him, only individuals had meaning; the state had none. In his view, "The community is a fictitious *body,* composed of the individual persons who are considered as constituting as it were its *members.* The interests of the community then is, what? The sum of the several members who compose it."

V

Bentham's political views accorded well with his economic liberalism. In 1787, he had published his first successful book, *A Defense of Usury,* which extended Smith's laissez-faire ideas to contracts concerning money, and won immediate approval from the author of the

15 Cf. Stephen, *op. cit.,* Vol. I, p. 288.
16 Mill, *op. cit.,* p. 403.

Wealth of Nations. Later, in a *Manual of Political Economy,* Bentham dealt especially with the subject of economic legislation. He declared that the most useful thing government could do in economic, as in most other, matters was to "be quiet." Government interference was either needless or obstructive: the individual knew best his own interest and would pursue it with the greatest ardor. It was, therefore, on basic principle, that Bentham opposed such seemingly humanitarian efforts as factory legislation.[17] His final advice to government was that "The requests which agriculture, manufactures, and commerce present to governments is modest and reasonable as that which Diogenes made to Alexander: 'Stand out of my sunshine!' "

Adam Smith had justified the free and uncontrolled play of self-interest by appealing to the guiding hand of Providence. Bentham defended laissez-faire economics on the grounds of utility. Similarly, instead of taking labor as a measure of value, he preferred to think of economic value as measured by pleasure. For this measurement, his "felicific calculus" was ready at hand. Thus, he built his laissez-faire economics on a basis of utilitarian hedonism and thereby made an original contribution to the doctrine of economic liberalism.[18]

It was this theory of economics, based on utilitarian hedonism, which led Karl Marx to call Bentham "the insipid, pedantic, leather-tongued oracle of the commonplace bourgeois intelligence of the nineteenth century" and to say that Bentham

> . . . assumes the modern petty bourgeois, and above all the modern English petty bourgeois, to be the normal man. Whatever seems useful to this queer sort of normal man, and to his world, is regarded as useful in and by itself. By this yardstick, Bentham proceeds to measure everything past, present, and to come. For instance, the Christian religion is "useful" because, in the name of religion, it forbids that which the penal law condemns. . . . Had I the pluck of my friend, Heinrich Heine, I should call Mr. Jeremy a genius in the way of bourgeois stupidity.[19]

[17] Nevertheless, the Utilitarians helped the worker by also opposing as government interference the Combination Acts forbidding the workers to combine in unions; the repeal of these acts, in 1824, left labor free to unite.

[18] Bentham's particular contribution to economics was to be further developed by Jevons.

[19] Karl Marx, *Capital,* Everyman's Library ed., 2 vols. (London and New York, 1951), Vol. II, p. 671.

In truth, Bentham was deficient in imagination. His inexperience of real life (the price, perhaps, of being the "Hermit of Queen's Square Place") woefully limited his knowledge of human nature. He ignored private morality and ethics and measured everything by what he considered to be an objective yardstick. He had no doubts and was aware of no limitations to his method. As his disciple, John Stuart Mill, said, "Self-consciousness, that demon of the men of genius of our time, from Wordsworth to Byron, from Goethe to Chateaubriand, and to which this age owes so much both of its cheerfulness and its mournful wisdom, never was awakened in him."[20]

But was not this limited perspective—this "bourgeois stupidity" —the price that Bentham had to pay for the concentrated depth of his vision? By his narrowness of life and thought, excluding the attractive vistas that lay on all sides about him, he was able, as he notes himself, "to cut a new road through the wilds of jurisprudence." He admitted that it was by dry and almost pedantic methods that he developed his "science" of legislation; as he explained, "truths that form the basis of political and moral science are not to be discovered but by investigations as severe as mathematical ones, and beyond all comparison more intricate and extensive. . . . They flourish not in the same soil as sentiment. They grow among thorns; and are not to be plucked, like daisies, by infants as they run."

VI

The real radicalism of Bentham was not so much his work in favor of Parliamentary reform and certainly not his economic liberalism; it was in the "new road" he cut. He was radical because he went to the roots, not only of the law, but of language. And because he went to the roots of language, therefore he went to the roots of all men's thoughts. His writings are filled with constant laments against the "tyranny of language" and the "ambiguity of language." He was always appealing for a "new road," a "new method," and a "new science" which would break through the "tyranny of language."

Bentham wished the language of politics and morality to be as rigorously scientific as was mathematics. His *Introduction,* to take a prominent example, is filled with mathematical images. He is con-

20 Mill, *op. cit.,* pp. 378–380.

stantly evoking measurement, proportion, and division; he is always "quadrating." After minutely dividing the possible consequences of a mischievous act, Bentham remarks: "There may be other points of view, according to which mischief might be divided, besides these: but this does not prevent the division here given from being an exhaustive one. A line may be divided in any one of an infinity of ways, and yet without leaving in any one of those cases any remainder." Fundamentally, he wished language to be mathematical. One feels that he would have been really at home only with Boolean algebra or symbolic logic.

Bentham, convinced that "the great difficulty lies in the nature of the words," was concerned with etymology and was constantly creating neologisms. Among many others, the following new words, or expressions, of his have now become part of the English language: international law, codify, codification, maximize, minimize, maximization, and minimization. We have already noted that he learned Latin at 4, and fluent French a few years later; he also knew Italian and Spanish, and had studied German, Russian, and Swedish. It is clear that he was in a position of some authority on the matter of language.

In fine, Bentham was a semanticist, who felt that "as far as legislation is concerned, it will depend upon the degree of perfection to which the arts of language may have been carried." So it is the confusion of language, he claimed, that misleads us into thinking that morality depends upon motives and not consequences. The same motive, of a desire for money, he pointed out, may lead a man to commit murder or to work industriously. Similarly, the motive of sexual desire may, in a bad sense, be termed lasciviousness; but, he asked, when exercised during the rights of marriage what is its name? So, with benevolence: it may serve as the motive to burn a man at the Inquisitional stake or to give him a free dinner.

On these grounds, Bentham defended Mandeville and others who asserted "Private Vices, Public Benefits":

> To this imperfection of language, and nothing more, are to be attributed in great measure, the violent clamours that have from time to time been raised against those ingenious moralists, who, travelling out of the beaten tract of speculation, have found more or less difficulty in disentangling themselves from the shackles of ordinary language: such as Rochefoucault, Mandeville and Helvétius.

Bentham held it to be a misuse of language even to say that it was real benevolence to extend charity to the poor or to pass Factory Acts. As violations of economic science, he contended, these would *actually* make worse the misery of the poor; they would be benevolent only in name.

Similarly, it was a notion about fictitious entities, borrowed from d'Alembert, that confirmed Bentham in his view that the Declaration of Rights was composed of meaningless sentiments. "Fictitious entities," i.e., names that do not "raise up in the mind any correspondent images," are metaphysical terms created by society. He contended that a "right" could not be created by merely asserting it; and it certainly was not a right by nature. A natural right was a mere "fictitious entity." In reality, rights were created only by the legislature prohibiting some act. "To know then how to expound a right," Bentham declared, "carry your eye to the act which, in the circumstances in question, would be a violation of that right; the law creates the right by prohibiting that act." A consequence of this is that "for every right which the law confers on one party, whether that party be an individual, a subordinate class of individuals, or the public, it thereby imposes on some other party a *duty* or *obligation*."[21]

We can see now that Bentham's investigations of language supplied him with much of his theory of legislation and morals. He punctured the "empty sounds," the "mere names," behind which sheltered the "sinister interests." He attempted to penetrate to the reality for which the names stood and, once in the temple of reality, to measure the worth of that reality by the standard of utility.

Words, and the logical arrangement of words—this was the crux of Bentham's method. He took as his litany the following chant: "O Logic!—born gatekeeper to the Temple of Science, victim of capri-

[21] On this view, property is not a natural right of all men but a *created right,* assigned to some men and involving all others in a duty to them. (Cf. *Introduction,* p. 235, n. 3, and p. 322.) What property ought to be created, Bentham added, "and to what persons, and in what cases they ought to be respectively assigned, are questions which cannot be settled here" (p. 236, n. 3). The answer to such "questions"—as to the legitimate assignation of property—is implicit in all of Bentham's theory: property should be assigned on the basis of utility to society. But this potentially revolutionary idea was not developed by Bentham, because he considered the individual best able to decide his own interest and that the addition of such selfish interests would equal the "greatest happiness of the greatest number."

cious destiny! doomed hitherto to be the drudge of pedants! come to the aid of thy master, Legislation."

The logical method Bentham used was that of classifying his subject matter by "bipartition," as illustrated by the so-called "Porphyrian Tree." Leslie Stephen has given a convenient summary of this method.

> Take any genus: divide it into two classes, one of which has and the other has not a certain mark. The two classes must be mutually exclusive and together exhaustive. Repeat the operation upon each of the classes and continue the process as long as desired. At every step you thus have a complete enumeration of all the species, varieties, and so on, each of which excludes all the others.[22]

This severe logical method did not make for light, pleasant reading; but it did provide a rational classification and codification of the material under consideration; in this case, the law. And if one of the fundamental traits of modern science is its use of classification schemes, such as the periodic table in chemistry or the species-genus connections in biology, then Bentham has a right to be considered a scientist in the field of morals and legislation. "If there be any thing new and original in this work [the *Introduction*]," he declared, "it is to the *exhaustive method* [our italics] so often aimed at that I am indebted for it."

Bentham applied the Cartesian method of doubt to morals and legislation. Certainly, Bentham was not interested in the historical origin of laws and had absolutely no respect for precedent. As he remarked, "In the days of Lord Coke, the light of utility can scarcely be said to have as yet shone upon the face of Common Law." If Bentham borrowed utility from Hume and laissez faire from Smith, he did not allow their historical method to affect him. Instead, he reconstructed the philosophy of legislation and of morals from the ground up, logically. This was the real radicalism of Bentham.

VII

Are we justified in seeing the beginnings of logical positivism in Bentham's escape from the "tyranny of language"? Can we number among his disciples, not only the Utilitarians and the Philosophical

22 Stephen, *op. cit.*, Vol. I, p. 246.

Radicals (the name given to his followers later in the nineteenth century), but also the logical positivists—those who assert that statements, such as "this is a beautiful painting," which are not "empirically verifiable," are merely nonsense, i.e., of no sense? It would seem that this was foreseen by John Stuart Mill, as early as 1838, when he wrote of Bentham:

He has thus, it is not too much to say, for the first time, introduced precision of thought into moral and political philosophy. Instead of taking up their opinions by intuition, or by ratiocination from premises adopted on a mere rough view, and couched in language so vague that it is impossible to say exactly whether they are true or false, philosophers are now forced to understand one another, to break down the generality of their propositions, and join a precise issue in every dispute. This is nothing less than a revolution in philosophy.[23]

There can be little question that Bentham attempted, through his method, to effect a "revolution." He wished to "codify" his materials rationally and to subject them to empirical verification; to measure them against an external and invariable standard. He worked resourcefully and at length to introduce into political and moral subjects the same scientific methods used in physics and mathematics.

In our judgment, Bentham failed to make a science out of morals and legislation. His scheme introduced no more certainty in these subjects than existed before. His failure, however, had the merit that it exhausted one set of possibilities of applying the methods of the natural sciences to the moral sciences. Thus, it was made clear that other methods (as we shall see when we study Hegel) might have to be resorted to in order to explore the human "sciences."

Nevertheless, having failed in creating a true science of politics, Bentham insisted that laws are the heart of government and should be logical constructs designed to produce the greatest happiness for the greatest number. Imprecise as that phrase may seem, it did make

[23] Mill, *op. cit.*, p. 372. Cf. Alfred Jules Ayer, *Language, Truth and Logic* (New York, Dover Publishers, no date given), p. 55, for a logical positivist's own tribute to Bentham.

We question, however, whether Bentham's achievement was a revolution in philosophy; rather, was it not a continuation of trends of thought which we have already encountered, in such figures as Descartes—with his method of doubt—and Locke—with his opposition to innate ideas?

On the general relation of Bentham to language—to semantics—see C. K. Ogden, *Bentham's Theory of Fictions* (London, 1932).

clear to the framers of new laws at what it was that government should aim. Thus, in England and other countries, in the 1820's, when things eased up a bit, the drafting of new laws was very much influenced by Bentham's work and thought. Having failed to create a science, Bentham had succeeded in developing an ideology for the rising middle classes.

VIII

Bentham did not restrict himself to political reform, however. His head was filled with other projects and plans. A consideration of these tells us much about his character as well as about the essential direction of his thought. He had an idea for a "frugality bank," for a "frigidarium" to keep perishable food from rotting during the off-season, and for a "Chrestomathic School" which was to give the middle-class children a Utilitarian training on new educational principles. He involved himself with such diverse projects as an "amphibious vehicle" for the British army (his brother, Samuel, supplied the mechanical plans), and a *Westminster Review* for the propagation of Utilitarian doctrines (started in 1824, with Bentham putting up the money).

His most famous project, conceived in collaboration with his brother, Samuel, was the Panopticon. In characteristic fashion, Bentham hoped both to solve a social problem and to make money. The social problem was crime. The profit was to come from using convict labor, in place of an engine, to make wood and metal products which Samuel had invented.[24] In Bentham's scheme, the Panopticon was supposed to supplant the prison.

The guiding principle of the Panopticon was constant supervision and oversight of the prisoners, ideally by one man. "Position, not form," Bentham asserted, "centrality of the keeper's lodge, with a commanding view of every part of the space into which a prisoner can introduce himself (by the help of peep-holes, blinds or any other contrivance which will enable the keepers to see upon occasion without being seen), such is the real characteristic principle." Circularity (although not the only possible solution) seemed to be the best architectural handling of this principle. The architectural part of the Panopticon had been invented by brother Samuel (an engineer and

24 See Stephen, *op. cit.*, Vol. I, p. 202.

officer who had served for many years in the Russian army, and whom Bentham visited in Russia for two years) for the muzhiks, or peasantry of Russia. "I thought it applicable to prisons," Bentham explained, "and adopted it. The inspection is universal, perpetual, all-comprehensive."

Bentham quickly perceived that the "universal, perpetual, all-comprehensive" features of the scheme might be extended to another social problem—pauperism. The Panopticon was magnificently adaptable to a workhouse or factory. Thus, the Panopticon "in both the prisoner and pauper branch" became one of Bentham's pet projects. He defined it as a "mill for grinding rogues honest, and idle men industrious."

Bentham suggested his scheme, around 1791, to George III and Parliament, where it at first received sympathetic attention. Indeed, Bentham invested considerable money in purchasing a plot and hiring workers for the construction. But the bill to enact the Panopticon, although passed, never got to the money stage. The affair dragged on until, eventually, in 1813, Parliament gave Bentham £23,000 in compensation for his efforts, and adopted someone else's plans for a penitentiary.[25]

The elaborate and extraordinary plan to build the Panopticon epitomizes much of Bentham: his "projecting" mind and his strange combination of philanthropy and profit-seeking; his inventive spirit and his lack of imagination. It demonstrates both the strengths and weaknesses of his ideology of reform and of his attempt at a science of legislation. Thus, it did not occur to Bentham that Parliament might be right when it gave as the reason for refusal, in 1813, the view that a prison should not be run for private profit; and that even if Bentham and his brother might be trusted, their successors might not.[26] His individualism was too strongly entrenched for him to realize that a social solution to social problems might be more useful than a laissez-faire project, which served primarily the self-interest of an individual (and of society only incidentally and problematically).

This element in Bentham comes into relief when we marshal him

[25] Bentham blamed the failure of the Panopticon on George III (who, he felt, was personally thwarting him) and unsuccessfully offered the plan to France and to any other country which would accept it.

[26] Cf. Stephen, *op. cit.*, Vol. I, p. 205.

against another reformer, Robert Owen. The two men actually became business partners. In 1813, Bentham was induced by a philanthropic Quaker, William Allen, to invest money in the New Lanark mills in order that Owen might buy out some recalcitrant partners and continue with his benevolent schemes for infant education and adult improvement. To Bentham, New Lanark at first probably seemed a version of the Panopticon. In any case, he made money on the deal. Eventually, however, he came to realize his ideological differences with Owen and scornfully gave as his opinion that the latter "begins in vapour, and ends in smoke. He is a great braggadoccio. His mind is a maze of confusion, and he avoids coming to particulars. He is always the same—says the same thing over and over again. He built some small houses; and people, who had no houses of their own, went to live in those houses—and he calls this success."

We shall soon be examining Owen and his "success" in detail. An apt contrast to Bentham, he points directly to the possible oversights in Bentham's theories. What if experience showed that laissez-faire economics were not really conducive to the "greatest happiness of the greatest number"? Worse, what if the masses of people, given power by Parliamentary Reform, exercised their "despotism of Public Opinion" in order to decide, in spite of the fact that experience might prove economic liberalism most useful, to embrace socialism?

These were problems which Bentham, with his powerful but limited perspective, never considered seriously; they formed no part of his science of morals and legislation. History, however, especially in the period after the peace of 1815, in England, insisted on posing these problems; and we must look at them through the eyes of Robert Owen in our next chapter.

CHAPTER 25

Robert Owen

I

ROBERT OWEN was born in 1771, the son of a Welsh saddler and ironmonger who was also the local postmaster in the small market town of Newton, Montgomeryshire, Wales. Owen's formal schooling ended at the age of 9. At this point, according to his own story, he had been serving as assistant to the village schoolmaster for two years! A voracious reader, he continued his education on his own; one result of his private reading was that, at the age of 10, he gave up all religious dogma.

At this same time, aged 10, faced with the fact that there were no prospects for a bright young lad in Wales, he left home. After a brief stay in London, he began work as a linen draper's assistant in Lincolnshire.[1] About four years later, he moved to a similar but better-paid position in Manchester, where he worked for another three years.

Owen's next endeavor was to start a business of his own. He borrowed £100 from his brother and went into partnership with a mechanic; their business consisted in making new spinning machines for the rising cotton industry of Manchester. When this enterprise fell through, Owen, who retained possession of some of the machines, set up for himself as a master cotton spinner. Soon, he abandoned his own small endeavor and became the manager of a large mill. Thus, before he was 21 years old, he found himself in charge of a factory employing 500 people.

[1] Owen went to London originally because his eldest brother was living and working there.

As a manager, Owen was extremely effective and extremely inventive. For example, he was the first person outside the United States, so far as is known, to spin Sea Island cotton (from the Southern States) and thus to realize the advantages inherent in the use of long-staple cotton. Rapidly, he became well known in the trade for the quality of his products.

Within a short time Owen was eager again for a partnership of his own. He found a large and well-equipped mill in Scotland, in the smallish village of Lanark, where he was able to become a partner-manager. As a result of this move, he also met and married the daughter of David Dale, the original owner, whom he and his partners had bought out.

At New Lanark mill, Owen was faced with conditions—and this in spite of his father-in-law's attempted philanthropy—prevalent in all factories at the time: child labor as a commonplace, and a great deal of immorality and crime. The first condition existed because the simple spinning machinery invented by Arkwright and others could be attended by children. Many of these children were paupers, brought from the poorhouses of Edinburgh and Glasgow, at the age of about 5 or 6, to remote mills like New Lanark, situated where water power was plentiful. Thus, of the 2000 hands at the New Lanark mill, 500 were children.

Within the cotton mills the second condition—immorality and crime—prevailed because here existed a form of society with no real order and no real ethics. Drunkenness and laziness cut down productivity. Thievery was widespread, and turbulent and degenerate conditions of life were rife. Adults and children alike existed in misery, working long hours for little money, constantly living in squalid surroundings. Uneducated and uncared for, they were given less consideration than the machines they tended.

Owen, a determined and very active young man, decided that an enterprise run under these conditions was not quite right. He resolved to clean up the factory, partly as a piece of efficiency, partly as a piece of social reform. This conception—that social reform and efficiency go hand in hand—runs through everything he did; it is a new conception, and we shall have occasion to return to it.

First, Owen insisted that the children should be educated. There was, of course, no legal necessity for this: compulsory education did not exist in any part of the world until, roughly speaking, the 1850's. Not only did he insist on the education of children; he devised new and progressive methods for their education. And as there were many women working who had children under 5, he started a school for these tiny offspring. This was one of the first infant schools, or kindergartens, in the world.

Further, he provided all his workers with better housing, food, and clothing, instituted a health fund and, skillfully, through his ability to handle workmen, eliminated the causes of drunkenness, theft, and fornication. After sixteen years of persevering in these measures, Owen had, as he himself claimed, "effected a complete change in the general character of the village."[2]

The upshot of this was that New Lanark came to be regarded not only as one of the most profitable and one of the most productive mills in the world but also as a sort of modern Utopia. Almost all of the philanthropists of the time visited it. Even the royal dukes of England and of Russia were interested.[3] They observed that the children (although, of course, they worked almost as hard as ever; nobody would do without child labor) had an air of being cared for which was different from that of the other mills.[4] They saw that everybody was clean and that the hands did not go about in a drunken stupor. All this was especially surprising because the Scottish mills, being farther

[2] Robert Owen, *A New View of Society and Other Writings,* Everyman's Library ed., introduction by G. D. H. Cole (London, 1949) (hereafter referred to as *New View*), p. 35. This book contains, besides the title piece, the *Address to New Lanark* and *Report to the County of Lanark* as well as a selection of other of Owen's writings; it is doubtful if anyone but a specialist need read more than is contained in this volume. As for his life, Owen's own *Life of Robert Owen* (1857) is the basic source for the period up to 1821. The standard biography is Frank Podmore, *Robert Owen,* 2 vols. (London, 1906), which reprints many excerpts from the *Life;* it also contains excellent illustrations. See, too, G. D. H. Cole, *Life of Robert Owen,* 2nd ed. (London, 1930), and *Socialist Thought—The Forerunners* (New York, 1953). Owen's son, Robert Dale Owen, sheds much light on his father in his autobiography, *Threading My Way* (1874). The best and most up-to-date study of Owen is the recent *Robert Owen of New Lanark* (London, 1953), by Margaret Cole, which also has a good bibliography.

[3] Cf. Margaret Cole, *op. cit.,* p. 86, for an account of the interest of the Grand Duke Nicholas of Russia in Owen's scheme. Owen even believed, as he tells us in his autobiography, that he had converted Napoleon at Elba to his new system.

[4] Owen, however, did refuse to employ children under 10 (other mill owners used children of 5 and 6) and employed them only for a 10¾-hour day. See Margaret Cole, *op. cit.,* p. 80.

removed from the center of civilization, tended to be behind the Manchester mills.

In spite of the success of New Lanark, Owen's partners pointed out to him that they were spending a great deal of money on taking care of the employees. Though it was, of course, Owen's contention that if they did not spend this money the factory would not be doing so well, his partners simply droned on that money spent is money spent. Owen finally got tired of this and, in 1809, decided to buy them out. Unfortunately, his new partners (for Owen could not swing the finances on his own) were much worse than the previous ones in their obstructionism. Faced once more with the necessity of buying out his associates, Owen, now in his early forties, joined forces in a partnership which included Jeremy Bentham and the well-known Quaker, William Allen. He offered his new partners no greater inducement than that they would get 5 percent on their money; all profit above and beyond that would be used to benefit the employees through communal services. Once again, the New Lanark mill was both financially successful and a model factory.

II

In the same year as the formation of his new partnership, 1813, Owen published what is really his only successful book, *A New View of Society*. It was based largely on his own experience, and his point was quite simple. He had become convinced that human character is formed in early childhood and is formed entirely by environment. This was the ground on which he based all his subsequent work. As a principle, it appeared on the title page of *A New View* in the following terms:

> Any character, from the best to the worst, from the most ignorant to the most enlightened, may be given to any community, even to the world at large, by applying certain means; which are to a great extent at the command and under the controul [*sic*], or easily made so, of those who possess the government of nations.

From this view Owen deduced two propositions which he made the center of his book. The first was that it is pointless to persecute people who commit crimes or, in general, to attach either praise or blame to what people do in life. As was the case with Bentham's Utilitarianism, this was not entirely an original point of view; but in the hands of Owen it came to have great influence. The modern theory of juvenile delinquency, for example, is essentially what Owen proposed in 1813; that the antisocial child behaves as he does because he has suffered great hardships, major emotional disturbances, and the like. It is this view which gradually won out against the antiquated eighteenth-century legal codes, which were based on the theory that the criminal of any age was essentially wicked and had to be violently and ferociously punished so as to set an example.[5]

Owen's second proposition was that in order to avoid their being criminals in later life, one must take hold of people in early childhood and see that they lead good lives. Thus, society has a responsibility—to itself and to the individual—to educate and to provide decent living conditions for all. The lessons Owen gave as to the formation of man's character made it "evident to the understanding, that by far the greater part of the misery with which man is encircled *may* be easily dissipated and removed; and that with mathematical precision he *may* be surrounded with those circumstances which must gradually increase his happiness."

III

On the basis of his propositions, Owen wished not only to make his own factory a model, but to reform other factories. His mind always went from the particular to the more and more general; and in this he was quite different from most socialist thinkers of the nineteenth century. He did not state, abstractly, that there should be laws to correct the sad state of a society in which children were dirty and uneducated. Instead, he began by showing in his own practice that the state of affairs could be corrected. Only then did he appeal for factory reform and, later, for social reform in a general way. His was typically the approach of the specific (and perhaps we ought to say practical and technical) mind.

[5] Before Owen's time, it was possible, in theory at least, for men in England to be hanged for about 200 offenses, some of them quite minor. In practice, the legal system was breaking down. Juries, faced with a statute saying that a criminal could be hanged for stealing anything more than 40 shillings, simply assessed the stolen objects (for example, a dozen silver spoons) at less than 40 shillings, whatever the shopkeeper might say.

Thus, shortly after he published his book, Owen started agitating for factory reform. The condition of the factory system in England (and, indeed, the world over) at this time cannot better be illustrated than by describing the bill which Robert Owen desired the British Parliament to pass. The bill, as introduced by Peel, had three provisions. The first was that no child should work in a factory until he (or she) was 10 years old (Owen originally proposed 12). The second was that no child should be put on night work (i.e., 9 P.M. through 5 A.M.) until he was 18 years old. And the third was that no child should work more than 10½ hours a day until his eighteenth birthday. These provisions speak for themselves as to existing conditions.

Owen's scheme, however, was considered extreme and the bill was mutilated during passage. Thus, although the first Factory Act—that of 1819—was due to his initiative, he disowned it as a mockery of what he had demanded. And even the watered-down Factory Act of 1819 was rendered useless by the fact that no mechanism of inspection was established to insure that the provisions of the bill were enforced.

Unhappily, at the moment that Owen became engaged in factory-reform agitation, the final defeat of Napoleon occurred. Ending the conflict in Europe, it ushered in a postwar world characterized—as many postwar worlds are—by disbanded armies, unemployment, and severe want and misery. In England, the problem was especially acute because machinery, which had hitherto been used only for spinning, began to be freely used for weaving. The result was widespread unemployment and starvation of the hand-loom weavers. We see this reflected in the fact that the price paid for weaving a piece of calico by hand was 6 shillings and 6 pence in 1814, and 1 shilling and 2 pence in 1829.[6]

The British Parliament cast around for a solution, and Robert Owen was asked to give evidence and to propose a plan. His proposals mark his emergence from a mere factory reformer to a social reformer in the large. Presenting his scheme to a special Committee for the Relief of the Manufacturing Poor, he made an analysis which was quite uncomplicated. He said, "The immediate cause of the present distress is the depreciation of human labour. This has been occasioned by the general introduction of mechanism into the manu-

[6] T. S. Ashton, *The Industrial Revolution*, p. 117.

factures of Europe and America." As a consequence, Great Britain possessed at the end of the Napoleonic Wars a new power which "exceeded the labour of *one hundred millions* of the most industrious human beings." When the war demand ceased, "it soon proved that mechanical power was much cheaper than human labour." The resultant discharge of human labor, Owen pointed out, worsened the downward swing of what we have now come to call the business cycle, by diminishing purchasing power.

Owen did not propose that machinery be eliminated. He repudiated entirely the Luddite solution to the problem: "Although such an act [the discontinuation of mechanical power] were possible, it would be a sure sign of barbarism, in those who should make the attempt." The problem, he insisted, was to rearrange the use of the new productive forces, not to destroy them. Some way of establishing a relation between men and machines which was profitable to all men had to be found.

Owen proposed, therefore, the setting up of what were, essentially, coöperative villages.[7] He suggested that communities of about 1000 to 1500 people be formed; that the occupants should live on about the same acreage of land (that is, a density of about one man per acre); that the villages be laid out as a parallelogram (formed by its buildings); that many activities, such as cooking and eating, take place communally; and that there be sufficient land set aside for agricultural purposes and enough factories for mechanical and manufacturing purposes. Conceiving of his village as a single, independent unit, which would make most of its own products, he held that a certain number of people—about 1500—would suffice to form a self-sufficient community.

This view of society strikes a new note. It involves an attitude very unlike that of most Utopias, because it bears constantly in mind that the community being designed is supposed to be an economic unit which can live. In Owen's words:

> The first object, then, of the political economist, in forming these arrangements, must be, to consider well *under what limitation of numbers, individuals should be associated to form the first nucleus or division of society.*

[7] This was a proposal for the Poor Law population only, not for the general society.

All his future proceedings will be materially influenced by the decision of this point, which is one of the most difficult problems in the science of political economy. It will affect essentially the future character of individuals and influence the general proceedings of mankind. It is, in fact, the cornerstone of the whole fabric of human society.[8]

Others, before Owen, had asked the question, what should be the optimum size of a community: Plato and Aristotle had discussed the "right" size of a polis, and Rousseau had insisted on smallness for a republican state. But all these discussions had revolved primarily around the problem of correct size for the *political* functioning of the community.[9] Owen was concerned with *economic* functioning. He insisted that the size of the social nucleus was important from the point of view of *production*—of goods and services—and that the political economist should take as his principle in determining this question "that it is the interest of all men . . . that there should be the largest amount of intrinsically valuable produce created, at the least expense of labour, and in a way the most advantageous to the producers and society."

The problem which Owen set himself was, so to speak: "What is the minimum number of people which can provide the necessities and luxuries of the community?" Let us take a far-fetched example. If the community is to contain, for instance, chess players, then it must be large enough to be able to afford at least one maker of chess sets. In short, the size of the community is finally determined by the amount of labor which is required to contribute the most peripheral luxury of the community, the one which least occupies the attention of the people. Undoubtedly influenced in his figure of 1000–1500 by the fact that this happened to be about the population of his mill in New Lanark, Owen emphasized the new way, of economic planning, in his projected Utopia.[10]

[8] *New View*, p. 264.

[9] Plato, of course, did erect his Republic on an economic basis, i.e., the existence of necessities and luxuries determined its "swollen" size; but his main interest was in the community as a *political* unit.

[10] Actually, Owen's proposed communities were very much like the *kibbutzim*, or communities, with which the state of Israel has been colonized in recent times. These individual Israelite communities tend to federate together and seem to represent a transition stage in the building of a country. Cf. Isaac Deutscher's very interesting article, "Israel's Spiritual Climate," *The Reporter* (April 27, 1954).

IV

The extraordinary thing is that Robert Owen initially got a great deal of support for his view. *The Times* wrote in favor of it; the Tory home secretary, Lord Sidmouth, gave it an interested hearing; and David Ricardo, the noted economist, sat on a committee to back the plan. But, as with Bentham's Panopticon, the government never took any real action on the matter. Instead, the Tory ministry resorted to repression as the solution for unemployment and distress.

In disillusion, Owen swung away from government and committees to an appeal to the people. This, coupled with his outspoken opposition to organized religion, lost him the future support of the established powers and turned him from a successful, philanthropic businessman into a platform agitator for various forms of coöperatism and socialism. In his new role, he went from one experiment to another.

The first "flyer" was in 1824–1825, when Owen took most of his money out of the New Lanark mills and bought about 20,000 to 30,000 acres of land in the New World. Here, in Indiana, he set up the famous Owenite community, called New Harmony, and invited "the industrious and well disposed of all nations" to join him. Those who came, however, were not always "well disposed"—many of them were actual scoundrels, for Owen had little choice in the sort of colonists he could recruit—and they were hardly as docile as his Lanark workers. Even his persuasive personality was unable to make the thing go (to be true to his own theory, of course, Owen should have started with a large number of small children). New Harmony lasted less than three years.

Owen returned to England in 1829 quite cheerful, although he had lost almost all the money that he had put into New Harmony, and immediately began to help shape a more elaborate social movement.[11] Supported by John Doherty, the great labor leader, and others, he formed the Grand National Consolidated Trades Union. The union, however, which may have reached a membership of half

[11] Owen finally sold much of the New Harmony land to individuals, keeping the rest for his family. In Indiana, today, remnants of the settlement still exist in private hands. For further details on New Harmony, see Cole, *op. cit.*, Chap. 17, and, especially, A. E. Bestor, *Backwoods Utopia* (Philadelphia, 1950).

a million in its brief existence, broke apart disastrously in the labor struggles of 1834. Thus, the chief merit of Owen's second experiment, which involved, essentially, the origins of the trade-union movement, was that it served as a prototype and, more immediately, as a warning for future labor efforts.

In the midst of this failure, however, Owen was already helping to start what has since become the coöperative movement. He set up nonprofit trading communities where, as in present-day coöperatives, a rebate was made at the end of the year on all purchases.[12] Several of these communities were founded in Scotland and Ireland, but the one most remembered was at Queenwood in Hampshire. Eventually, however, the coöperative colony at Queenwood itself was, like New Harmony, a failure.

In the course of his experiments, Owen had cross-fertilized the trade-union movement with the coöperative movement. He induced many of the unions to set up their own producer's coöps; and frequently, in these coöps, goods were produced for a price which was calculated entirely on their labor content—in accord with an idea devised by Owen in his *Report to the County of Lanark* (1820). The money used consisted of pieces of paper which said, for example, "Equal in value to three working hours," and all goods could be exchanged on the basis of this labor value. Owen even founded a National Equitable Labour Exchange, in 1832, to facilitate this new form of exchange on a country-wide scale.[13]

Clearly, all these schemes of Owen were addressed to the working class movement, where he found his main disciples. Long before the time, in 1858, when he went back to his birthplace in Wales to die, he had completely traversed the path from a successful and wealthy cotton entrepreneur to an eccentric and poor projector of social reforms. He had turned from addressing and petitioning the rulers to teaching and organizing the ruled.

12 The basis of Owen's idea for the coöperatives was the belief (which in some respects was shared by Adam Smith) that the value of an object is created solely by the laborer who makes it. The rebate was an effort to return to the buyer what might otherwise be a "profit" extracted from his labor.

13 There is a reproduction of one such note in Podmore, *op. cit.*, Vol. I, p. 230. The National Equitable Labour Exchange itself failed in 1834. Cf. Margaret Cole, *op. cit.*, p. 184.

V

Owen's life can be considered as something of a capsule within which the economic-social ideas of the European world developed under the impact of the Industrial Revolution. He is the Lunar Society realizing its limitations; becoming aware of the misuse of laissez-faire economics when considered in the light of humanitarian interests and the facts. Thus, although a characteristic Lunar Society type, a successful entrepreneur in the rising cotton industry, Owen came to the conclusion that the new economic system, instead of freeing the individual, had made a slave out of him. It was to this aspect of laissez faire that he called attention.

Hitherto, legislators have appeared to regard manufactures only in one point of view, as a source of national wealth. The other mighty consequences which proceed from extended manufactures *when left to their natural progress,* have never yet engaged the attention of any legislature. Yet the political and moral effects to which we allude, well deserve to occupy the best faculties of the greatest and wisest statesmen.

Without directly naming Bentham, Owen attacked those "whose profound investigations have been about words only." The one-time factory manager paid attention to facts and not words, and the facts of the laissez-faire system showed, he claimed, that there was "increasing misery and demoralization of the working classes." Not only had the introduction of manufacturing and laissez-faire economics *not* bettered the worker's lot, but it had worsened it to a point below feudal conditions. In fact, Owen suggested, the industrial slavery was worse than Negro slavery. Not only economic, but moral degradation marked the new system, and none of the "social affections" were left. Instead, as Owen described the situation, the worker labors today "for one master, to-morrow for a second, then for a third, and a fourth, until all ties between employers and employed are frittered down to the consideration of what immediate gain each can derive from the other."[14]

[14] *New View,* p. 124. Karl Marx, who had read Owen carefully, made the theory of the worker's increasing degradation and misery under capitalism a fundamental part of his analysis. (And it was exactly over this point that the German Social Democrats, led by Bernstein, broke with orthodox Marxism.) Further, it is of interest to compare Owen's theory—of the "frittering" away of any other connec-

It was not that Owen lacked the very reasonable desire to combine justice with efficiency which is the essence of the Benthamite faith, or that he rejected Bentham's utilitarian and hedonistic principles; it was rather that on their basis he came to very different conclusions. Owen agreed that "that government, then, is the best, which in practice produces the greatest happiness to the greatest number." And he also accepted, without reservations, the theory that men are governed in their actions either by immediate or future self-interest. But, on these very grounds, he insisted that men turn to the "social" schemes devised by him rather than to the individualist world of Bentham. Happiness would never come, Owen claimed, "by any arrangements that it is possible to make by individualizing man in his proceedings, either in a cottage or in a palace; for while his character shall be so formed, and while the circumstances around him shall be, as they then must be, in unison with that character, he cannot but be an enemy to all men, and all men must be in enmity and opposed to him."

Thus, the major difference between Owen and Bentham is this. Bentham had elaborated a Utilitarian doctrine whose end was society's good but whose motive force was found in the individual's self-interest. Owen, one of the earliest liberals to recognize publicly the evils of the Industrial Revolution, rejected such a doctrine and projected a plan emphasizing the achievement of society's general welfare through social means.

VI

Let us try now to assess Owen's overall influence. What, in sum, is the significance of his life and ideas? Perhaps we can subsume his achievement under three principles—all three provoked by the conditions facing England as a result of the Industrial and the French Revolutions.

Starting with a remedy for the rapidly declining economic and moral situation of the period around 1813, Owen turned first to

---

tion between employer and worker than that of "immediate gain"—with Marx's famous statement, in the *Communist Manifesto,* that the new bourgeois system "has pitilessly torn asunder the motley feudal ties that bound man to his 'natural superiors,' and has left remaining no other nexus between man and man than naked self-interest, than callous 'cash payment.' "

"The Principle of the Formation of the Human Character," to the study of psychology, which he called his "New View." Then, upon this theory of character formation, he raised the edifice of a socialistic society; or, at least, he tried to expound its principles and to experiment with models of socialistic communities. To these two fundamental contributions to the ideas and practice of the modern world, he added a third: the conception of the factory as being not only an economic unit but a "social structure," i.e., a community.[15] Each of these three strands deserves separate and detailed treatment.

When we discussed Hobbes and Locke, we said that Newtonian and Galilean physics "had supplied the model of society as an artificial construction, made up of individual atoms of humanity. But it had left unanswered the basic question: the nature of the fundamental building block—man." It was within this tradition—of attempting to obtain, as he put it, "a correct knowledge of human nature"—that Owen operated. Indeed, his "New View," as we have said, was not very original. The central idea, about the individual's character being shaped for him by society, was to be found in Helvétius, Rousseau, and others. What was new was the intensity with which Owen turned his vision upon the subject.

Owen marks, in an important sense, the culmination of Locke's epistemology. If man is a blank tablet, as Locke implied, any characteristics can be stamped upon him. A new twist to Locke's ideas is given, however, by Owen's insistence that the child's exposure to experience be consciously guided by society and that only certain impressions be implanted in his mind. Where Locke saw man formed by nature, Owen perceived the influence of society; this was the "correct knowledge of human nature" that Owen had sought.

The consequences of this simple principle were profound. Among other things, it meant, as we have already suggested in referring to the treatment of juvenile delinquency, the end of ethics. With this "correct knowledge," men would realize that "they are not subjects for praise or blame, reward or punishment." The only "value" which remained was happiness.

Perhaps Owen did not realize to what sort of world this might lead. He did not see that evil as well as good might come of the idea

[15] We have touched on this point before, when we discussed the optimum size of the community as being determined by economic needs.

of going, so to speak, "beyond good and evil" to a world of unalloyed happiness.[16] The *reductio ad absurdum* of his idea, however, is depicted in Aldous Huxley's wonderful satire on Utopia, *Brave New World.* In this version of the "New View," character is not merely formed from infancy by Pavlovian conditioned reflexes, but is partially determined from birth by chemical formulas; babies in Huxley's Utopia are "bottled babies." Unhappiness is unknown in the Brave New World, where a wonder-drug, "soma," and an improved version of the movies, the "feelies," supplement any deficiencies of the chemical and psychological conditioning. The hero of the book, who wants the freedom to be "unhappy," is an accidental misfit.

Now, it is true that Owen did not have this sort of world in mind. For one thing, he strongly opposed the specialization of Huxley's new world. Everyone in Owen's villages is to take his turn at various occupations. Thus, "Instead of the unhealthy pointer of a pin,—header of a nail,—piecer of a thread—or clod-hopper, senselessly gazing at the soil or around him without understanding or rational reflection, there would spring up a working class full of activity and useful knowledge."

For another thing, Owen believed in equality, whereas Huxley's new world is characterized by inbred inequality. Owen derived the belief in equality in a new fashion from his theory of character formation. He claimed that all men are equal on the ground that they are all born with equally plastic natures. "Human nature, save the minute differences which are ever found in all the compounds of the creation, is one and the same in all; it is without exception universally plastic, and by judicious training, *the infants of any one class in the world may be readily formed into men of any other class.*"

It is clear, then, that the Brave New World can only be thought of as an unintended consequence of Owen's new science of character formation. Nevertheless, it *is* a possible consequence of Owen's principle. Like the totalitarian implications of Rousseau's ideas, there is a dangerous strain which lurks within the amoral hedonism of Owen's "New View."

We come now to Owen's second principle. According to Owen, the science of character formation would mold a new man. And,

[16] Our use here of "beyond good and evil" is not, of course, in the sense in which Nietzsche used the phrase.

gradually, Owen came to realize that the new man could only exist in a new state and society. With an ecstatic voice, he described "the fundamental differences between man in the OLD and man in the NEW state of society. In the first he *has been* a wretched, credulous, superstitious hypocrite: in the *last* he *must become* rational, intelligent, wise, sincere and good. In the OLD, the earth has been the residence of poverty, luxury, vice, crime and misery;—in the NEW, it will become the abode of health, temperance, wisdom, virtue, and happiness." The NEW society, of course, was coöperative and socialistic; and we have already described how Owen arrived at his socialism and what his concrete suggestions were.

The next important consideration for Owen was how the new society would come into effect. He believed that there was absolutely no need for revolution or violence. All that was needed was the dissemination of the "correct knowledge" which he had discovered. The removal of ignorance would be gradual but inevitable, for Owen believed fervently that the "Truth would conquer." In his writings, there is a steady millennial note. He was convinced that mankind was on the threshold of a "new era."

For this reason, Owen has been called a "Utopian socialist." In the book *Die Entwicklung des Socialismus von der Utopia zur Wissenschaft* (*The Development of Socialism from Utopia to Science*), Friedrich Engels lumped Owen with such reformers as Morelly, Saint-Simon, and Fourier and accused him of sharing the delusion of the eighteenth-century *philosophes* that society could be reorganized simply by the power of enlightenment. Engels and Marx did not believe that change would occur because men were suddenly presented with certain abstract ideas; change would occur because it was a dialectic necessity, and the mechanism by which this dialectic worked itself out was the class struggle.

How much justice is there in Engels' charge: that because Owen had not linked his ideas to a social force working for socialism, he depended, finally, on the power of conversion? First, of course, we ought to remark that the power of conversion, as the example of Christianity shows, is not to be underestimated. And Owen himself increasingly viewed his system as a NEW RELIGION, "the most glorious the world has seen, the RELIGION OF CHARITY, UNCONNECTED WITH FAITH," which would supplant all former moral and religious instruction. Thus, his writings resound with tones from the Bible and are filled with religious images.

The socialist movement in Great Britain in fact derives from men whose inspiration was in the main religious. There is to this day a large element of nonconformity and evangelism in the English, and particularly the Welsh, labor movement—and Robert Owen was Welsh. The background of British socialism is the Bible and the Sermon on the Mount, not the writings of Marx and Engels.

If, however, we leave aside his evangelical power, how Utopian is Owen in comparison with the "scientific" Marxists? We obtain some clue to the answer if we examine the dedications Owen made of the four essays which compose his *New View:* the first is to William Wilberforce; the second to the BRITISH PUBLIC; the third to the "Superintendents of Manufactories"; and the fourth to HIS ROYAL HIGHNESS, THE PRINCE REGENT OF THE BRITISH EMPIRE. In short, Owen was prepared to address anyone who would listen to him. His idea was:

> Let truth unaccompanied with error be placed before them; give them time to examine it and to see that it is in unison with all previously ascertained truths; and conviction and acknowledgment of it will follow of course. It is weakness itself to require assent *before* conviction; and *afterwards* it will not be withheld. To endeavour to force conclusions without making the subject clear to the understanding, is most unjustifiable and irrational, and must prove useless or injurious to the mental faculties.[17]

Gradually, however, as we have seen, Owen lost faith in his power to persuade philanthropists, like Wilberforce and Bentham, or politicians, like Lord Sidmouth and the Prince Regent. Was this an error of impatience on his part? Basically, we believe, there was nothing wrong in Owen's expectation of reform from the state bureaucracy. We suggest the reading of a few pages from Joseph Schumpeter's stimulating book, *Capitalism, Socialism and Democracy,* on this very point: Schumpeter maintains, correctly we think, that the appeal to the state was not so Utopian as Marx and Engels thought; that intellectuals and bureaucrats are also moved by a dialectic necessity. In Schumpeter's words: "For the state, its bureaucracy and the groups that man the political engine are quite promising prospects for the

[17] *New View,* p. 24.

socialist looking for his source of social power. As should be evident by now, they are likely to move in the desired direction, with no less 'dialectical' necessity than the masses."[18] What prevented Owen from applying his own belief in gradual change in relation to the ruling class was the "millennial" note in his character; the Fabian Socialists, however, knew better than he.

In principle, therefore, Owen was not being Utopian. In practice, however, he was. What he needed, really, was to capture the machinery of reform developed by the Benthamites. Instead, he allowed himself to be quickly rebuffed by the upper classes and turned toward the lower classes. Here, as we have seen, he met with somewhat more rapid success; and the coöperative and trade-union movements bear the strong stamp of Owen's ideas upon them. But ultimately the development of the English movement was in a direction quite different from Owen's: the trade-union destiny lay either in the day-to-day contest for higher wages and improved working conditions or in the political struggle for the nationalization of industry. These formed no part of Owen's new society, which was to have as its basis small voluntary coöperative associations.

Owen's ultimate failure, therefore, to link his ideas effectively with either the state bureaucracy or the working class stems from the essential nature of his ideas. He wanted a total reorganization of the fundamental elements of society. He wished a "new man" in a "new society." It is this notion we had in mind when we said a few pages back that Owen's third fundamental contribution was "the conception of the factory as being not only an economic unit but a 'social structure,' i.e., a community." It is this which for one brief moment captured the imagination of Karl Marx. As he admitted:

> We can learn in detail from a study of the life work of Robert Owen, the germs of the education of the future are to be found in the factory system. This will be an education which, in the case of every child over a certain age, will combine productive labour with instruction and physical culture, not only as a means for increasing social production, but as the only way of producing fully developed human beings.[19]

[18] Joseph Schumpeter, *Capitalism, Socialism and Democracy* (New York, 1947), pp. 310–311.

[19] Karl Marx, *Capital*, Everyman's Library ed., 2 vols. (London and New York, 1951), Vol. I, p. 522.

Owen's third distinctive feature was, then, that he conceived the function of the factory system to be the production of "fully developed human beings." It was a social, more than economic, purpose which he had in mind when he talked of the factory. He accepted the Industrial Revolution, with its wide-scale introduction of factories and machines, but he wished to convert man's new technical and economic powers into social benefits. If the factory is to dominate man's life, he believed, then it must be made so that men may live in it.

In many ways, therefore, Owen anticipated the doctrine of what we may call "corporatism." This doctrine, unfortunately, at least in name, has been mainly connected with fascist ideology, but it is certainly capable of development in other directions. It is, in fact, developing differently in America, where the *social* function of capital and work is coming into new and greater prominence. With the tremendous shift in American industry from control by owners to control by managers, the new form of "corporatism" emerges more clearly.[20] In this new manifestation, corporations set up their factories as total communities rather than as simple plants. The corporation's social duties in relation to its workers, the public, and the community at large is stressed. Within the plant, consideration is given to the psychological as well as to the physical needs of the workers, and status systems are taken into account. Externally, the "community" aspect of the plant is so strong that much of present city and regional planning in the United States is due to private corporations rather than to public authorities.[21]

At the root of these developments are again the ideas and work of Robert Owen. On the simplest level, he anticipated what is today called "scientific management." In his earliest writings, he bluntly talked of the worker as a part of the factory mechanism and declared that "from the commencement of my management I viewed the population, with the mechanism and every other part of the

20 For details of this shift, cf. A. A. Berle and G. C. Means, *The Modern Corporation and Private Property* (New York, 1933), and James Burnham, *The Managerial Revolution* (New York, 1941).

21 In fact, so strong is this drift of the corporation into noneconomic roles that a recent writer (John Knox Jessup, in an article called "A Political Role for the Corporation," *Fortune*, August, 1952) has suggested, although perhaps with some exaggeration, that a plurality of welfare communities, i.e., the corporations, rather than a single national welfare state is emerging in the United States.

establishment, as a system composed of many parts, and which it was my duty and interest so to combine, as that every hand, as well as every spring, lever, and wheel, should effectively co-operate to produce the greatest pecuniary gain to the proprietors." Candidly, he asked his fellow manufacturers: if "due care as to the state of your inanimate machines can produce such beneficial results, what may not be expected if you devote equal attention to your vital machines, which are far more wonderfully constructed?" And Owen's own success demonstrated the strictly economic advantages to be derived from "scientific management." Thus, there is little in Elton Mayo, the supposed pioneer sociologist of industry, that cannot be found, in germ, in the English cotton entrepreneur.[22]

Owen quickly abandoned this narrow approach. Occasionally, he would still talk of his planned villages as a "machine" for living, but this was only a verbal throwback to his factory origins. Increasingly, he centered all his attention on the social benefits of his "scientific management" or "correct knowledge of human nature." Whereas New Lanark was a factory turned into a community, New Harmony was planned as a community with a factory attached.

How was the shift, however, from the dismal England of Owen's day to the "new era" in which the factory would be a social community to be made? In Owen's mind, there loomed no difficulty. He had made the change at New Lanark, and from the experience thus gained he had formulated the principles whereby others might accomplish the same thing. He declared, "The principles developed in this 'New View of Society' will point out a remedy which is almost simplicity itself, possessing no more practical difficulties than many of the common employments of life; and such as are readily overcome by men of very ordinary practical talents." In fact, Owen continued: "It is so easy, that it may be put into practice with less ability and exertion than are necessary to establish a new manufacture in a new situation. Many individuals of ordinary talent have formed establishments which possess combinations much more complex."

Owen's idea is a profound one. He was substituting his Captains

22 Owen, of course, was overlooking the fact that his fellow manufacturers were more concerned with their machines than with their workers because the former represented a heavy capital investment whereas the oversupply of unskilled labor presented no replacement problem. Only when industry must spend money in training its personnel do they become economically as important as machines.

of Industry for Plato's Philosopher-Kings; he was saying that the organizational skills necessary for entrepreneurship in industry are also those needed in undertakings for the reconstruction of society. Society's basis, Owen insisted, is economic even though the economic is not its end. Thus, the "good life" cannot be lived independently of the economic life, and it is only men of the Lunar Society type who can lay the correct foundations of the latter.

Owen was, therefore, advising society to place its organization in the hands of technicians and businessmen.[23] He repeatedly stated that it is not the role of the workers to effect change. He cautioned them to obedience and the acceptance of gradual improvement. It is, instead, the Captains of Industry who must accept *social* responsibilities and take over the leadership not only of economics but of society. The Wedgwoods, Boultons, and Watts were to lay aside their particular self-interest and, altruistically, like Owen, devote their talents to the running of society. This, in essence, was Owen's final message.

VII

Was the failure of New Harmony, however, an indication that something was wrong with Owen's "entrepreneurial" view of society? Had he overlooked some factor which had not been present in New Lanark? We believe this to be the case. In Owen's factory, there had been no political problem; he simply exercised a paternal despotism. In New Harmony, all the difficulties of village government arose. And Owen had no concept of politics to offer as a solution. In the sketch of an ideal community which he drew in *Report to Lanark,* he suggested only that the villagers govern themselves

> . . . upon principles that will *prevent* divisions, opposition of interests, jealousies, or any of the common and vulgar passions which a contention for power is certain to generate. Their affairs should be conducted by a committee, composed of all members of the association between certain ages—for instance, of those between thirty-five and forty-five, or between forty and fifty. . . . By this equitable and natural arrangement all the numberless evils of elections and electioneering will be avoided.[24]

[23] This was the same message being propounded by the contemporary French thinker Saint-Simon. For the latter, see *Henri Comte de Saint-Simon, Selected Writings,* ed. and tr. with an introduction by F. M. H. Markham (New York, 1952).

[24] *New View,* p. 287.

But it is exactly the "common and vulgar passions" which do intrude themselves into political arrangements. As to the "principles" that will prevent divisions, Owen tells us almost nothing about them. We know only that he did not believe in political parties: his anathema is upon all of them—Radicals, Whigs, or Tories—and he "solicits no favour from any party." He wished to do away with "those localized beings of country, sect, class, and party who now compose the population of the earth." And he believed that this could be done once the cause of man's selfishness was removed: "Society has been hitherto so constituted that all parties are afraid of being over-reached by others, and, without great care to secure their individual interests, of being deprived of the means of existence. This feeling has created a universal selfishness."[25]

Such was Owen's simple, and negative, approach to political problems. In our judgment, it overlooks all the real factors involved in politics. In its naïvete it is of a piece with his belief that the *sole* cause of misery is from "the ignorance of our forefathers." According to Owen, this cause of misery was on the point of disappearing. Mankind was emerging "from the abodes of mental darkness to the intellectual light" provided by the New View. Whereas before all was error, now there had been formulated "a system which shall train its children of twelve years old to surpass, in true wisdom and knowledge, the boasted acquirements of modern learning, of the sages of antiquity, of the founders of all those systems which hitherto have only confused and distracted the world, and which have been the immediate cause of almost all the miseries we now deplore."

In this quotation, we suggest, the basic weakness of Owen's approach is highlighted. It is not only that he had no political philosophy; worse, he refused to recognize any intellectual forerunners. There was a strong anti-intellectual strain in him: he declaimed against "closet theorists" and impractical, learned men, and he opposed "book learning." He saw the entire past as "insanity" (now, as a result of his work, to give way to "rationality"). In short, he was

[25] *Ibid.*, p. 288. Are we being too critical if we say that it is a poor "knowledge of human nature" which sees the "contention for power" as arising solely from particular economic arrangements? Owen's idea is akin to the Marxist theory that with the elimination of the capitalist system all evil will disappear. This would seem to carry Rousseau's placing of the theodicy problem in society to its ultimate extreme.

repudiating the entire intellectual tradition of the West which we have been tracing until this moment.

Here is the real "utopian" element in Robert Owen. He was "out of time" rather than out of place. Although he recognized, empirically, the problems thrown up by his time, he was really against history in an even more fundamental sense than Bentham. Although Owen made the profound discovery that man has no nature independent of his "social nature," he did not discover that man has no nature that is not of an historical character.

For this insight, we must turn to our next figure, Georg Hegel. But before doing so, let us stress, not the gaps in Owen's work, but the extent of his achievements. He contributed to a theory of character formation, played a major part in announcing coöperative and socialist doctrines, and preached and practiced a form of corporatism in which the factory became a community. It was a great deal for one man to do; and perhaps we can forgive him a lack of historical consciousness when we think of these "by-products" of his New Lanark factory.

CHAPTER 26

Kant and Hegel: The Emergence of History

AT THE time of the Renaissance, the intellectual tradition of Europe was re-created by Mediterranean thinkers, many of them from northern Italy. Now, more than 300 years later, at the beginning of the nineteenth century, the Western tradition had widened. The ideas of the French Revolution were influential; the English Utilitarians were working out a reasoned system; and American thinkers had actually had the chance to begin a new society. At the same time, the small German states reëntered the intellectual ferment of the West, which Martin Luther had once so strongly provoked; and their contribution had a subtle influence on the course of history.

One sign of the times in Germany was a new birth of romantic poets—who, incidentally, did much to fix her still fluid language. The greatest of these poets, and the most powerful creative mind in Europe since Newton, was Johann Wolfgang von Goethe. Goethe was a scientist as well as a poet, and a highly individual scientist. He was interested particularly in the growth and form of plants, in which (like Erasmus Darwin in England) he sensed the underlying unity in the development of nature which Charles Darwin later expressed in the theory of evolution. Goethe also rediscovered the Renaissance and the antiquities of Rome; he was a successful public servant, particularly in education; and his wide and vivid interests fired the minds of men far beyond Germany.

Another development went on in German thought at this time

which now seems most characteristic: a tradition of philosophy was begun. Its founder was Immanuel Kant, and he remains the most important theoretical philosopher that Germany has produced. But those who followed Kant had a practical influence which went beyond philosophy, and which has shaped the political outlook of men ever since. The philosopher who has had most political effect was Georg Wilhelm Friedrich Hegel.

We have to see this practical influence against its philosophical background. Hegel was, after all, a professor of philosophy, and his life was devoted to problems with which other philosophers had left him. It is therefore remarkable that out of his academic work should grow a way of looking at men and states which has overturned empires—remarkable, and worth our study.

Immanuel Kant

I

The problems of philosophy in the seventeenth and eighteenth centuries were related to the advances of science then. Much work in philosophy was an attempt to find foundations for the new science, and many philosophers were scientists. Kant was among these, and his philosophy was such an attempt to close a gap in the foundations of science which had been opened unexpectedly in his boyhood.

An early basis for science had been found by Thomas Hobbes in the principle that every effect has a material cause—and the same cause always has the same effect. Since Hobbes' interest in science had sprung from his reading of Euclid, he thought of the connection between cause and effect as having roughly the same force as a proof in geometry. We do not need experience to teach us that the propositions of geometry follow from the axioms—we only need logic. So, thought Hobbes, effects flow from their causes logically, by inner necessity, and we should be able to grasp this necessity intellectually.[1]

The view that cause and effect are linked by a necessity which transcends our experience was held by scientific philosophers well into the eighteenth century. True, the Irish philosopher George Berkeley challenged the assumption that what our senses tell us entitles us to infer the behavior—and even the existence—of physical

[1] See Chap. 11, "Hobbes and Locke," pp. 195–199.

objects outside ourselves. But this was so radical a piece of skepticism that no one could take it seriously—certainly no one could suggest in what way it ought to change our behavior toward our sense experience.[2]

The effective challenge to the outlook of Hobbes came in 1739 from a young Scottish philosopher, David Hume. With searching clarity he pointed out that the sequence from cause to effect seems necessary to us only because it is familiar; we have no reason to expect this sequence to occur again except our habits.

> We are determined by CUSTOM alone to suppose the future conformable to the past. When I see a billiard-ball moving towards another, my mind is immediately carry'd by habit to the usual effect, and anticipates my sight by conceiving the second ball in motion. There is nothing in these objects, abstractly considered, and independent of experience, which leads me to form any such conclusion: and even after I have had experience of many repeated effects of this kind, there is no argument, which determines me to suppose, that the effect will be conformable to past experience. The powers, by which bodies operate, are entirely unknown. We perceive only their sensible qualities: and what *reason* have we to think, that the same powers will always be conjoined with the same sensible qualities?
>
> 'Tis not, therefore, reason, which is the guide of life, but custom. That alone determines the mind, in all instances, to suppose the future conformable to the past. However easy this step may seem, reason would never, to all eternity, be able to make it.
>
> This is a very curious discovery.[3]

The curious discovery did not at once impress Hume's contemporaries, and he was disappointed. But the verdict of the following centuries has been with Hume. No one has refuted his contention that the step by which we make an induction, and conclude from one experience that we shall have a similar experience next time, has no

[2] The matter of this and the following paragraphs has been discussed briefly in Chap. 12, "The Method of Descartes," pp. 224–225.

[3] The quotation is taken from *An Abstract of a Treatise of Human Nature,* which Hume published anonymously in 1740 in order to draw attention to his large *Treatise of Human Nature* (1739) which had made little impression. The *Abstract* has often been ascribed to Adam Smith, but this is a mistake, as J. M. Keynes and P. Sraffa showed in their facsimile edition of it (Cambridge, 1938). The *Abstract* restricts itself to the most original part of Hume's work—the analysis of the relation of cause and effect—and is strongly recommended to all students.

logical sanction. The connection between cause and effect is empirical and not necessary.[4]

Hume had thus given body to some of the assertions of Locke. He had shown that even our most carefully formulated knowledge—that is, even science—is built up empirically. The pattern of experience is not, as mathematics is, held together by logical and necessary relations. Experience has to be lived, it cannot be imagined.

II

As always when there is a swing of opinion, the new view was taken too far. Hume had shown that cause and effect are not connected by logical necessity.[5] Soon it came to be assumed that there are no necessary connections in nature at all, and no guides to experience or to conduct except empirical success. This is, of course, the mental climate of the English Utilitarians and reformers—of Jeremy Bentham and Robert Owen. It is, for example, the climate which made Bentham feel that the Declaration of Rights was high-flown verbiage.

Kant and the German philosophers after him examined what Hume had proved with greater subtlety. They agreed that some of the connections which we find in nature have no other sanction than experience, repeated until it becomes habit. But Hume had not shown that *all* the relations that we find in nature are of this kind. Now Immanuel Kant asked whether there are not *some* connections in nature which are necessary because they underlie experience. Are there relations which are inescapable because without them our experience would be impossible? Is there some framework on which experience is founded, and without which the mind cannot grasp the external world at all?

[4] Important work has been done in recent years by psychologists in studying the conditions which prompt us to believe that a sequence of events is causally connected. See A. Michotte, *La Perception de la Causalité* (Louvain, 1946).

[5] It is not essential to Hume's argument, or to the discussion in these sections, that the same cause must always have the same *unique* effect. In essence, what is being discussed is the belief that there is a necessary connection between an event and some event or events which have been observed to follow it regularly in the past. But "regularly" need not mean "uniquely"; it need only mean that there is a determined pattern or distribution of events, one or other of which is always observed to follow the first event. Thus, Hume's argument, and the discussion in these sections, is equally relevant to the pattern of events in quantum physics.

These are the penetrating—and still unanswered—questions to which Kant devoted his difficult works. He concluded that there are some necessary relations in nature, which we feel as necessary modes of thought. In particular, Kant claimed to show that space and time are a basic framework of this kind: we know them a priori. Whatever the empirical laws of nature, Kant held that they must conform to the necessities of space and time as we know them. In his view, we cannot imagine a world in which either could be different.

We know now that Kant's claim, in the form that he made it, is false. The discovery, soon after he wrote, that there are consistent geometries which are not Euclidean[6] (as our everyday experience is) raised the first doubt. And in this century, special relativity has shown that different observers may see the same sequence of events in a different order of time, and has thereby destroyed the Newtonian picture of a uniform time which Kant had.

Nevertheless, we must not dismiss Kant's thought outright. He asserted that *Euclidean* space and *Newtonian* time are given to the mind a priori. Kant was mistaken in these examples. But behind the examples stands a larger assertion: that some things in the world could not be different, and that our minds are attuned to these things. This assertion still remains open.[7] For example, Kant held also that there is an underlying framework in nature, and in our thought, to which the physical behavior of material bodies must conform in any possible universe. Such categories as substance itself must be, in Kant's view, a priori; and so must such relations as the vexed relation between cause and effect.

In this way, Kant had consciously turned the argument of Hume inside out. When Hume had said, "Necessity is something that exists in the mind, not in objects," he had thought himself to be attacking

[6] This discovery was made independently by Wolfgang and Johann Bolyai, by Karl Friedrich Gauss, and by Nikolai Ivanovitch Lobachevsky, all early in the nineteenth century. See also the reference to John Playfair in Chap. 12, "Descartes."

[7] For example, one of the pioneers of relativity, the distinguished astronomer Arthur Stanley Eddington, believed that some of the fundamental constants of nature can be calculated simply by considering *how* we observe them, and without actually observing them. Among the constants which Eddington believed to have these necessary values are the number of dimensions of space and of time: see his book, *Fundamental Theory* (Cambridge, 1946).

causality, by demonstrating that nature gave no evidence for it. But to Kant, this phrase meant that nature must conform to causality, because causality is the only way in which the mind can grasp her workings. Hume had shown that there is no empirical evidence for causality; and Kant concluded from this that empiricism is not enough, and that nature can only be understood if we see that under empirical experience lies a framework of a priori knowledge.

We have interpreted Kant's philosophy in scientific terms. This is not the only way of interpreting it, and the philosophy has influenced more nonscientists than scientists.[8] But it is the best way of seeing Kant's mind, because Kant himself was trained as a mathematician and physicist, and for much of his life he earned his living as a lecturer in physics. He made an original contribution to science in 1775, when he put forward for the first time the theory that the planets have been condensed from a mass of gas, which Laplace formulated more accurately in 1796. Only at the age of 45, in 1769, did Kant begin to trouble himself with the philosophical difficulties in the foundations of science which Hume had thrown up. In that year Kant had the great revelation, that some knowledge must be a priori in order to make empirical science possible at all, which turned his career to philosophy. In the following year, in 1770, Kant was elected to the chair of logic and metaphysics in his native university of Königsberg in East Prussia, and he outlined his approach in his inaugural lecture and published it fully in his book *The Critique of Pure Reason* in 1781.

[8] Kant had a deep influence on German scientists in the nineteenth century, such as H. Helmholtz: see, for example, Helmholtz' own notes (written in 1881) on the assumptions which underlie the famous fundamental essay, *Ueber die Erhaltung der Kraft,* which he had written in 1847. Outside Germany, Kant's main influence was on nonscientists. For a lively and very different exposition of the nonscientific content of this whole sequence of philosophical debates, see (as an extreme example) R. G. Collingwood, *The Principles of Art* (Oxford, 1947), Chap. 9, and *Sensation and Imagination* (Oxford, 1938), pp. 172–194.

On Kant, in general, see Edward Caird, *The Critical Philosophy of Immanuel Kant,* 2nd ed., 2 vols. (London and New York, 1909); Friedrich Paulsen, *Immanuel Kant: His Life and Doctrine,* tr. by J. E. Creighton and Albert Lefevre (New York, 1902); Alexander D. Lindsay, *Kant* (London, 1934); and Richard Kroner, *Kant's Weltanschauung,* tr. by John E. Smith (Chicago, 1956). On Kant's ethical theory, see C. D. Broad, *Five Types of Ethical Theory* (New York, 1930), Chap. 5, "Kant"; and H. J. Paton, *The Categorical Imperative: A Study in Kant's Moral Philosophy* (Chicago, 1948).

III

Kant lived a life so exquisitely regular and uneventful that it seems to come straight out of Jules Verne. He was born in Königsberg in 1724, and there he lived, taught—and finally died, at the age of 80, in 1804. He was small, ascetic, and slightly misshapen; he never married. He did everything at the same hour every day; the inhabitants of Königsberg are said to have set their watches by the walk that he took from four o'clock to five o'clock every afternoon, wet or fine. Kant had no taste for music, for imaginative literature, or even for natural beauty; and he seems not to have been able to call up visual images in his mind, so that he lacked a gift which is probably important to experimental scientists.

Yet even the academic life and theoretical work of a philosopher of science can alarm authority in an age of revolutions. It was the point of Kant's approach to the problems of knowledge that man and nature are profoundly in accord—and in this accord, there was no mention of God. Some rational a priori knowledge of nature was, in Kant's philosophy, inborn in man; and there was an implication that some rational a priori knowledge of morality is also inborn—an implication which Kant later set out explicitly (for example, in his *Critique of Practical Reason*). The church was alarmed, and in 1792 (when Louis XVI had just been condemned, and the war against the French Revolution had begun) it prevented the publication in Berlin of a work by Kant which dealt with aspects of religion. Kant got the work published in Königsberg instead, and was promptly forbidden by the King of Prussia to lecture or publish on religious matters. He obeyed this command so long as the King of Prussia lived, but he virtually gave up teaching and withdrew into a melancholy protest.

In short, Kant's contemporaries were more interested in the moral than in the scientific applications of his work. Kant argued that there are moral necessities in nature as well as scientific ones. His ground was the same as that on which he based his scientific argument. The behavior of nature, he had said in his scientific argument, presupposes some underlying necessities, such as time and space. The behavior of men, he argued in his later works, presupposes the existence of underlying necessities without which their conduct would be meaningless; and these necessities are pieces of a priori morality.

They are imperatives which our moral nature is formed to obey of itself.

Georg Wilhelm Friedrich Hegel

I

The question which drove Kant was, "How does it come about that the human mind so naturally understands what goes on outside it?" To Kant, man and nature were no longer separate; so that his philosophy seemed to his contemporaries to give a new status to man. The knower and what he knows had moved closer, and were felt to be in natural accord. Man's reason appeared as an instrument in tune with the workings of the universe, and man's morality appeared as an instrument inborn in man, which needed no guide but its own dignity. Kant had given a new sense of dignity to men, in which the limitations of nature were not obstacles but the natural conditions for human freedom. Free will in man was a necessity of nature. (Precisely, Kant's point was that the conception of morality necessitates free will as a postulate just as much as the conception of natural science requires necessity as a postulate.)

These do not seem to us revolutionary doctrines. We have come to take for granted the dignity of man and the reach of his reason. And the subtle relation between the knower and what he knows, between the observer and the laws that he finds, have become evident to us in relativity and quantum physics.

But in the setting of the times from 1789 onward, these were revolutionary doctrines in Germany, whose scattered states had lagged behind the intellectual march of America, England, and France. The men of the new Romantic Movement in Germany may not have understood Kant well, but they felt that his philosophy gave a sanction to their faith in man's independence. Hume had been a materialist, now Kant was an idealist—this is how philosophy strikes forthright men, who do not stop to ask the technical meaning of "idealism" in philosophy. Many of these men had passed through the university of Jena, a small town in the state of Weimar where Goethe was minister of education. The philosopher Johann Gottlieb Fichte lectured at Jena enthusiastically on behalf of Kant and the French

Revolution. The Romantic poets Novalis and Johann Friedrich Hölderlin spent some time there, and there the poet Friedrich Schiller met Goethe in 1794—characteristically, at a session of the Jena Scientific Society.

In this constellation of men who passed through Jena, the philosopher Georg Wilhelm Friedrich Hegel was influenced and in turn became influential.[9] Hegel was born in 1770, a serious, rather drab boy who all his life had difficulty in expressing himself, and even at school was nicknamed "the old man." He had a university career so average that (as his biographers have noted with mischievous glee) the only comment his tutors could offer was that he was below average in philosophy.

An unlucky awkwardness of this kind dogged Hegel all his life. In 1781, William Herschel had caused a sensation in astronomy by his discovery of the planet Uranus, which had brought the number of planets (now that the moon was no longer counted among them) back to the classical figure of seven. In 1800, Hegel wrote a dissertation to prove, on lines which Kant had inspired in him, that the number of planets is seven not by accident but by a necessity of nature. It was unfortunate that, as soon as the dissertation was written, on January 1, 1801, a working astronomer found an eighth planet, Ceres.[10]

Hegel was a private tutor until, at the age of 31, he found a minor post at Jena in 1801. He made no special mark there, but in 1806 he himself was profoundly impressed by an experience outside his academic work. The Emperor Napoleon defeated the armies ranged against him at a great battle near Jena, and Hegel saw him enter Jena. He wrote to a friend: "I saw the Emperor, that world-soul,

[9] Among friends with whom Hegel corresponded were Hölderlin and the philosopher Friedrich Wilhelm Joseph von Schelling, whose theory of the imagination influenced the English Romantic poet Samuel Taylor Coleridge. For Hegel's early development, see the stimulating work by Wilhelm Dilthey, *Die Jugendgeschichte Hegels* (Berlin, 1905). In English, Edward Caird's *Hegel* (Philadelphia, 1883) is a good short introduction; G. R. G. Mure, *An Introduction to Hegel* (Oxford, 1940), is not simple, although it is short. *Early Theological Writings*, tr. by T. M. Knox, with an introduction and fragments tr. by Richard Kroner (Chicago, 1948), is of great interest for an understanding of the young Hegel; it includes a bibliography.

[10] The astronomer was Guiseppe Piazzi. For the whole episode, see George Sarton, *Guide to the History of Science* (Waltham, Mass., 1952).

riding through the city to reconnoitre. It is in truth a strange feeling to see such an individual before one, who here, from one point, as he rides on his horse, is reaching over the world, and remoulding it."

The rise of Napoleon had affected liberals in England and on the Continent of Europe differently. In England, those who had stood by the French Revolution finally broke with it when Napoleon made himself emperor in 1804. (For example, the Whig leader Charles Fox, after years of bitter opposition to the war against France, entered the government soon after.) On the Continent, where Napoleon's armies actually arrived and overthrew governments, many aging liberals felt that the chaotic spirit of the French Revolution had found solid and responsible form in him. So Goethe (who was then nearly 60) admired Napoleon and was impressed by him at Jena. And so Hegel felt him to be the "world-soul," whose autocratic person somehow fulfilled the ideas of rational liberty in which the French Revolution had begun.

II

The entry of Napoleon into Jena in 1806 disrupted university life, and Hegel had to make his living again in minor jobs. Nevertheless, he established a philosophic reputation, and he wrote *The Science of Logic,* which contains the essentials of his method. In 1817, Hegel was elected to the chair of philosophy at Berlin, which Fichte had held, and he dominated philosophy throughout Europe until he died in the great cholera epidemic of 1831.

Hegel's philosophy begins from the question which had engaged Kant: What is the accord between the mind and the world outside it? How is it that the one naturally understands the other? Like Kant, Hegel felt that there must be a profound unity between the knower and what he knows, and that knowledge would be impossible without such a unity.

Hegel thought of this as a unity of opposites, and the force of his method lies in his insistence that such opposites must be united at each step in human progress. The method is the dialectic, and in essence it goes back to Socrates. In the dialectic we begin with a thesis—say, with man as a person who seeks to know. To this thesis, nature presents an antithesis: the impersonal world resists the knower. There is a conflict between the thesis and the antithesis, and this is resolved only by a step of synthesis, which fuses the two: the knower and what is to be known generate a higher synthesis—generate knowledge itself.[11]

This strange but lively procedure had been applied by earlier thinkers to the activities of the mind. Hegel's originality lay in applying it to the concrete realities of life. It is not only in our thoughts, said Hegel, that thesis and antithesis have to be synthesized to give a higher understanding. Every process in life calls out its contradictory process—and life takes its important steps only when it synthesizes these two into a higher form. Life is not merely being, and death is not merely nonbeing; the essential step of progress is the synthesis of the two—is, becoming.

> The bud disappears when the blossom breaks through, and we might say that the former is refuted by the latter; in the same way when the fruit comes, the blossom may be explained to be a false form of the plant's existence, for the fruit appears as its true nature in place of the blossom. These stages are not merely differentiated; they supplant one another as being incompatible with one another. But the ceaseless activity of their own inherent nature makes them at the same time moments of an organic unity, where they not merely do not contradict one another, but where one is necessary as the other; and this equal necessity of all moments constitutes from the outset the life of the whole.[12]

Hegel's example here is interesting because it concerns the forms of plants, which also preoccupied Goethe, and from which Goethe derived an idea of evolution. But in Goethe's idea evolution is a simple progress. Hegel, for the first time, saw all progress, whether in the history of man or in the evolution of life, as a succession of revolutionary steps.

III

There are two implications of the dialectic method as Hegel used it which are important. One concerns the nature of reality. The other

[11] See, however, Gustave E. Mueller, "The Hegel Legend of 'Thesis-Antithesis-Synthesis,'" *Journal of the History of Ideas* (June, 1958), where it is argued that Hegel made only qualified use of this triad, and that its rigid use was later foisted on Hegel by the Marxists.

[12] *Hegel Selections*, ed. by J. Loewenberg (New York, 1929), pp. 2–3. This is a useful little volume containing some of Hegel's work in English translations.

concerns the progress and development of the real world. We shall follow these separately.

The nature of reality had already been seen by Kant to be more subtle than the empiricist philosophers of England and France had pictured it. Empiricist philosophers had simply thought of two worlds—the public world outside a man's head, and a private world inside it. For the empiricists, these were loosely connected by the man's senses; but the two were as concrete, and as separate, as a color film and a black-and-white copy.

Kant had seen that this simple picture will not do. We can postulate the real world only because we know it, and the part that the senses play in our knowing has to be analyzed with care. Man is not simply a passive receiver on whom the outside world prints a set of impressions. The knower and what he knows influence one another; what is known is in part imposed by the knower; so that the knower is active, is creative,[13] and thereby becomes what Kant called a self or ego.

Yet to Kant, this personal ego was still a part of a universal and transcendental ego, because it shares the a priori concepts which are common to all men. And equally, Kant believed that there is a reality which is independent of men; behind the thing as it is known, there is what Kant called a thing-in-itself.

Now Hegel's dialectic broke down even this distinction between the knower and what he knows. He denied that there is any thing-in-itself. To Hegel, there is no reality until we know it. We exist by virtue of knowing the outside world—but the world also exists only by virtue of our knowing it. "The real is the rational and the rational is the real," said Hegel: and he meant that things exist only because the mind thinks of them, just as the mind exists only because it thinks. Descartes had said long ago, *Cogito ergo sum*. Now Hegel went further. He said that my thinking does more than prove my existence: it creates it—in Hegel's phrase, "Being is thought"—and it also proves and creates the existence of the outside world. In the dialectic, Hegel claimed, "the opposition between being and knowing" is ended. The knower and what he knows, thesis and antithesis, are fused in a single synthesis of experience.

[13] For a fuller exposition of this important point of view, see J. Bronowski, *Science and Human Values* (New York, 1958), and Michael Polanyi, *Personal Knowledge* (London, 1958).

This is a bold conception. It asserts the importance of the individual even more forcefully than Kant did. As Hegel wrote to a friend, "I hold it one of the best signs of the times, that humanity has been presented to its own eyes as worthy of reverence. It is a proof that the nimbus is vanishing from the heads of the oppressors and gods of the earth. Philosophers are now proving the dignity of man."

But when this has been acknowledged, man in Hegel's conception becomes a curiously vague and amorphous creature. If the outside world owes its reality to the mind, then all matter becomes somewhat unreal; and the human body itself, and the human senses with it, become as thin, as much a construct of the mind, as the rest of the world. This leaves us with no center and no anchor other than an intangible spirit. And Hegel built up an elaborate and mystical devotion to the spirit, which is made more elusive by the ambivalence of all his terms. At the drop of a hat, Hegel's mysterious spirit can turn into something harsh and dictatorial, because after all the spirit is what expresses itself in the concrete world. "Out of this dialectic rises the universal spirit, the unlimited world-spirit, pronouncing its judgment—and its judgment is the highest—upon the finite nations of the world's history; for the history of the world is the world's court of justice." A few quick steps, and spirit becomes the world-spirit, the world-spirit becomes the world-soul, and the world-soul becomes the Emperor Napoleon passing judgment on the petty world by violating it. Like some other sensitive men (like Leonardo), Hegel was fascinated by violent authority, and in his most mystical moments longed to be dominated by it.

IV

We have heard Hegel speak of the judgment of history, in which the universal spirit reveals itself. This is the other implication of his dialectic, and it has dominated the Western outlook ever since Hegel wrote: the preoccupation with history.

It is the essence of the dialectic method that every thesis generates its antithesis, and that the conflict between them can be resolved only by a synthesis which goes beyond both. In the realm of ideas, this is a simple progression: for example, the psychology of Freud is full of it. But Hegel claimed that this was not only a description of ideas, nor a picture of the human mind. He turned it into a practical account of the everyday progress of the world. By this step, he elevated the passing of time to the rank of a creative force.

In the Middle Ages, it had occurred to no one to think that the future would be better than the present; no one thought that it would even be different. To us, this is a flat view of the world, but it remained general well into the eighteenth century. Aquinas gave no attention to the passage of time, and Descartes and Locke gave it little attention. Their systems were static, and the solutions which they offered to the problems of man and state seemed to them to be permanent.

All this began to change late in the eighteenth century.[14] We have seen history entering the work of Adam Smith and filling the mind of Edmund Burke. We have seen progress as a natural idea in the minds of Franklin and Jefferson, as well as of the English manufacturers. And we have seen evolution begin to enter the minds of Goethe and Erasmus Darwin. In Hegel's dialectic, we see this new outlook turned into a philosophical system.

If the everyday world is subject to a dialectic process, then it is always in a state of change. But more than this: because the dialectic process always moves to a higher synthesis, the changes in the world are changes for the better—at least, they are changes in the direction of more complexity, integration, of greater fullness. The dialectic process is not merely a progression; by its very nature, it is a progress.

The dominant and the most original theme in Hegel is his preoccupation with history. History to him is the great transformer, the great mover; she is the justification for every accident of existence; she is the working out, the physical realization of the spirit of all men. And if men make history, it is in order that history shall make states; history becomes a goddess in whom Hegel finds the fulfillment of all dreams—the conservative dreams of old age as well as the radical dreams of youth. But more than this—and it is Hegel's contribution to a developing science of man—he is suggesting that man *is* his history; and that only an understanding of history can enable man to understand himself.

[14] The Italian writer Giovanni Battista Vico had published a book in 1725, *Scienza Nuova,* which stressed the importance of the historical study of man; but it was overlooked by his contemporaries, and came into its own only in the nineteenth century.

History is always of great importance for a people; since by means of that it becomes conscious of the path of development taken by its own spirit, which expresses itself in laws, manners, customs, and deeds. History presents a people with their own image in a condition which thereby becomes objective to them.

Hegel's major work was called *The Philosophy of History,* and was delivered as a series of lectures in the nine years from 1822 to 1831.

As so often in the field of ideas, it is hard to say how far Hegel initiated the preoccupation with history, and how far he voiced a feeling which many of his contemporaries held. Goethe and Schiller both wrote works of history, and a school of historians had begun to grow in Germany. Hegel took from Schiller the phrase, *Die Weltgeschichte ist das Weltgericht* ("The history of the world is the world's court of justice"), which he uses as a motto of his own thought.[15] The nineteenth century was discovering history, which hitherto had been used mainly to point moral maxims, but which now became an interest in itself. Since Hegel, we think it the most natural thing in the world to look at any subject—whether art or science, whether culture or technology—by way of its history.[16] We see history as an unfolding; we believe that we shall grasp the present better when we understand how it grew from the past; and above both, we have our eye on the future.

History in this sense of Hegel's is not merely a record of the past. It is a progress, an evolution; and any scheme which disregards history (such as Robert Owen's) is thought Utopian. Charles Darwin's enunciation in 1858 of the theory of evolution is in fact one expression of the new concern with history, on the cosmic scale.[17] And just

[15] See Bruce Mazlish, "History and Morality," *Journal of Philosophy* (March 13, 1958), for the origin of the viewpoint that history determines morality, and for a discussion of the value of such a moral position.

[16] Special emphasis has been given to the historical and cultural nature of man in recent years by José Ortega y Gasset and by Wilhelm Dilthey: see the former's essay, "History as a System," in *Toward a Philosophy of History* (New York, 1941), and, for the latter, H. A. Hodges, *The Philosophy of Wilhelm Dilthey* (London, 1952).

[17] The theory of evolution by natural selection was proposed independently and at the same time by Charles Darwin and by Alfred Russel Wallace, and they thereupon agreed to communicate it in a joint paper to the Linnaean Society in London in 1858. Darwin went on to publish his massive evidence, which he had long accumulated (and which he had kept back for fear of giving offense to religious opinion), in 1859 in the classic *On the Origin of Species by Means of Natural Selection.*

as evolution is most interesting when it accounts for the development of man, so history for Hegel is above all an account of the development of states. In both developments, the struggle for survival has been a shaping force. "War has the deep meaning that by it the ethical health of the nations is preserved and their finite aims uprooted. And as the winds which sweep over the ocean prevent the decay that would result from its perpetual calm, so war protects the people from the corruption which an everlasting peace would bring upon it."

V

Hegel thought of history as the working of a universal spirit. The spirit is reason, and cannot be wrong; and the institution in which this spirit expresses itself is the state. The state has to be accepted as the practical realization of cosmic reason; it is secular but it is not therefore evil. "That harmony which has resulted from the painful struggles of History, involves the recognition of the Secular as capable of being an embodiment of Truth; whereas it had been formerly regarded as evil only, as incapable of Good—the latter being considered essentially other-worldly." More than this: the state becomes the fulfillment of the mystical spirit of a people, and history works providentially to make it so. Hegel is thus giving to the state the same romantic homage which Edmund Burke gave it. When Hegel wrote that a constitution is not something that can be thought out overnight, but "is the work of centuries, the idea and the consciousness of what is rational, in so far as it is developed in a people," he might have been echoing Burke. So he might have been when he wrote:

> It is nothing but a modern folly to try to alter a corrupt moral organisation by altering its political constitution and code of laws without changing the religion—to make a revolution without having made a reformation, to suppose that a political constitution opposed to the old religion could live in peace and harmony with it and its sanctities, and that stability could be procured for the laws by external guarantees, e.g., so-called 'chambers,' and the power given them to fix the budget, &c.[18]

[18] *Hegel Selections,* p. 276. It should be noted, however, that Hegel's eulogy of the state must be qualified by an awareness that he placed the state, as part of what he called objective mind, *below* art, religion, and philosophy, or absolute mind, in his system. See further Herbert Marcuse, *Reason and Revolution,* 2nd ed.

"A revolution without having made a reformation" is almost word for word the phrase which Burke used repeatedly to criticize the French Revolution.[19]

Hegel had begun by advocating a philosophy of history because history expresses the dialectic process of change. Now he seems to find these changes resolved forever in the political state at the beginning of the nineteenth century, as an expression of the spirit of a people. This is no longer a revolutionary but an authoritarian doctrine, and sounds much like the justification by Adolf Hitler of the destiny of Germany. Even before his death, Hegel was accused, with some justice, of using his philosophy to glorify the state of Prussia. In this aspect of his work, he had traveled the road of Burke: the romantic ardor of youth had become the romantic conservatism of old age. "Right-wing" Hegelians could shelter under this part of Hegel's work in defense of the status quo.

What of the other, the "left-wing" or revolutionary interpretation of Hegel? How could the spirit of a people change, and find a new expression in a new state? It was very fine for Hegel to write,

> A Nation is moral—virtuous—vigorous—while it is engaged in realising its grand objects, and defends its work against external violence during the process of giving to its purposes an objective existence. The contradiction between its potential, subjective being—its inner aim and life—and its *actual* being is removed; it has attained full reality, has itself objectively present to it. But this having been attained, the activity displayed by the Spirit of the people is no longer needed; it has its desire. The nation can still accomplish much in war and peace at home and abroad; but the living substantial soul itself may be said to have ceased its activity.[20]

(New York, 1954), p. 87. Marcuse's book is a stimulating piece of work; the second edition cited has an excellent bibliography. A violent attack on Hegel is to be found in Karl Popper, *The Open Society and Its Enemies,* 2 vols. (London, 1945); Also on Hegel are: Benedetto Croce, *What Is Living and What Is Dead of the Philosophy of Hegel,* tr. from the 3rd Italian ed. of 1912 by Douglas Ainslie (London, 1915), a highly stimulating and independent piece of work, as might be expected from its author; and Jean Hyppolite, *Etudes sur Marx et Hegel* (Paris, 1955), which has an essay on the significance of the French Revolution in relation to Hegel.

[19] Both Burke and Hegel also shared some understanding of the new economic thought which Adam Smith had formulated; and Hegel, in his discussion of Luther, stresses the economic effect of Protestant thought in a remarkably modern way which anticipates Max Weber and R. H. Tawney. For an appreciation of Hegel's economic outlook, see Marcuse, *op. cit., passim.*

[20] G. W. F. Hegel, *Philosophy of History,* tr. by J. Sibree (New York, 1900), p. 74.

It was very fine to write like this, but how could decay be avoided? How could an antithesis be generated? How could a synthesis be found?

To these questions, Hegel now gave the answer which has tempted political philosophers ever since Machiavelli. The spirit of a people expresses itself in its heroes. History is made, says Hegel, by "world-historical individuals." These are the men who personify a nation when it is changing; yet they have "no consciousness of the general idea they are unfolding while prosecuting their own private aims. On the contrary, they are practical political men." And when such men express the dialectic of history, they break with the old without scruple: they are above morality. "Such conduct is indeed morally reprehensible, but so mighty a form must trample down many an innocent flower, and crush to pieces many an object in its path."

The ambition of these supermen is the stuff of history, and serves its ends; and Hegel is contemptuous of the pedagogues who criticize them. "What pedagogue has not demonstrated of Alexander the Great—of Julius Caesar—that they were instigated by such passions, and were consequently immoral men? whence the conclusion immediately follows that he, the pedagogue, is a better man than they, because he has not such passions; a proof of which lies in the fact that he does not conquer Asia."

Hegel, of course, was a pedagogue himself, with a difference. He could feel himself taking part in the conquest of Asia, as an adviser to Alexander. He would give Machiavelli's advice that the hero—that is, the dictator—is the state; but with a new moral twist. "The State is the actually existing, realised moral life. For it is the Unity of the universal, essential Will with that of the individual; and this is 'Morality.' "

VI

Hegel's ideas are often contradictory in detail, but their large pattern is characteristic of political philosophy in the last 150 years. For example, his dialectic was given a materialist form by Ludwig Feuerbach, and in this form Karl Marx made it the foundation of Communist philosophy. A belief in the development of states by a repeated process of contradiction and revolution was, of course, welcome to Marx.

It is therefore natural that, in the long run, Communism—and Fascism—has been faced with the same difficulties that faced Hegel. There is no final form of society; and the man who feels this finds himself at odds with Communist or Fascist society, however much his philosophers (from Rousseau to Hegel and Marx) insist that the individual will and the collective will are one. The state as an end becomes a monster. And whoever rides the monster—Napoleon or the King of Prussia, Stalin or Hitler—becomes not a hero but a tyrant.

The state, we believe, is a means which can help its members to greater freedom. Even the recognition, which the state imposes, that laws are necessary, can help to free men; just as we act more freely, because more rationally, when we understand the limitations which the laws of nature impose on us. But the state is not itself the expression of freedom. The error of Hegel and his followers, idealists and materialists, right-wing and left-wing Hegelians, lies here—in regarding the state as an expression instead of an instrument, an end instead of a means. As this book shows, history is not made by states or for states; it is made by people and for people. Least of all is history made entirely or even mainly by world-historical individuals; it is made, more strangely, by men like Hegel.

The elevation of history to an impersonal goddess has remained with us since Hegel undertook it. Since Hegel, revolutionaries and progressives have tended to claim that history marches on their side. Perhaps it is true of all progressive ideas that their adherents must feel themselves to be going with the stream of history. On the other hand, traditionalists have often seemed to believe that history has reached its climax in them.

Whatever hopes for the future we hold, progressive or traditionalist, we derive them from our sense of history. Our sense of history, therefore, gives us our personal sense of mission. Since the beginning of the nineteenth century, this sense of mission has been formed by the images of progress and of evolution; we search for the movement of history—the play of ideas and the tempo of events. This is the search that should inspire our interest in history today; for history is not so much a book as a movement, not a story but a direction, and not a reverie in the past but a sense of the future.

Conclusion

I

IS THERE a lesson to be learned from history? The question has much the same force as if we were to ask: Is there a lesson to be learned from evolution? The study of evolution tells us that, as men, we stand at the end of a long development, which has endowed us with physical and mental gifts that mean more to us as we understand them better. The study of history makes us equally the heirs of a long development, and elucidates for us the cultural gifts which we owe to those who lived and struggled and thought before us.

The past, then, has not simply disappeared in the wastes of time. The ideas which the past has evolved are alive: the empirical way to truth, the insistence on reasoned explanations, the conviction that men have a claim to liberty and justice. Ideas have their roots in the minds of men, have developed out of their conflicts, and have shown their strength by survival.

So much of ideas in general. Now, at the end of this book, we ought to summarise the specific movements of ideas which we have traced. It remains for us to draw the threads together, and to ask: What, concretely, does a study of the changing thought and modes of life from the time of Leonardo to Hegel tell us? What are the positive ideas which have been created in the 500 years since the Renaissance?

II

A dominant trend of the period which this book covers is the rise of the scientific method, both in the natural and in the human sciences. Of the many ideas which flow together into the scientific method, one is fundamental: the idea that nature—physical nature and human nature—follows consistent and permanent laws. To the medieval mind, nature had been in some sense a constant miracle: it was sustained and renewed from each moment to moment by a new divine intervention. The Renaissance mind and the modern mind also find nature wonderful, but for a different reason: they find it wonderful that nature consistently follows the same laws. The rise of the scientific method rests on our conviction that nature is not arbitrary, but is profoundly lawful.

When men say that nature is not arbitrary but follows consistent laws, they mean implicitly that her laws are intelligible to the human mind. (If men cannot understand how a law of nature works, they do not recognize it as a law; they think it arbitrary.) Therefore the conviction that nature is lawful has stimulated men since the Renaissance to do two things to discover her laws; and both these are essential to the scientific method.

On the one hand, the conviction that nature is lawful has stimulated men to keep a close watch on natural phenomena, to observe and to experiment, in order to see in what manner she repeats herself—to find, practically, the pattern of her consistency. And on the other hand, it has stimulated men to think behind the practical pattern, to analyze and to reason, in order to find its simple and rational organization—in order to find intelligible laws. The combination of the empirical and the rational makes up the scientific method.

We have traced this interplay of empirical experiment and of rational inquiry, which makes up scientific progress, in this book. In our view, the successful combination of the two is prefigured first in the person of Leonardo. In the century after his death, the combination of these two modes inspired the pioneers of the Scientific Revolution in the physical sciences: Copernicus, Kepler, and above all, Galileo. Their work reached its fulfillment in another age of history, the age of the Royal Society and the towering achievement of Isaac Newton.

The empirical approach and the rational approach have continued to be necessary jointly to the development of science. Whenever one has been fostered at the expense of the other, progress has taken a one-sided turn. Thus the dominance of rational inquiry, too little supported by experiment, in Descartes and his followers was a handicap to French science almost to the time of the French Revolution. Descartes' method of doubt was effective philosophically in destroying traditional prejudices about the workings of nature, but it could not create a positive and practical scheme of natural laws.

By contrast, the decay of the Royal Society in eighteenth-century England caused rational inquiry to be undervalued. It gave to the rising intellects of England and of America about 1760 a predominantly practical bias and an empirical view of science. As a result, the technical achievements of men of the stamp of James Watt and Benjamin Franklin were too easily detached from their human and ethical context, and allowed to proliferate into the smug inhumanity of early industrial society.

III

The fundamental movement of the Western mind since the Renaissance toward a vision of nature governed by consistent laws has also been felt in the human sciences. Human nature has been believed to follow intelligible laws, just as physical nature has. In particular, men have believed since the Renaissance that the way in which human beings think and feel must somehow shape the structure of human societies. Somehow, the laws of society cannot be arbitrary: they must derive from, they must conform to, fundamentally they must satisfy the needs and the aspirations of individuals. The very use of the same word "law" in government and in science implies that government must be like science, and must learn to conform to the nature of its material. In this view, which has penetrated Western thought since the Renaissance, legislation is not a matter of edict but, at bottom, of research; and the state, if it is to survive, must not impose its laws but must discover them in the nature of human relations.

There is, however, one respect in which the evolution of the human sciences since the Renaissance has differed from that of the physical sciences. The human sciences have not integrated the empirical approach and the rational approach. Those who have practiced, or at least have advocated, the empirical study of human societies have, on the whole, divorced it from rational analysis, often on the grounds that the latter tends to be biased by a priori and by moral judgments. And on the other side, those who have tried to base a theory of society on the rational analysis of the motives of individuals have tended to dismiss the workings of their own society as irrelevant or distorted versions of their ideal Utopia. We can ignore here the necessary qualifications, which we have made in the detailed treatment of our chapters, and embrace the following broad classification. Thus Machiavelli, whose "new route" initiated the empirical study of power politics, is contemptuous of those who look for rational motives beyond such politics. Bayle and, in some ways, Montesquieu revived the empirical tradition, and it was then established in its modern form by Adam Smith. Again, there is a hint of Machiavelli's contempt for the accepted rational motives in Burke, and more than a hint in Hegel: in both, the empirical workings of history are given a mystical admiration which would ultimately negate a scientific study.

The rational analysis of society as a construction to serve and satisfy human needs also begins at the Renaissance, in Thomas More and his fellow humanists, among them Erasmus. We have traced it through Hobbes and Locke, whose different political deductions sprang equally from the analogy of the physical sciences in the age of Newton. From there the analysis of society was carried forward by Voltaire to the *philosophes* who prepared the climate of ideas for the French Revolution; and equally it flowed more directly into the minds of men like Franklin and Jefferson, which expressed the climate of the American Revolution. Its final rationalization is reached in Jeremy Bentham, who was so contemptuous of all existing forms of society and yet so anxious to be practical in his rationalism. In Bentham, the apparatus is still dominated by Newtonian science. Only after Kant had given a more subtle dialectic of the relation of the scientific observer to what he observes, of the knower to the known, could Hegel attempt a dialectic of history which tried to introduce a more refined rationalism.

There are many reasons why the human sciences have failed to give as satisfying an account of their field of inquiry as the physical sciences have given of theirs. One important reason is that to which we have pointed. The human sciences have failed to find as coherent a method as the natural sciences, because they have not fused together the two modes of inquiry, the empirical and the rational. This is still a major shortcoming in the human sciences: that the attempt to construct a reasoned analysis of society from the motives of individuals is divorced from the practical study of the large-scale functioning of states and communities.

This continued separation of the rational approach from the empirical in the human sciences may be due to their failure to evolve a logical link, a way of reasoning from motive to action, which is subtle enough to fit their human material. The simple reasoning of cause and effect which gave coherence to the physical sciences from the time of Hobbes onward is too crude to give an account of the interplay of human motives in a large society. It may be that the new concept which the human sciences require is a statistical one, and will turn out to involve nothing more radical than the application of large computing machines to social and economic problems. Or it may be that some more profound conception is required to marry the rational and the empirical approaches in the human sciences.

IV

In their confident search for the laws of physical and human nature, the sciences have displayed another drive. This is the secularization of thought, and its emancipation from absolute edicts which are not open to inquiry. Machiavelli and Galileo, Bayle and Locke, Franklin and Adam Smith, and Robert Owen prove that this impulse has remained powerful and unexhausted ever since the Renaissance. It has been powerful in literature and art and philosophy also.

There is a sense, paradoxical as it may seem, in which this movement of secularization has also been powerful in the religious changes of the period covered by this book. Certainly the humanist movement of Erasmus and his contemporaries was a movement of secularization, for they tried to show that the religious virtues were of themselves natural human virtues. In this sense, humanism was and is precisely what its name implies: a wish to find the source and the criterion of what is good, just and beautiful in the human gifts. This wish implies a belief that, in the end, every man must judge for himself, in matters of truth and right as well as in matters of taste.

The belief that men cannot accept authority but must make their own judgments inspired the Reformation and its different expressions, including its extreme expression in the Puritan Revolution in England. In this belief, revolutions of faith and social revolutions necessarily go hand in hand, and so we have seen them do. Even the later Industrial Revolution was powered by an ethic of its own, which changed the character of the Christian virtues in the direction of thrift, frugality, and resignation. In this respect, John Wesley in the eighteenth century was as much an innovator as Luther and Calvin were in the sixteenth.

Secularization, then, is one facet of that advancing humanism which the Renaissance introduced and to which, in some way, every personage in this book is a witness. It is not an affront to traditional values, but a desire by the human spirit to examine itself, and to find a necessary relation between its gifts and its values. When they painted men and women as beautiful, the artists of the Renaissance said that the secular attitude is a form of pride in the gifts which make us human. And the work of philosophers and scientists and historians says today that the relation of man to nature and to himself is not predetermined, but is for man to study, to form, and thereby to enrich.

V

The new scientific and secular attitudes to life have had their effect on every part of man's activity. The subject matter of philosophy, of painting, and of literature has changed through the period of this book. It is obvious that no medieval writer could have written a modern thesis on electronics; but it is equally important to remember that no medieval writer could have written a modern novel.

Just as the changes in outlook since the Renaissance have affected the content of what men think and write, so they have affected the manner in which they think and write. The style of a man or of an age mirrors the thought, and we catch that thought most effectively, its color and its unconscious personality, when we read at first hand how the man or the age expressed themselves in their own words.

The splendid and passionate writing of the Elizabethans is in a sense the last expression of the high Renaissance; its mode is classical rather than modern. Thereafter, the Puritan Revolution ushers in the persons and the manner of the Royal Society, who regarded it as a matter of principle that they should write simply and practically about simple and practical facts. The style of Pascal as much as the style of Jefferson derives from these strict models, who were suspicious of any form of words that might underline the truth (which they regarded as carrying conviction by the mere statement) or might seem to persuade instead of to assert. And it is notable that the new style can display itself as effectively in the pointed satire of Voltaire as in the homely quips of Franklin.

The new style, in fact, had a particular leaning to satire, and was used in this way almost at the beginning of the Scientific Revolution by Galileo in his *Dialogues.* The dialogue form also lends itself particularly, as Socrates had shown many centuries before, to that method of questioning the accepted view of the nature of things which Descartes consolidated into the philosophical method of doubt. In satire and in dialogue, the method of doubt expressed itself in the masterpieces of French literature before the French Revolution, in the hands of Pascal, Voltaire, Montesquieu, and even Beaumarchais. Just as the *philosophes* put the new ideas of science and secularization into the content of the *Encyclopedia,* so the writers displayed them by their very style.

VI

The changes in man's view of the world became concrete in action: in the overthrow of old institutions and in the development of new social institutions which might better clothe man's new aspirations. Indeed, the first expression of the new ideas was usually a criticism, not of the traditional values of Christianity, but of the existing Christian institutions. Erasmus and the humanists did not only attack the ideals of monastic virtue; they criticised the way in which monasteries were run in practice. Luther did not attack Roman Catholicism; he opposed what he regarded as its abuses. Characteristically, the Reformation began in an argument about a peripheral abuse of papal power, a device for raising money—the sale of indulgences.

More directly, the rise of the scientific attitude led to the formation of new loci for the discussion of ideas. These widened, step by step, from the courts of princes in the city-states of Italy, to the royal scientific societies of England and France, to the fashionable salons of enlightened ladies in French society, and so to the dissenting academies of England and the debating clubs of America in the late eighteenth century. By these steps there was created that tradition, in which information and discussion are compounded, which now expresses itself naturally in all Western institutions: parliaments, universities, clubs, and newspapers.

There were also created institutions which were more directly geared to scientific development: institutions so precise that they may be called social inventions, and which were designed to exploit the technical inventions of science. Among these institutions are the taxation and insurance systems of the Italian city-states, the stock exchanges of Holland and England in the seventeenth century, the joint-stock companies of the eighteenth century, and the large-scale factories which were the most important invention of the Industrial Revolution. These economic institutions in their turn set going new social institutions, such as the Lunar Society and others like it, the associations of working men, and the schools of Robert Owen. All these were human creations, formed in order to express the changing personality of men; yet inevitably they themselves changed the men who met in them.

VII

In each transformation in the period covered by this book, men have had before them a single vision: the vision of their own humanity. This above all is what the Renaissance did—to inspire men with a feeling that there is a picture of man, the essential man, to which they themselves aspire. The Renaissance made ideas a new prime mover which could shape men and their societies, and the men then went on to reshape the ideas. In this historical circulation, one of the most important ideas has been the idea which man has had of himself, in each generation.

The Renaissance ideal of man had an element of condottiere brutality, which has lingered on in Western thought; it is perhaps inseparable from the Western admiration for power, over nature and over men. Something of this sense of power, of mastering the techniques and desires of the earthly life, is present in the ideal men of the Reformation—in Calvin's "new man," for example, and in the Puritan soldiery. A different direction was set by the Tudor ideal of a gentleman, which has remained alive from that day to this, and by the seventeenth-century ideal of the virtuoso. This direction leads from the humanist of the sixteenth century to the *philosophe* of the eighteenth, and is seen at its best in the tolerant, rational, free, and yet convinced and single-minded men of English dissent and of the American Revolution.

The religious dissent of the churches and the reasoned dissent of humanism have both been formative in these ideals, for example, in fostering religious and political individualism. Other institutions have threatened to deform these ideals, for example, the herding of men into factories which has made it too easy to treat first the labor of man, and ultimately man himself, as a commodity. The totalitarian distortions of modern industrial nations, in Germany and in Russia for example, have no doubt derived in part from this confusion of man with the machines in the factory, and from the image of the state as a kind of factory.

VIII

In the 500 years since Leonardo, two ideas about man have been especially important. The first is the emphasis on the full development of the human personality. The individual is prized for himself. His creative powers are seen as the core of his being. The unfettered development of individual personality is praised as the ideal, from the Renaissance artists through the Elizabethans, and through Locke and Voltaire and Rousseau. This vision of the freely developing man, happy in the unfolding of his own gifts, is shared by men as different in their conceptions as Thomas Jefferson and Edmund Burke. It is the picture which Hegel had of the heroes of the age of Napoleon, and—though this will be a surprise to those who have not read him—which Karl Marx had of the artisans and workers whom he idealized.

Thus the fulfillment of man has been one of the two most formative grand ideas throughout the period of this book, and it has remained so to our day. Men have seen themselves as entering the world with a potential of many gifts, and they have hoped to fulfill these gifts in the development of their own lives. This has come to be the unexpressed purpose of the life of individuals: fulfilling the special gifts with which a man is endowed.

The self-fulfillment of the individual has itself become part of a larger, more embracing idea, the self-fulfillment of man. We think of man as a species with special gifts, which are the human gifts. Some of these gifts, the physical and mental gifts, are elucidated for us explicitly by science; some of them, the aesthetic and ethical gifts, we feel and struggle to express in our own minds; and some of them, the cultural gifts, are unfolded for us by the study of history. The total of these gifts is man as a type or species, and the aspiration of man as a species has become the fulfillment of what is most human in these gifts.

This idea of human self-fulfillment has also inspired scientific and technical progress. We sometimes think that progress is illusory, and that the devices and gadgets which became indispensable to civilized men in the last 500 years are only a self-propagating accumulation of idle luxuries. But this has not been the purpose in the minds of scientists and technicians, nor has it been the true effect of these inventions on human society. The purpose and the effect has been to liberate men from the exhausting drudgeries of earning their living, in order to give them the opportunity to live. From Leonardo to Franklin, the inventor has wanted to give, and has succeeded in giving, more and more people the ease and leisure to find the best in themselves which was once the monopoly of princes.

Only rarely has a thinker in the last 500 years gone back from this ideal of human potential and fulfillment. Calvin was perhaps such a thinker who went back, and believed as the Middle Ages did, that man comes into this world as a complete entity, incapable of any worthwhile development. And it is characteristic that the state which Calvin organized was, as a result, a totalitarian state. For if men cannot develop, and have nothing in them which is personal and creative, there is no point in giving them freedom.

IX

The second of the two grand formative ideas which this history displays is the idea of freedom. We see in fact that human fulfillment is unattainable without freedom, so that these two main ideas are linked together. There could be no development of the personality of individuals, no fulfillment of those gifts in which one man differs from another, without the freedom for each man to grow in his own direction.

What is true of individuals is true of human groups. A state or a society cannot change unless its members are given freedom to judge, to criticize, and to search for a new status for themselves. Therefore the pressure of ideas throughout the period of this book has been toward freedom as an expression of individuality. Sometimes men have tried to find freedom along quiet paths of change, as the humanists did on the eve of the Reformation, and as the dissenting manufacturers of the eighteenth century did. At other times, the drive for freedom has been explosive: intellectually explosive in the Elizabethan age and the Scientific Revolution, economically explosive in the Industrial Revolution, and politically explosive in the other great revolutions of our period, from Puritan times to the age of Napoleon.

Yet our study shows that freedom is a supple and elusive idea, whose advocates can at times delude themselves that obedience to tyranny is a form of freedom. Such a delusion ensnared men as diverse as Luther and Rousseau, and Hegel and Marx. Philosophically, there is indeed no unlimited freedom. But we have seen that there is one freedom which can be defined without contradiction, and which can therefore be an end in itself. This is freedom of thought and speech: the right to dissent.

X

The evidence of history is strong, that those societies are most creative and progressive which safeguard the expression of new ideas. Societies appear to remain vigorous only so long as they are organized to receive novel and unexpected—and sometimes unpleasant—thoughts. It may seem odd that the government of some countries gives a special status to the opposition; Canada, like Great Britain, actually pays a salary to the leader of the opposition. Yet this legalization of opposition, this balance between power and dissent, is the heart of the Western tradition.

The great creative ages have tended to be those in which reasoned dissent was welcome. One lesson that history teaches here is that such dissent is creative in all fields at once. The art of the Renaissance went hand in hand, in time and in place, with its science. The Elizabethan poets and adventurers were contemporaries of the first great scientists who spoke English—of Francis Bacon, of William Gilbert and William Harvey. The Romantic Movement in poetry at the end of the eighteenth century coincided with the uprush of inventions then, and both took strength from and gave strength to the revolutions of the age.

Here we have reached an odd conclusion: that there is a tradition of Western thought since the Renaissance, which is a tradition of dissent—that is, a tradition of questioning what is traditional. Yet, this conclusion is not as paradoxical as it seems. For we have found that ideas do not follow a simple progression, and their history is not a featureless avenue. History is made in conflict, and the history of ideas is a conflict of minds.

The conflict of minds is still an abstract phrase—as abstract as the history of ideas. Behind the minds are men: ideas are made, are held, and are fought for by men. To read the history of ideas out of its context of men and events is to violate it. Ideas are as human as emotions, as powerful, as conflicting and as indestructible. The aim of this book has been to show ideas in their full setting: of men, of groups of men, and of events. The men whom we have chosen were not singular; what they said was said by others also. Yet it is important that it was said by the particular men whom we have singled out for study; they have left the stamp of their style on the ideas they expressed. As a result, our history is, in part, a tribute to the men in its pages. But it is also more and wider than that. It is a tribute to all men who have lived and who live by ideas, and who, in their turn, create and give life to other ideas.

Indexes

Index of Names

Index of Subjects